DESIGNING
HUMAN INTERFACE
IN
SPEECH TECHNOLOGY

DESIGNING
HUMAN INTERFACE
IN
SPEECH TECHNOLOGY

FANG CHEN
Interaction Design,
Department of Computer Science
Chalmers University of Technology
Göteborg University
Sweden

 Springer

Fang Chen
Interaction Design,
Department of Computer Science
Chalmers University of Technology
Göteborg University
Sweden

Designing Human Interface in Speech Technology

e-ISBN-10: 0-387-24156-6

ISBN-13: 978-1-4419-3697-4

e-ISBN-13: 978-0-387-24156-2

Printed on acid-free paper.

Printed in the United States of America.

9 8 7 6 5 4 3 2 1   SPIN 11052654

springeronline.com

Dedication

This book is dedicated to my
two lovely boys

Henry and David

Contents

Preface..xvii
Abbreviation.. xxi
Acknowledgement... xxiv

1.   INTRODUCTION...1
   1.1   NEW TIME WITH NEW REQUIREMENT...1
   1.2   THE ASR APPLICATIONS...2
      1.2.1   The History...3
      1.2.2   Analysis on the Consumer Applications.........................5
      1.2.3   Using ASR to Drive Ship –One Application Example...................7
   1.3   INTERFACE DESIGN 9
      1.3.1   Neuroscience and Cognitive Psychology.........................9
      1.3.2   Speech Interaction Studies...10
      1.3.3   Multidisciplinary Aspects...11
      1.3.4   Task Orientated or Domain Orientated System...................12
   1.4   USABILITY AND USER-CENTERED DESIGN.................................13
      1.4.1   Usability Issue...13
      1.4.2   Understand the User...14
      1.4.3   About Design Guidelines...15
   1.5   RESEARCH PHILOSOPHY..17

  1.5   RESEARCH PHILOSOPHY 17
         I Ching Philosophy
         Research and Design Methodology
         Research Validity
  1.6   CONCLUSION 25

2.   **BASIC NEUROPSYCHOLOGY**... 27

  2.1   INTRODUCTION 27
  2.2   GENERAL ASPECTS OF NEUROSCIENCE... 28
         Basic Neuron Structure and Signal Transfer
         The Principles of Brain Functioning
         Methodology of Neuropsychological Studies
  2.3   PERCEPTION 33
         The Sensory System
         Sensory Modality
         Cortical Connection and Function
         Perception and Recognition Theories
  2.4   LANGUAGE.. 39
         Neurology of Language
         Speech Perception
         Word Recognition
         Language Comprehension
  2.5   LEARNING AND MEMORY 45
         Learning Theory
         Memory
         The Mechanism of Memory
         Working Memory
         Forgetting
  2.6   CONCLUSION 50

3.   **ATTENTION, WORKLOAD AND STRESS**... 53

  3.1   INTRODUCTION.. 53
  3.2   ATTENTION 54
         Neuroscience Perspective
         Focusing Attention
         Dividing Attention
  3.3   MULTIPLE-TASK PERFORMANCE 58
         The Resource Concept
         The Efficiency of Multiple Task Performance
         Individual Differences
  3.4   STRESS AND WORKLOAD 61
         Stress
         Workload

3.5   THE RELATIONSHIP BETWEEN STRESS AND WORKLOAD 63
         Stressors Classification
         Stress: Its Cognitive Effects
         Stress: Its Physiological Effects
         Stress: Its Effects on Speech
         Fatigue
3.6   THE MEASUREMENT OF STRESS AND WORKLOAD 68
         The Measurement of Stress
         Workload Assessment
         Performance Measurement
         Subjective Rating Measures
3.7   PSYCHOPHYSIOLOGICAL MEASURES 72
         Psychological Function Test
         Physiological Function Test
3.8   ENVIRONMENTAL STRESS 78
         Acceleration
         Vibration
         Noise
         Auditory Distraction
3.9   WORKLOAD AND THE PRODUCTION OF SPEECH 83
3.10   ANALYSIS OF SPEECH UNDER STRESS 85
         Emotion and Cognition
         Speech Measures Indicating Workload
         Acoustic Analysis of Speech
         Improving ASR Performance
3.11   RESEARCH PROBLEMS 91
3.12   CONCLUSION 94

4.   **DESIGN ANALYSIS** 95

4.1   INTRODUCTION 95
4.2   INFORMATION PROCESSING THEORY 97
4.3   THE ECOLOGICAL PERSPECTIVE 98
         Ecological View of the Interface design
4.4   DISTRIBUTED COGNITION 103
4.5   COGNITIVE SYSTEM ENGINEERING 104
4.6   WORK AND TASK ANALYSIS 106
         Task Analysis
         Cognitive Task Analysis
         GOMS—A Cognitive Model
         Cognitive Work Analysis
4.7   INTERACTION DESIGN PROCESS 115
         Human-System Interface
         Design Process

Interaction Design

4.8   SCENARIO-BASED DESIGN 119
4.9   DISCUSSION.. 120
4.10   CONCLUSION... 122

5.   USABILITY DESIGN AND EVALUATION................................ 123

5.1   INTRODUCTION.. 123
5.2   DIFFERENT DESIGN APPROACHES 123
5.3   THE CONCEPT OF USABILITY 125
       Definitions
       Usability Design
       Usability Evaluation
5.4   HUMAN NEEDS AND SATISFACTION 135
       Trust
       Pleasure
5.5   USER-CENTERED DESIGN 141
       What is UCD Process?
       Planning the UCD Process
       Specifying the Context of Use
       User Analysis
       User Partnership
       Usability Requirements Analysis
       UCD Process in Practices
5.6   SOCIAL TECHNICAL ISSUE 152
5.7   ADAPTIVE AND INTUITIVE USER INTERFACE........................... 153
5.8   USAGE-CENTERED DESIGN... 155
       Creative Design
       Design Criteria and Process
       Ecological Approach
5.9   UNIVERSAL ACCESS 160
5.10   ETHNOGRAPHY METHOD FOR CONTEXTUAL INTERFACE DESIGN . 163
5.11   CONCLUSION... 164

6.   HUMAN FACTORS IN SPEECH INTERFACE DESIGN 167

6.1   INTRODUCTION.. 167
6.2   THE UNIQUE CHARACTERISTICS OF HUMAN SPEECH................ 169
6.3   HUMAN SPEECH RECOGNITION SYSTEM................................ 171
       Automatic Speech Recognition System
       Speech Feature Analysis and Pattern Matching
       Speech Synthesis
       National Language Processing
       Language Modeling
6.4   HUMAN FACTORS IN SPEECH TECHNOLOGY............................ 179

6.5   SPEECH INPUT 180
       Human Factors and NLP
       Flexibility of Vocabulary
       Vocabulary Design
       Accent
       Emotion
6.6   ASSESSMENT AND EVALUATION 186
6.7   FEEDBACK DESIGN 188
       Classification of Feedback
       Modality of Feedback
       Textual Feedback or Symbolic Feedback
6.8   SYNTHESIZED SPEECH OUTPUT 192
       Cognitive Factors
       Intelligibility
       Comprehension
       Emotion
       Social Aspects
       Evaluation
6.9   ERROR CORRECTION 199
       Speech Recognition Error
       User Errors
       User Error Correction
6.10   SYNTAX 205
6.11   BACK-UP AND REVERSION 206
6.12   HUMAN VERBAL BEHAVIOR IN SPEECH INPUT SYSTEMS 208
       Expertise and Experience of the User
       The Evolutionary Aspects of Human Speech
6.13   MULTIMODAL INTERACTION SYSTEM 212
       Definitions
       Advantages of Multimodal Interface
       Design Questions
       Selection and Combination of Modalities
       Modality Interaction
       Modality for Error Correction
       Evaluation
6.14   CONCLUSION 224

7.   THE USABILITY OF SPOKEN DIALOGUE SYSTEM DESIGN225

7.1   INTRODUCTION 225
7.2   THE ATTRACTIVE BUSINESS 226
7.3   ERGONOMIC AND SOCIO-TECHNICAL ISSUES 228
       The User Analysis
       The Variance of Human Speech

           Dialogue Strategy
    7.4   SPEECH RECOGNITION ERROR..232
           Error Correction for Dialogue Systems

    7.5   COGNITIVE AND EMOTIONAL ISSUE..234
           Short-Term Memory
           Verbal/Spatial Cognition
           Speech and Persistence
           Emotion, Prosody and Register
    7.6   AFFECTIVE COMMUNICATION 237
    7.7   LIMITATIONS OF SUI..238
           Speech Synthesis
           Interface Design
    7.8   USABILITY EVALUATION..241
           Functionality Evaluation
           Who Will Carry Out Usability Evaluation Work?
           The Usability Design Criteria
           Evaluation Methods
    7.9   CONCLUSION 249

8.   **IN-VEHICLE COMMUNICATION SYSTEM DESIGN 251**

    8.1   INTRODUCTION...251
           Intelligent Transport System
           Design of ITS
    8.2   IN-VEHICLE SPEECH INTERACTION SYSTEMS 256
           Design Spoken Input
           Multimodal Interface
           ETUDE Dialogue Manager
           DARPA Communicator Architecture
           SENECs
           SmartKom Mobile
    8.3   THE COGNITIVE ASPECTS..266
           Driver Distraction
           Driver's Information Processing
           Interface Design
           Human Factors
    8.4   USE OF CELLULAR PHONES AND DRIVING..272
           Accident Study
           Types of Task and Circumstances
           Human Factors Study Results
    8.5   USE OF IN-VEHICLE NAVIGATION..276
           Accident Analysis
           Types of Tasks

Navigation system-Related Risk

8.6　SYSTEM DESIGN AND EVALUATION 281
　　　System Design
　　　System Evaluation/Assessment

8.7　FUTURE WORKS 286
8.8　CONCLUSION 287

9.　　SPEECH TECHNOLOGY IN MILITARY APPLICATION......... 289

9.1　INTRODUCTION.. 289
9.2　THE CATEGORIES IN MILITARY APPLICATIONS 291
　　　Command and Control
　　　Computers and Information Access
　　　Training
　　　Joint Force at Multinational Level
9.3　APPLICATION ANALYSIS 294
9.4　COMPARISON BETWEEN SPEECH INPUT AND MANUAL INPUT 297
　　　The Argumentation between Poock and Damper
　　　Effects of concurrent tasks on Direct Voice Input
　　　Voice Input and Concurrent Tracking Tasks
9.5　AVIATION APPLICATION 301
　　　The Effects from Stress
　　　Compare Pictorial and Speech Display
　　　Eye/Voice Mission Planning Interface (EVMPI) Model
　　　Application in Cockpit Fast Jet
　　　Battle Management System
　　　UAV Control Stations
9.6　ARMY APPLICATION 312
　　　Command and Control on Move (C2OTM)
　　　ASR Application in AFVs
　　　The Soldier's Computer
　　　Applications in Helicopter
9.7　AIR TRAFFIC CONTROL APPLICATION............................... 316
　　　Training of Air Traffic Controllers
　　　Real Time Speech Gisting for ATC Application
9.8　NAVY APPLICATION.. 320
　　　Aircraft Carrier Flight Deck Control
9.9　SPACE APPLICATION 321
9.10　OTHER APPLICATIONS 322
　　　Computer Aid Training
　　　Aviation Weather Information
　　　Interface Design for Military Datasets
9.11　INTEGRATING SPEECH TECHNOLOGY INTO MILITARY SYSTEMS... 324
　　　The Selection of Suitable Function for Speech technology

Recognition Error, Coverage and Speed
Interface Design
Alternative and Parallel Control Interface
Innovative Spoken Dialogue Interface
9.12   CONCLUSION.. 330

References.. **331**
Index... **377**

Preface

There is no question of the value of applying automatic speech recognition technology as one of the interaction tools between humans and different computational systems. There are many books on design standards and guidelines for different practical issues, such as Gibbon's book *Handbook of Standards and Resources for Spoken Language System* (1997) and *Handbook of Multimodal and Spoken Dialogue Systems: Resources, Terminology and Product Evaluation*" (2000), Jurafsky's *Speech and Language Processing* (2000), Bernsen, et al.'s book *Designing Interactive Speech System* (1998), and Balentine and Morgan's book *How to Build a Speech Recognition Application – A style Guide for Telephony Dialogues* (2001), etc. Most of these books focus on the design of the voice itself. They provide certain solutions for specific dialogue design problems. Because humans are notoriously varied, human perception and human performance are complicated and difficult to predict.

It is not possible to separate speech behavior from their cognitive behavior in their daily. What humans want, their needs and the requirements from the interacting systems are changing all the time. Unfortunately, most of the research related to speech interaction system design focus on the voice itself and forgets about the human brain, the rest of the body, the human needs and the environment that will affect the way of their thinking, feeling and behavior. Differing from other books, this book focuses on understanding the user, the user's requirements and behavior limitation. It focuses on the user's perspective in the application context and environment. It considers on social-technical issues and working environment issues.

In this book, I will a give brief introduction of fundamental neuron science, cognitive theories on human mental behavior and cognitive engineering studies and try to integrate the knowledge from different cognitive and human factors disciplines into the speech interaction system design. I will discuss research methodologies based on the research results from human behavior in complicated systems, human-computer interaction and interface design. The usability concept and user-centered design process with its advantages and disadvantages will be discussed systematically.

During the preparation of this book, I strongly realized that there is very little research work in the human factor aspects of speech technology application studies. This book will emphasize the application, providing human behavior and cognitive knowledge and theories for a designer to analyze his tasks before, during and after the design and examples of human factors research methods, tests and evaluation enterprises across a range of systems and environments. Its orientation is toward breadth of coverage rather than in-depth treatment of a few issues or techniques. This book is not intended to provide any design guidelines or recommendations.

This book shall be very helpful for those who are interested in doing research work on speech-related human computer interaction design. It can also be very helpful for those who are going to design a speech interaction system, but get lost from many different guidelines and recommendations or do not know how to select and adjust those guidelines and recommendations into specific design, as this book provides the basic knowledge for the designers to reach any goals in design. For those who would like to make some innovative design with speech technology, but do not know how to start, this book may be helpful. This book can be useful for those who are interested in Human-Computer Interaction (HCI) in general and would like to have advanced knowledge in the area. This book can also be a good reference or course book for postgraduate students, researchers or engineers who have special interests in usability issues of speech-related interaction system design.

The present book will be divided into two parts. In part one, the basic knowledge and theories about human neuroscience and cognitive science, human behavior and different research ideas, theories and methodologies will be presented. It will provide design methodologies from a user-centered point of view to guide the design process to increase the usability of the products. It will cover the tests and evaluation methodologies from a usability point of view in variances of design stages. The human-factors related speech interface design studies in the literature are reviewed systematically in this part.

In part two, the book will focus on the application issue and interface design of speech technology in various domains such as telecommunication, in-cars information systems and military applications. The purpose of this part is to give the guides for the application of cognitive theories and research methodologies to the speech-related interface design.

Abbreviations

ACC.. Adaptive Cruise Control
ACT... Alternative Control Technology
AFVs .... Armored Fight Vehicles
AH .... abstraction hierarchy
AMS .... Avionics Management System
ANN .... artificial neural network
ASR.. automatic speech recognition
ASW.. anti-submarine warfare
ATC.. air traffic control
ATT .... Advanced and Transport Telemetric
AUIS... Adaptive User Interfaces
BP .... blood pressure
C2OTM .... Command and Control on Move
C3I.. command, control, communications and intelligent
CAI.. Computer-Assisted Instruction
CAVS .... Center for Advanced Vehicular Systems
CFF.. critical flicker frequency
CFG .... context-free grammar
CLID... Cluster-Identification-Test
CNI.. communication, navigation and identification
CNS .... central nervous system
CSE.. Cognitive System Engineering
CTA.. cognitive task analysis
CUA.. context of use analysis
CWA.. cognitive work analysis
CWS .... collision-warning system
DM .... dialogue manager
DRT.. diagnostic rhyme test
DVI.. direct voice input
EEG .... electroencephalogram
EID .... ecological interface design

EOG..electrooculogram
ERP..event-related potential
EVMPI...Eye/Voice Mission Planning Interface
FAA..Federal Aviation Administration
fMRI...function MRI
GA...global array
GCS..ground control system
GPS..Global Position System
GIDS...generic intelligent driver support
GOMS..goals, operators, methods selection roles
GUI..graphical user interface
HCI..human-computer interaction
HMM...Hidden Markov Model
HR...heart rate
HRV..heart rate variability
HTA..Hierarchical Task Analysis
HUD..head-up display
IBI...inter-beat interval
ICE..in-car entertainment
IP..information processing
ITS..Intelligent Transport Systems
IVCS...in-vehicle computer system
IVHS...Intelligent Vehicle Highway Systems
IVR..interactive voice response
KBB..knowledge-based behavior
LCD..liquid crystal display
LVCSR...large-vocabulary speaker-independent continuous
MCDS..Multifunction Cathode ray tube Display System
MEG..magnetonencephalogram
MHP..Model Human Processor
MRI..magnetic resonance imaging
MRT..modified rhyme test
NASA-TLX...NASA Task Load Index
NL...Natural Language
NLU..Natural Language Understanding
PET..positron emission tomography
PNS..peripheral nervous system
POG..point-of-gaze
PRF..performance-resources function
PRP..psychological refractory period
RBB..rule-based behavior
RSA..respiratory sinus arrhythmia
RTI..Road Transport Informatics
SACL...Stress Arousal Checklist
SBB..skill-based behavior
S-C-R..stimulus-central processor-response
SIAM...speech interface assessment method
SQUID...superconducting quantum interference device
SRK..skill, rule, knowledge
SSM..soft-system methodology

SUI .... speech user interface
SWAT.. subjective workload assessment techniques
TEO .... teager energy operator
TICS .... Transport Information and Control Systems
TOT .... Time on Task
TTS... text-to-speech
UAVs.. Unmanned Aerial Vehicles
UCD.. user –centered design
UWOT .... User Words on Task
WER .... Word Error Rate

Acknowledgements

I would like to thank Professor Rinzou Ebukuro, Certified Ergonomist, JES, Japan, helped me prepare the material and the texts for Chapter 1. Lay Lin Pow, Zhan Fu and Krishna Prasad from the School of Mechanical and Production Engineering Nanyang Technological University, Singapore, helped me prepare the material and the texts for Chapters 7 and 8.

Chapter 1

INTRODUCTION

1.1 NEW TIME WITH NEW REQUIREMENT

The fast-developing information technology with its quickly growing volume of globally available network-based information systems has made a strong impact on the changes of an individual's life and working environment and the organization and social structure in many different aspects. The highly flexible, customer-oriented and network-based modern life requires the human-system interaction to be suitable for a broad range of users to carry out the knowledge-intensive and creative work under different contexts and various cultural backgrounds. Thus the requirement for the interface design is changing due to the differences of environment, organization, users, tasks and technology development. Nowadays, users are facing an overwhelming complexity and diversity of content, function, and interaction methods. Talking to the machine is long cherished human desire, and it is regarded as one of the most natural human-system interaction tools in the socio-technical systems context as repeatedly told in stories of science fiction.

In the later '90s, speech technology made its quick development. This is due to the extensive use of statistical learning algorithm, and the availability of a number of large collections of speech and text samples. The possibility of the high accuracy rate of automatic speech recognition (ASR) technology and the relative robust, speaker-independent, spontaneous (or continuous) spoken dialogue system, and humanized synthetic speech, together with the dramatically increase of computer speed, storage and bandwidth capacity, offered the new possibility for human beings to communicate with computer information systems through speech.

The development of speech technology creates the new challenge to the interface designer. As Bernsen (2000; 2001) pointed out, the speech field is no longer separate from many other fields of research, when it works for a broad range of application purpose, the human factor skills as well as system integration skills, among other skills are needed. According to his opinion, in the coming few years, the application of speech technology will still be task-orientated. So "the system must be carefully crafted to fit human behavior in the task domain in order to work at all" (2000). This requires deep understanding of humans, human behavior, the task and the system that are going to perform under the context of society, organization and performance environment.

The application of speech technology to different industrial fields has over a 30 years' history. However, there are more cases of failure than success in the past years. Even up to now, users are still hesitating to use speech interaction systems; very few of them can reach the users' demands of the system performance. People often blame the failure of the technology itself, because the automatic speech recognition (ASR) technology has insufficient accuracy. Researchers and engineers are still looking for better solutions to increase the recognition accuracy. Based on the accumulated application experienced from the past years, we should realize that the failure of the recognition accuracy probably is just part of the reason of the unsuccessful application of the technology. Some of the problems have originated mainly from lack of the understanding of the end users, their tasks, their working environments and their requirements. There is little systematic investigation about the reasons of failures on the applications, other than the misrecognition that has been reported in this field. Actually, the ASR system, based on the statistical calculation of voice spectrum matching, can hardly reach the human-like speech recognition performance, as there are so many factors that can affect human speech, therefore making varied voice from any possible voice samples. To be able to have the successful design of a speech-related interaction system, one needs to understand the constraints of the technology, to find appropriate and adequate applications and to make the proper interaction design that fits the user's nature in cognition and performance. These are typically the weak points, and very few experts could explain it to the prospects.

1.2   THE ASR APPLICATIONS

In this section, we take a brief view of automatic speech recognition technology applications and outline the progress in the US and Japan in history. It is found that there are other factors besides the technology itself that contribute to either the success or failure of the application. These factors can be: the problems of finding appropriate speech applications; lack of communications among cognitive researchers, application engineers, management and the end users; and further solid human factor research and development for applications.

1.2.1 The History

Dr. T., B. Martin, RCA Corporation originated the application history of speech input in the United States in 1969 with the installation of a speech recognition system at the Pittsburgh Bulk Mail Center of USPS for destination input to the mailbag-sorting system. In May 1970, Dr. Martin established the well-known Threshold Technology Inc. (TTI) for speech business development. TTI was listed in the stock market NASDAQ in 1972. Then he established EMI Threshold Limited in the United Kingdom in 1973 for development of its European market.

A TTI-installed voice input for sorting destinations and inspection data at United Air Lines, a luggage sorting system at Trans World Air Line in 1973, and after that at least an 11-systems operation were confirmed in the market by a well-known speech researcher (NEC's internal report, internally issued by Mr. Y. KATO 1974, not published). Since then the application field has been enlarged to different areas such as the inspection data input of Pull-Ring Can, automatic assembly line, inspections at delivery reception site, physical distribution centers, automobile final-inspection data input for quality controls at General Motors, cash registers at convenience stores, warehouse storage controls, cockpit simulations, air traffic control, supporting equipment for physically handicapped persons, map drawings, depth soundings and so forth (Martin, 1976). Most of them were for industrial applications. There were over 950 installations in the U.S. market around 1977, according to the sales experts of TTI.

NEC Corporation started up the speech business in Japan in 1972. It developed the well-known DP matching method (Sakoe, 1971). The system allows continuously spoken word input instead of discrete input that was commonplace for industrial field applications at that time. The first experimental DP model for field applications was completed in 1976. A trend of field introduction of the speech technology began with the sorting input in 1977 and 1978 (Abe, 1979; Suzuki, 1980) and then spread out to the other fields, such as meat inspection data input, steel sheet inspections, steel coil field storage controls, crane remote control input, cathode ray tube inspections, automobile final inspections and so forth, just as the application development in the US. There is an interesting survey result on DP-100 installations and inquiry from prospects for introduction in 1982 as shown in

Figure 1-1 (Ebukuro, 1984 a; 1984b). Many other manufacturers entered into this market one after another in that age but many of them disappeared soon after, due to the fact that the business scale was too small and the required investment for the research and development was too much. There was very little systematic investigation about why manufacturers gave up the already installed voice input system. This probably was regarded as not only the factory's secret, but also the lack of research and developments on the application basics.

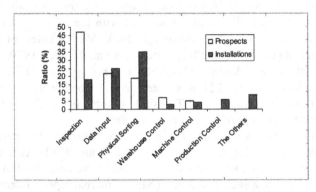

Figure 1-1. DP-100 Speech input applications in 1982

Speech recognition for PC input was introduced to the Japanese market by the Japanese-affiliated company of a well-known U.S. manufacturer in around 1997 and was immediately followed by some Japanese manufacturers. Many of the main suppliers have withdrawn their products from this consumer market in Japan. The interesting phenomenon is that the sales talks in trite expressions this time were almost the same as the early '80s. Examples of sales talks from for vendors are summarized in Table 1-1.

Table 1-1. Group of examples of sales talks in trite expressions

- Japanese word recognition has improved input accuracy.
- You can use it by only 1-min. adjustments.
- Simple/Impressive.
- Voice input makes PC input easier.
- New century by voice.
- You can do chat with PC in a dialog sense.

1.2.2 Analysis of the Consumer Applications

The report of the analysis of American market status was published by the ZD Net News December 22, 2001[1]. The main points in this report can be summarized as:

a) The real reason of why industrial and market statuses did not fit with the expectation of the application possibilities was not clear.
b) The real market scale was not clear.
c) Speech experts can rarely explain real state-of-speech application technology; e.g., product adaptability and adjustability to the applications and suitably applicable areas and site to the prospects.
d) Hand-free speech application developments for consumer applications may still require some more time.
e) Popularization of speech input technology may take time.
f) PC users (consumers) may abandon the speech products.
g) Only call-centers and supply chains look to gain a profit.

The situation of the Japanese market was not the exception as well. Why had those brilliant installations like the sorting centers eventually disappeared? What would happen to the other potential applications such as to consumer products and system solutions application? Although there are numerous reasons for disappearing and thriving, one of the main reasons from the view of the application basics is that we ignore the human factors study related to human speech in these working settings. The managements were interested in decreasing the workloads and improving the working environments, while the designers had too little knowledge about the human speech performance under the task behavior nature since many of them have disregarded the importance of human factors on the human behavior in the system operations.

The speech input is simple and impressive; voice input makes input tasks easier since the speech is human nature. These are true but not always and not everywhere. A comparison between the disappearing and thriving of ASR Systems in industrial applications is given in Table 1-2. It is summarized from the ergonomic point of view and shows that the advantages and disadvantages of the application depend on the characteristics of the tasks and the users. This historical fact showed us again the truth that an advanced technology can be hazardous if we do not apply it properly.

[1] http://www.zdnet.co.jp/news/0010/20voice

Researchers and development engineers may tell lofty ideas on speech input prospective aspects with their deep philosophical and practical experiences standing firm on the rolling deck. These may be based on a lot of premises that cannot tell at once and are sometimes implicitly hidden in explanations. We, vendors and the users, including the managements and the end users, must know and understand this nature of engineers. The application study may fill up this type of communication gap. The lack of communications and understanding among designers, vendors and users could make the situation worse.

Table 1-2. Differences between disappearing and thriving

Difference	Disappearing (Luggage Sorting Input)	Thriving (Meat Inspection)
Operators Vocabulary Vocabulary	Unskilled workers Some tens of words for sorting destinations	Qualified personnel Not so many words for inspection results
Inputs	Destinations that they are obliged to read	Familiar input words that are spoken under judgment
Work load and working environment	Requirement of higher input speed Facing the objects in finding destinations that requires effort to find and read out the destinations Fixed position Noises from higher-speed running machine	The speed follows on the auctioning speed Automatically fed hanged meat that requires little effort to facing and finding. Can walk around Noises from slower running Machine

How should we improve the communications among different partners in the business chain such as researchers, technology engineers, speech analysts, vendors, industrial managements and the end users? The principle that a better understanding of the user, the task characteristics, user requirements, the working environment and the organization may bring all the partners in the business chain together to communicate and understand each other.

A well-known successful case of this kind of sophisticated technology in Japan is the OCR (Optical Character Reader) applications for postal code readings. The engineers and business developers have worked closely together with the postal administration, which followed the introduction of the principle of management on the errors and took the socio-technical systems approach. They have had complete understanding of the different aspects related to the application of the ASR technology and have developed a smart method of system introductions to the user. The printing bureau of finance administration Japan has had applied this principle and had the successful installations of a sophisticated printing quality assurances system.

1.2.3 Using ASR to Drive Ship, One Application Example

The Ship Building Division developed the speech input sailing system called Advanced Navigation Support System (ANSS) with Voice Control and Guidance by Mitsubishi Heavy Industries Ltd. in September 1997 in applying NEC Semi-syllable speech input system DS-2000 and a voice response system (Fukuto, 1998). The system allows a diagonal marine ship operation applying a speech response system under the assistance of the Navigation Support System. Since then, nearly ten marine ships have been equipped with this system for their one-man bridge operations.

The first system was successfully installed on the coastal service LPG carrier called the SIN-PUROPAN (propane) MARU (Gross Tonnage, 749t, Deadweight, 849t) owned by Kyowa Marin Transportation. The first technical meeting on the successful completion of the trial operation of the SIN-PROPAN MARU was held in Tokyo on July 1, 1998. It was reported that the error rate of voice input was first 8% to 9% or more, and finally it was reduced to 2% (by applying noise-canceling method and fine tunings). The training period for three navigators who had not been preliminary acquainted with the ship operation was only three days at the primary stage of ship operation, including speech input trainings. Two captains, who were firstly very skeptical about the speech navigations, made the following comments about speech navigations:

a) The voice operation was very easy even in the congested waters and narrower route. It was enjoyable, just like playing games.
b) We can have extra time to do the things other than the formal operations, e.g., give an earlier operational direction to the controller than the past navigations.
c) We have had the feeling of solidarity with the ship controller.
d) The navigator can concentrate his effort to watching the route for sailings, acquisition of information for navigation, radar range switching, and so forth, as the eyes and hands were free from machine operation.

It is very interesting to look at the implementation process of this successful achievement. It is again telling us how important it is to maintain the ergonomic and socio-technical systems concept in mind, such as the deep understanding of the operators, behavioral psychology, on their tasks requirement and good communication among different partners in the chain.

The statistics of collision on Marine ship sailing (Report on Marine Accident 2002, by Japan marine accident Inquiry Agency) showed that collision from insufficient watching was 56.8% in 2002 in Japan. The remainders are illegal sailing (18.8%), ignorance of signal (9.5%) and the others (14.9%). It shows the significant influence of safety sailing by watching at the bridge. Continuously observing traffic situations and various displays working with the navigation systems at the bridge are heavy mental and physical burdens for a captain. Speech technology introduced in the communication by the ANSS in operations has effectively reduced the burdens by keeping eyes and hands free. The further feelings of intimations to the machine system are confirmed. The expected effectiveness for the protection of collisions (Fukuto, 1998) is also confirmed.

Safety issues caused by the loss of situation awareness and the potential collision due to the performance constraints and the mental workload of operators can be very sensitive. Unlike the parcel sorting errors, a trivial error in operation may sometimes lead to significant marine disasters if the captain cannot take countermeasures to the critical or dangerous situations when an error has occurred.

In the development process, the designer and the ship owner were very aware of the safety issues and ergonomic matters, and they explored every avenue in studying all the possible items relating to safety and human behaviors in the operational environment and had technical meetings between the engineer and ship owner (Shimono, 2000). The designers followed the ergonomics principles and user-centered design process. The project started with a solid task analysis to identify what were the functions that speech navigation could be used. They performed the user analysis to identify the operators' behavior, their needs and design requirements and even the possible vocabulary. The end users were involved in the design process and setting up the design criteria in detail, such as a) the navigator must answer back surely to the spoken command for navigation, b) the system must issue the alert messages such as collision, strike or running on a rock or ashore, deviate or diverge the course and so forth by artificial voice, and c) the system must have a function as the speech memorandum registrations and automatically arranging announcement and so forth. By these considerations, they introduced an adaptable and adjustable speech input system; they limited the vocabulary number, using normal and stereotyped words for navigation, control and management to make the speech input performances sure; they ensured reliable resources of speech technology for maintaining long-term system performance to increase the usability of the system. Introducing the ANSS into ship operation makes team working among the captain, the mate and the engineer change, especially in safety considerations. The socio-technical systems considerations have been incorporated into the sailing systems design and management. Assessments were conducted from various fields, such as ergonomics and safety, technology, reliability, resources and regulations, and various matters regarding environmental issues.

The success of the integration of speech technology into the navigation system in the Marine ship is mainly due to the factor that the designers put the ergonomics and safety considerations as the first priority and applied the socio-technical systems principle in designing the speech navigation system.

1.3   INTERFACE DESIGN

John Flach once pointed out that "an important foundation for the design of interfaces is a basic theory of human cognition. The assumptions (explicit or implicit) that are made about the basic properties of the human cognitive system play a critical role in shaping designers' intuitions about the most effective ways to communicate with humans and about the most effective ways to support human problem solving and decision making" (1998). This is one of the significant points in the ergonomics, human factors and the socio-technical systems approaches. As outlined in the recent progress of speech applications technology, if the interface design does not accommodate human cognitive demands, fit into the user's application environment, or fulfill social and organizational demands, the product itself will not easily be accepted by the user.

1.3.1   Neuroscience and Cognitive Psychology

A human being is a biological entity. The cognition that happens in the brain has its biological ground. Cognitive psychology tries to discover and predict human cognitive behavior under different environmental conditions while neuroscience tries to explain this behavior from the neural structure and biological function of the brain. In cognitive science, people talk about perception, long term memory and short term memory; about speech and language comprehension; about workload and cognitive limitation; about the process from perceptions to action while neuroscience tries to find evidence on what happens on different perceptions and where such perceptions happen in the brain; how different perceptions connect to each other in the brain; what memory is; and where we have the distinction of short-term memory and long-term memory in the brain; etc. It is very common that people apply different theories to explain certain human cognitive behavior. The best evidence to prove these theories is to find explanation in neuroscience.

The application of cognitive psychology to the system design is what we call "cognitive engineering" and human factors studies. Human factors engineering, socio-technical systems researchers and cognitive sciences have accumulated knowledge regarding efficient and effective user interface design in general. But there is relatively little study regarding the interface design in interactive speech communication systems. There are very limited publications related to human factors and speech interaction design. These limited results of studies will be summarized in different chapters in this book.

1.3.2   Speech Interaction Studies

Basically, the appropriate mode of voice-based man-machine communication contains three different systems: a *speech recognition system—a speech understanding and dialogue generation system* and a *speech synthesis system.*

The *speech recognition system* is the basic part for any voice-based man-machine interaction system. A speech recognition system identifies the words and sentences that have been spoken. The requirement for the recognition accuracy of such a system is different for the different interaction system it is applied. For example, for spoken dialogue systems, to permit for natural communication with humans, the speech recognition system should be able to recognize any spoken utterances in any context, independent of speaker and environmental conditions, with excellent accuracy, reliability and speed.

The *speech understanding system and dialogue generation system* infer the meaning of the spoken sentences and initiate adequate actions to perform the tasks from which that the system is designed. The difficulty of understanding language depends very much on the structure and the complexity of the sentences and the context in which it has been spoken.

A *speech synthesis system* generates an acoustic response. The requirement for the acoustic effect of a generated response can be very different depending on the applications as well. For example, for most dialogue systems, it must feel as natural as native speaking to carry dialogue between humans.

The actual state of technology on the speech interfaces may be required to add additional input/output modalities for compensation in order to make the interaction between human and systems seem as natural as possible. This is due to two different reasons: one is because human speech and the automatic speech systems make errors easily; another reason is because human communications are multimodal in nature. As it was demonstrated in the speech sailing system, manual input modality was introduced as the compromise to the performance limitation of speech recognition systems. Voice portal systems, dialogue through the Internet, are also not the exception of using speech input only. Even though these three principal components of a speech interface are working perfectly, the careful and proper design of the interaction is still important.

1.3.3 Multidisciplinary Aspects

Similar to many other human-system interaction designs, the design of a speech interaction system is multidisciplinary in nature. Knowledge from different disciplines is needed for the design. Sometimes multidisciplinary team collaboration is necessary for a usable product design, and this point has often been ignored. The possible knowledge that is needed for such a design is shown in Figure 1-2.

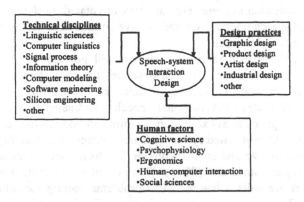

Figure 1-2. The multidisciplinary aspects of speech-system interaction design

This figure may not cover all the possible disciplines that may be involved in the speech technology related interface design. If I consider the technical disciplines and design practices in Figure 1-2 as part of system design technologies, then different disciplines under the human factor category would be another important part of the design of the speech interaction system. The knowledge of applied psychology can contribute to the speech technology system since it tells us how to understand human behavior when interacting with the system. The studies on cognition and human factors of speech technology applications together with the socio-technical systems approaches would guide the development of the speech recognition technology in a proper and adequate direction. It provided the guarantee of the usability of the designed system. Very often, people misunderstand the function of cognitive and human factor knowledge and believe that it takes place only when we need to evaluate the usability of the final system. Many practical cases have proved that the human factor experts should be involved in the design process as early as possible. In the design procedure of the speech sailing system, the case we described in the earlier session, the fact that made it successful was due to the incorporation of multi-disciplinary professions especially the human factor experts who were involved in the early project plan and entire design process.

1.3.4   Task-Orientated or Domain-Orientated Systems

Nowadays, most of the speech interaction systems that are on the market or under development are task-orientated, which means they are designed for performing specific task(s). All the applications in industry for certain controlling functions, in military systems, in public communication systems, in-vehicle communication systems, etc., are task-orientated. Different kinds of task-orientated speech interactive systems have continued to be developed for new and more complex tasks.

In the future, the speech interactive system will go into the domain-oriented natural interactive age. Bernsen describes the characteristics of a domain-orientated natural interactive system as "when the system understands and generates conversational speech coordinated in real time with practically adequate facial expression, gesture and bodily posture in all domains." The domain-orientated systems are user-independent and they can "understand free-initiative and spontaneous or free-form natural interactive communicative input from users" (2001). There is not such a system on the market yet. There are quite a few technical problems waiting for solutions (Bernsen, 2001). There are not any human factor studies that can be found in this area either.

The task-orientated systems are highly restricted or specialized forms of domain-orientated systems. In the near future, the most popularly used system will still be the task-orientated system. In this book, we will focus mainly on the studies of task-orientated speech interaction systems. The demands for the interaction of domain-orientated systems should be very different from the task-orientated system, but human beings and human nature do not change. The knowledge of human cognition and different design principles may still be useful for the future study of new interaction systems.

1.4 USABILITY AND USER-CENTERED DESIGN

The development of speech communication technology has set a new challenge for the human-system interaction design. Designers are looking for new design methods to fulfill the new requirements. As more and more applications come up, the arguments for the usability of the interaction systems are loudly heard. The traditional user-centered design (UCD) process is also getting interest from the designers. At the same time, the UCD or the Human-Centre Design (HCD) would look for new design methodologies for designing the products/software that are suitable for different application context.

1.4.1 Usability Issues

The concept of intuitive user interface, the possibility of speech-related interaction systems and multimodal interface design are getting more and more interest from the designers who are curious to make the products highly usable. Consequently, the usability evaluations of the modern IT products/software are also facing new way of measuring effectiveness, efficiency and satisfaction. Especially, "satisfaction" may not just be the synonym of the "comfortable", but also fulfill the high-level needs and pleasure of the potential users.

Some statistical data from a IT system development study (Smith, 1997) showed that roughly overall 30% of software products are fully successful and that some 70% either fail or only produce some marginal gain to the organization. Besides the technical failure, the failures of utility and usability are the common cause. The review of the speech technology in the industrial application history, as in section 1.2, strongly indicates the importance of the usability consideration. The utility failure means the design system failed to perform the functions for which it was devised. It may not only fail to meet the real, or full, task-related needs of the organization but also fail to meet the individual's requirement. These facts strongly suggest the importance of the concept of UCD in the application basics when implementing adequate system design using speech technologies.

Comparing the speed of the development of the speech technology, there are relatively few cases of the successful applications of the technologies. Throughout history, most of the speech technology companies did not have any profit from their products, and some of them had to close down their business in the area. Bernsen (2001) named three main reasons for the lack of usability of the interface design: a) The far distance between HCI research and system development; b) the lack of usability theory for the respective interface designer; and c) even with the theory, to develop the design support tools for usability tends to be unfeasible due to its time-consuming learning process.

Implementation of the technology into the application system and making the new interaction system useful is the problem. The implementation of new information technology systems in the workplace is always a troublesome process in which the record of success is ingloriously low. Wastell, et al., (1996) pointed out that the critical factors for successful outcomes lie in product factors and process factors. The product factors refer to the technical quality of the artifact such as the degree to which the computer-based system satisfies the user's technical needs and the usability of the products. The process factors refer to the organization, management of the system and the role of the job. The technical system itself is not enough to guarantee successful implementation. Failure to involve end-users in the development process, for instance, may engender alienation, resistance and ultimately rejection. Thus the incorporation of the concept of UCD into the speech system design must be encountered at the stage of system planning.

Wastell's arguments also strongly suggest incorporation of the concept of socio-technical systems approaches into the system design and the needs of application technology developments on the application basics. When considering integration of a new technology into the production system, the optimal integration will take consideration of the user (human) and the machine to perform a task within the constraints imposed by the physical environment. At the same time, the social environment in which the task is to be undertaken, together with the management system and organization, must be considered.

1.4.2   Understanding the User

Where any interface design has to be usable, the UCD principle and a concept of usability evaluation should provide the leading method for any interface design in which the speech interaction system is the exception. In speech interaction system design, bridging between cognitive science and the design engineering is one of the significant issues from the human factor view.

In many cases, the design concept could be considered the designer's implicit theory about aspects of human perception and cognition. The knowledge transfer from cognitive scientists to the interface designer is not so simple (Gillan, 2001). One of the main reasons is that many of the cognitive studies do not come from the engineering and designers' perspective. For designers, it is the idealistic situation to have some handy tools to formulate human cognitions and behavior so they can apply them directly into the design, while human mental behavior far more complicated than any mathematical model can describe. It naturally comes to the situation that the study results of cognitive science are difficult to use directly by the designers. In the absence of this transfer, designers tend to create designs based on their own assumptions about how people perceive and think, rather than based on a set of empirically validated principles.

How to bridge the gap between cognitive science and design engineering? The effort shall come from both sides: the design engineers shall learn to understand human behavior and the cognitive engineers shall try to understand designer's perspective and needs.

The key issue in the user/human-centered interface design is basically the understanding of the user's behaviors on the task implementations through the interface system operations. The user's mental model reflects the user's perspectives on the tasks of the interface activities that are designed for the system operations. To be able to develop the user's mental model, it is important to understand the human cognitive process. Consequently, we must understand that the human being is the biological entity, so the human cognitive process is happening inside the human body and dynamically interacting with the environmental conditions and statuses.

1.4.3   About Design Guidelines

In many user interface design books, authors try to provide their own "guidelines" for the engineers and believe that this should be the easiest and useful way, "out of the shelf," for the engineers to use it for their design. Different user interface design guidelines for speech systems can be found in the literature (Baber, 1996b; 2001; Jones, 1989). These guidelines vary in the extent to which they are derived from their specific research findings. Their scope is rarely made explicit, and it remains for the designer to judge the applicability of a guideline to a particular user interface and to apply it accordingly. The body of the guidelines is also known to be incomplete— many design issues are just not covered by guidelines (Life, 1994). For example, some guidelines for speech interface design are provided by Baber and Noyes (2001) in Figure 1-3. How do such "guidelines" help designers in their designs? Are there many unsolved problems behind the lines such as how to match the technology and the type of the work? What are the criteria of such a "match?" How does one measure the match? The answers to these questions are design-context dependent. Without a deep understanding of human cognitive behavior and user characteristics, together with the deep knowledge of application domain, the designers would not find it possible to answer these questions. If the designer has the necessary knowledge, he would not need such a "guideline!" One of the main problems for such guidelines is that they are always at the qualitative, not the quantitative and practical level.

Guidelines:
- Match the type of work the recognizer intends to perform with the characteristics of the available technology
- Get to know the hardware and the software environment in which the speech-control application is intended to function—the recognition system needs to work in conjunction with other parts of the application
- Because of the problems of end-point recognition, isolated item recognition should be used in safety-critical application
- A key feature of the successful, operational speech-control applications is that they are not safety-critical

---

Figure 1-3. Examples of design guidelines for speech-system interaction design

The worse cases can be that different books may provide different kinds of guidelines with different detail levels. Some of them do not state the application conditions clearly; some of them may even be contradictory. Guidelines and cognitive theories often have a very large "external validity" statement, or try to pretend have a high "concept level," while any design is working in context-based creativity. How does one guide the engineer to find the "right" guideline and right "theory" for his specific design?

Life and Long (1994) has suggested that if the guideline could be presented as "IF (condition) THEN (system performance consequence) BECAUSE (interaction model constraint), HENCE (guideline, expressed as a system design prescription)," then it can be very handy. This is almost an impossible dream, however, because nobody can cover all the possible application conditions where a speech interaction system may be applied. Many of the application conditions are unpredictable in the present situation; the situation and application context may change due to the development of the technology. At the same time, there are almost unlimited consequences and constraints one can identify according to the application context. The matrix of the three entities (IF, THEN, BECAUSE), with an enormous amount of variability in each, can come out with millions of detail guidelines, which may be difficult for the designers to find out the proper ones while there still may be the danger of not covering the situation where the designer is needed.

Through a survey and examination of numerous successful and false cases with the traditional method, the way of empirical advice on the speech system design can be derived from general principles and known facts. Such advices are more for the design process, rather than detail guidelines. The theory of ergonomics may be helpful to carry out this procedure. Some implicit suggestions on the firmness level may be read out from Table 1-2, that composed from ergonomic view, if has ample experiences, either successful or false, on the physical speech application problems together with the metaphysical concepts. Even though the design principles cannot be simply composed from the laboratory studies and particular experimental results, it is essential to disclose each time the result has come in order to share the experiences that originate those particular concepts and guidelines among the researchers in the field of speech application basics.

1.5   RESEARCH PHILOSOPHY

How does one transfer the knowledge from cognitive science to the interface designer? How does one conduct the cognitive and human factor research work to have a better understanding of the human perception-cognition-action in the performance under the complex environment? How does one let the design engineers, who used to think the way "IF—THEN—BECAUSE—HENCE" to understand the complicated human mental behavior? As the interface design is a multidisciplinary issue, how can we set up the communication between different disciplinary partners? Certainly, we need some useful research philosophy and methodologies to guide us.

1.5.1   I Ching Philosophy

Maybe we can borrow some ideas from the world's oldest philosophy—"I Ching"—that probably has more than 5,000 years' history. I Ching uses binary code to predict and explain the changes, including the immutability of the phenomena of natural, political, social, personal and mental process, and even the physiological changes inside the body. Confucius, the most famous educator and philosopher in Chinese history, had spent many years studying I Ching and developed many theories from it more than 2,500 years ago. Shuen Zhi, the famous militarist who lived 2,500 years ago, developed his war strategies theory also from I Ching philosophy. Chinese medicine, even Chinese Gongfu and the martial arts, also have their roots in this philosophy. It is regarded as the philosophy of philosophy, the philosophy of the universal.

I Ching used the binary modal as its coding system, the note "--" symbolizes "Yin (negative)" and the note "—" symbolizes "yang (positive)." Taking three notes arbitrarily out of these two fundamental codes, as one symbol in which three different combinations of "—" and "--" are formed, built up eight trigrams. One trigram balanced against the other to create the sixty-four hexagrams, each of which describes a unique dynamic process of the phenomena. Nowadays, most of the computing systems are still using the binary modal. Astoundingly, the sixty-four DNA swatches were also found and can even be cross-coded in a binary way with the genetic code. It is probably not just a simple coincidence that the number of I Ching hexagrams and the number of DNA swatches are the same. Probably, here is an underlying root that brings mind and body together. To answer this question is outside of the interests of this book.

Can we learn something from I Ching? Can we, cognitive science and engineers, use its philosophy to guide our research and design work? The answer is very sure, but the question is how. Let's start the discussion by understanding the basic philosophy of I Ching.

There are three basic principles that I Ching tried to explain and to explore: The first one is *Change*. Everything in this world, in the universe, including a human's mind, and behavior and feeling is constantly in change, and it is not possible to find the absolute static moment. The second is the *Regulation*. Any changes follow certain regulation. The regulations behind the complicated changes may be very simple. The third is *Constant*. Behind all the changes, there is constancy. It is this "constancy" that keeps the world changing. There are three essential elements one can use to describe any changes: reason, phenomena and quantity. Anything that happens in this world has its reason, has certain rules to follow and shows certain phenomena. The phenomena can be described not only qualitatively, but also quantitatively.

I Ching philosophy demonstrates to us that even if there is only a single change that happens in the system (such as one of the hexagrams changes from yin to yang), it might affect the entire function of the system. How much the effects can be depended on where and when the change takes place and how the situation of the patterns is before the changes. So in other words, the effects of the changes and the results of the changes are always context dependent. Even though the change is context dependent, there are some simple regulations to follow, so it is possible to predict the results of the changes.

In the highly abstract level of the I Ching coding system, any change is just the matter of changing the quantity of negative Yin and positive Yang inside the study objects, or the entity, or the movement, etc. If we say Yang is the positive side of the object, then Yin is the negative side of the object, as has been described. Once the old balance of the Yin and Yang changes, a new situation happen, and the consequence of the changes will happen. Even after certain series of the changes, the Yin and Yang will turn back to the similar balance again. The situation is already different, compared to what it was before. Actually, there never has been a stable balance between Yin and Yang relation in anything in this world. The unbalance of the opposite force makes the initial energy change. The regulation of the changes can be studied from at least nine different directions. To make this idea more clear, I would like to use the original I Ching hexagram to explain the different directions shown in Figure 1-4.

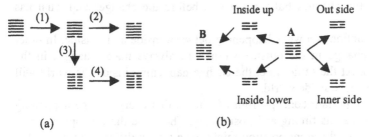

(a)   (b)

Figure 1-4. Study of changes in I Ching

In I Ching, every hexagram has its meaning. As shown in Figure 1-4(a), when one of the six lines changed from Yang to Yin, the function changed and the phenomena changes (as seen in (1) in (a)). If we turned the changed one upside down (as (2) in (a)), it becomes another hexagram, and the meaning of the hexagram is very different. If all the six lines of Yin and Yang are exchanged to the opposite one (as (3) in (a)), then it becomes another hexagram. This new hexagram can also be seen upside down (as (4) in a). Here, these four changes and the new hexagram from these four changes describe four different regulations of the changes. Now we look at Figure 1-4(b): every hexagram can have four different trigrams and every trigram provides the essential meaning for the hexagram. The top one is called the outside trigram and the bottom one the inner-side trigram. These symbolize the changes that happen inside the entity (for example, the human body or mind) and outside the entity (for example, the environment or organization). As the changes from inside are very important, if we take off the top line and bottom line, then the lower three lines and top three lines can also be seen as two different trigrams. If we put these two trigrams together, it becomes another hexagram (as B in (b)).

What do these changes tell us? According to I Ching philosophy, Yin and Yang are the two sides of anything in this world, as big as the universe, or as small as the molecular. They are opposites of each other, but they are always together. They contain each other. Any changes that happen in this world are just a matter of changes between the Yin and Yang. In the sixty four hexagrams of I Ching, the orders of the Yin and Yang lines are different. They symbolize different change patterns of the world. So every hexagram itself can explain certain phenomena and change patterns. But it is not enough to understand all the aspects within the changes. So the interpretation of the nine changing directions is also equally important. The changes demonstrated in Figure 1-4 tell us that when any changes happen, we not only look at it from its surface, from one direction, but we should study it from several directions. We should pay attention to the changes not only happening from outside, but also from inside. We should not only look at the outcome of the changes but what it was before the changes and what has changed.

The interaction is always happening between inside and outside. In other words, the change in the outside world will always make changes in the inside world. At the same time, the change happening inside the world will also affect to the outside world.

Another important concept from I Ching is that every change happening in this world has its timing and positioning. The same thing happening at a different time or different position might have a very different effect on the entire entity. The changes of the world not do follow linear regulation, but are analogical without end-stop. That is why the layout of the total sixty-four hexagrams in I Ching is in a circle.

The above discussion can be summarized as:

a) Everything changes constantly, but follows certain regulations.
b) The change has its reasons, its phenomena and quantity.
c) The changes follow certain regulations, which can be simple.
d) The changes can be studied from many different directions, from inside and outside.
e) Anything that happens has two sides, the positive side and the negative side.
f) Timing and positioning are very important for studying the changes.
g) The changes normally do not follow the linear model, but analogical without end-stop.

Human behavior and cognition can be considered as the most complicated, constantly changing phenomena. Cognitive science is trying to explain the reasons and the principle/regulations behind the changes. There are many different theories that have been developed or are on the way to being developed. It is hard to say which one is better than the other, or which one is right and which one is wrong. Most likely, different theories try to explain different phenomena in the changing process, or from different directions to explain the same changes, or explain the changes on different timings.

We cannot explain all the phenomena happening in different situations. It is not because the change is irrational, but we do not know the reason and the regulation yet! The cause of changes can be internal and external, or the interaction of both. In most of the cases, the theory or the outcome of the research results has its valid condition, most probably that a few cognitive theories can apply together, even sometimes looking contradictory to each other. For example, for children, it is easier for them to understand speech than reading the text of a similar story. One cannot say in general that speech is easier to understand than reading. Comparing with a machine reading a long document to them, most adults will prefer to read the text of a document, because the perception of speech requires higher demands on working memory when compared with reading the written text. Here the situation for children and for adults becomes very different. This is probably the most difficult part for a technical engineer—to understand human cognitive behavior and apply the knowledge into his design. For example, if he read from somewhere that "people find it easier to understand reading text than to understand speech," and he uses this piece of knowledge in his design without understanding why and in what situation, and he may be totally wrong in the application.

For the designers and researchers, it is important to have the right attitude and understanding toward different studying results, theories and different design guidelines and recommendations. It is critical to choose the right cognitive knowledge, theories and even the design guidelines. Even if the proper ones were found, it is a potential danger to apply it directly the design process without certain modifications based on the design context and the understanding of the changes from the inner operator's factors and environmental factors.

1.5.2 Research and Design Methodology

By examining the philosophy of I Ching, we can see that it is it is not possible to have very handy "guidelines" that the designer would like to use directly without any adjustment, as the condition when the guidelines were developed may not be the same as the designer's environment. The question is how to make such adjustment. The designers need to learn the principles and the regulation of the changes and understand its phenomena under their design context. For every interface design, it is the designer's "experimental results" based on the knowledge he has about human cognition, human behavior and the study methodology he developed based on previous experiences and the material and the equipment that are available for his "experiment," including the time available and the budget available.

The world is immensely complex, but is highly patterned, because the change from one pattern to another pattern should follow some simple regulation. Human cognitive systems must cope with this world by finding these patterns and regulations. Cognitive processes intend to find patterns of the interaction between humans and the world. How shall we study the reason and the regulation of the changes? The regulation of the changes should be simple and universal. Actually, the entire research work should be to find out the regulation of the changes and understand how to apply these regulations to explain and predict the changes under the real context.

The knowledge we have about human cognition and behaviors mostly come from experimental studies. It is from the long historical and research tradition that we tried to isolate the causes of changes into one, two or three factors and tried to manipulate the causing factors to change in a couple of levels and then to measure the effects on different causing levels. At the same time, we tried to ignore the other factors that may produce the influence or give co-effects to the results. The other ignored factors are considered "randomized factors," and it was assumed that they have equal effects on the final results. This principle has dominated the research work of different areas.

The advantage of such research methods is to have a deep understanding of the functions and the effects from those controlled manipulated factors. But we should understand that in almost all situations, the key factors did not affect the results independently; they always work together with other uncontrolled factors (some of them we know and some of them we don't know). Since we do not know the real interactions between different factors and how it affects the results, most of the studies turned out to have very little external validity. Unfortunately, most of our knowledge of human behavior and cognition gained from such studies. For the designer, he cannot ignore those "random" factors because the interface will always be applied in a very different timing and positioning compared with the laboratory settings. That is why the knowledge transfer becomes so difficult.

Why is it hard to apply the research results the design practices? We forget about the changes that are happening constantly, and on the context base, or that are the consequences caused by the changes difficult to predict. Change seldom is caused by a single factor but by the interaction of multiple factors. When the cause affects the changes happening in different timing and positioning, which means the environmental factors and "uncontrolled random factors" are different, the results most likely be different. Only studying the phenomena that at each moment of change is far from enough; we shall study the reasons of causing the changes. Only studying the main factors that are believed to be the main factors for the changes is also far from enough. We shall, at the same time, understand that the changes never happen just because of one or two key factors; even the other factors considered as "random" factors may also affect the changes. I Ching told us that the changes are always context dependent. At different timings, different environmental conditions and internal conditions, the same causing factor may have different outcomes. But the regulation of changes can be the same. So if we only study the key causing factors and the phenomena of the changes, we are always far from the truth. That is why there are so many theories that try to explain the same phenomena, because even with the same causing factors, just the timing different the position different, and outer or internal conditions different, the phenomena of the changes are different.

In most of the research work, we often focus on the positive changes and forget about the negative effects. Actually, the positive and negative effects are always hand-in-hand, just like the Yin and Yang in I Ching. Any new technology that is introduced to the society, any new solution that is discovered, may improve the performance or solve certain problems, but at the same time, they may also introduce new problems. Nothing can be perfect!

Most of the human cognitive and behavior theories have tried to explain the phenomena and their connection with the causing factors, but not to understand the reasons and regulations of the changes. People are most likely from one direction to study the phenomena of the changes and believe that they have found the universal regulations. When the others studied the phenomena from another direction under different conditions and developed a new theory to explain the phenomena, they tried to criticize the earlier ones or argued who was right and who was wrong. Actually, all of them could be correct; it is just a matter of from which direction people study the phenomena and under what kind of condition (the timing and the positioning).

Since the constraints come from the research methodologies, the theories developed by these methodologies have their limitation of application. When it comes to different situations, it needs to be adjusted before the application, and the adjustment can be developed into a lot of other theories and those theories will further need to be adjusted before the application in the design. The most difficult problems for design engineers are that it is hard for them to judge in what situation, at what time and which position, which theories can be applied and what kind of adjustment they will make and how to make the adjustment.

1.5.3 Research Validity

There is one question every study has to answer (it is often difficult to get the right answer): the validity of the results. As we pointed out earlier, almost any empirical research work is context dependent. How can we know that our research results are applicable, in what condition, under what limitation? The difference between the context specific results and the concept-dependent regulation is a matter of different levels of abstraction, as Figure 1-5 indicates.

To evaluate the research work, the validity of the work is always considered. The validity refers to the approximate truth of an inference (Shadish, 2002). If the research work is valid, it means that the judgment is made about the extent to which relevant evidence supports that inference as being true or correct. As Shadish, et al., (2002) indicates that "validity is a property of inferences. It is not a property of design or methods, for the same design may contribute to more or less valid inferences under different circumstances." The validity of a study reflects how much we can generate from the particular study.

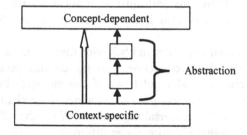

Figure 1-5. The level of analysis

Normally, there are three distinguishing validities in any study: First, the *Internal validity* is defined as the validity of inferences about whether observed covariation between A (the presumed treatment) and B (the presumed outcome) reflects a causal relationship from A to B as those variables were manipulated or measured (Cook, 1979). Internal validity indicates if the research work is correct or not. Sometimes, people can find a certain correlation between A and B by observation, or statistical analysis, but it does not mean A is the cause of B. Based on earlier studies, or others work, or different theories and knowledge in the area, the researchers should be able to explain why A is the cause of B. Similar to internal validity, people may specify it in certain details such as concurrent validation involves observing the relationship between the test and other criteria that are measured at the same time. More often than not this involves a correlation between the test in question and one or more other measures for which a hypothesized relationship is posited (Hammond, 2000). The second, *construct validity* is defined as the validity of inferences about the higher-order constructs that represent sampling particulars (Shadish, 2002). This is more related to the research design, for example, the outcome from a given tests measurement described as intelligence or as achievement. In other words, do we really measure the thing we are supposed to measure? Sometimes, people call it "content validation" as well (Hammond, 2000). Content validation simply asks the question: Is the content of the test relevant to the characteristics being measured?

The third, *external validity,* is defined as the validity of inferences about whether the cause-effect relationship holds over variations in persons, settings, treatment variables and measurement variables (Shadish, 2002). In other words, external validity indicates in what degree we can generalize the result from the study. If we look at Figure 1-5, internal validity and construct validity are related to the research work itself, and if we use the right research method, we select the important and reasonable factors to study and carry out the right measurement in the study. These are still in the context-specific level. The external validity or the generalization of the study indicates how high we can reach the concept-dependent level by the study.

Sometimes in the literature, one can find different names of validities that are similar to the external validity in different application aspects. For example, *Ecological validity,* in which the lack of ecological validity means findings generated in laboratory conditions where behavior must be tightly controlled may not tell us anything interesting about life outside the laboratory (Davis, 2000). *Predictive validity* asks the question: Does the test predict later behavior? It is a very important feature of a psychometric test since the choice of a psychometric test in research is commonly informed by assumptions of its predictive quality (Hammond, 2000).

In any research work, it is essential to hold its internal and constructed validity. The external validity is what the research work is looking for; the higher the external validity, the better. In research reports, people often discuss the internal validity of the study, but the most interesting is the discussion about what the research outcome is and where this outcome can be applied. Often we can find that people tried to "extend" their findings too much! Quite a lot of these "design guidelines" came from this "extended" external validity. It is extremely dangerous for a technical engineer to follow a guideline for his design without the deep understanding of why.

1.6   CONCLUSION

Speech technology has developed quickly in the past few years. It is time to discover its possible application areas. Many task-orientated speech interaction systems are now on real-time application, or under development. To design a usable speech interaction system, only making the technology work is far from enough. It requires multiple collaborations among different disciplines.

To increase the usability of a speech interaction system is an urgent requirement. It challenges new design philosophy and new design methodology. Regarding the usability issue, one of the key questions is to understand the user, the user's mental behavior and limitations, the user's needs and user's preference of interacting with the system. The theory of human cognition is the basic foundation for any interface design. Human cognition is the important knowledge for interaction designers who need to have a deep understanding of the user.

Technical engineers normally lack the solid human cognition knowledge and therefore would like to look for design "guidelines" for their interaction design. In most of the cases, these design guidelines were developed from specific research context under certain conditions. The designers need to select the right guidelines and then need to adjust them to specific design. The question is: a) how to select the right one? And b) how to adjust it to the design context? Often, these guidelines do not tell the deep cognitive knowledge behind them and the application limitations.

Chapter 2

BASIC NEUROPSYCHOLOGY

2.1   INTRODUCTION

A human being is a biological entity. Any stimulation from the outside world has to have the effects on the human body before any action can take place. People come from different perspectives to study human mental behavior; therefore, there are neuroscience, human neuropsychology, cognitive psychology and engineering psychology. Neuropsychology is the science of the relationship between brain function and behavior (Kolb, 2000). Cognitive psychology deals with how people perceive, learn, remember and think about information (Sternberg, 1999). Engineering psychology specifies the capacities and limitations of the human cognition for the purpose of guiding a better design (Wickens, 2000). These three research areas are closely connected to each other, but their emphasis and purposes of the study are different. In history, cognitive engineering, or cognitive psychology looked at the brain as a "black box." By observing human behavior when they are given different stimulations/tasks (input) under certain conditions and by observing human subjects' reactions (output) to the stimulation, the researchers try to guess what is going on inside the brain. Neuroscience studies the structure of the brain and tries to allocate different functions in the brain, including the connection of different functions structurally. The purpose is to have better understanding of how a brain works biologically. Up to the '80s we knew too little about how the brain worked biologically, so the knowledge from this area provided very little help for the cognitive engineers for their design purpose.

As far as the technical development, neuroscientists have many useful tools to study the brain function. The research methodologies have been constantly improved in this area. It is probably time again to look at the development in the neuroscience and apply the knowledge to the cognitive engineering studies and to the design. Besides, to have a better understanding of human behavior, especially mental behavior, some basic knowledge from neuroscience and neuropsychology is necessary. In this chapter, I will give some brief summaries of the main theories and findings from neuroscience and psychology regarding the human sensation and perception, human language process, learning and memory.

The anatomy of the human brain is very complicated. It is difficult to describe it in a few sentences. Since the detail of the brain structure and the location of a specific function in the brain is not our interest, if any readers are interested in it, Bryan Klob and Ian Q. Whishaw's book *"An introduction to brain and behavior"* (2001) can be recommended. To prevent from the unnecessary "information overloading," I will try to avoid, as much as I can, using anatomical vocabularies in this chapter.

2.2   GENERAL ASPECTS OF NEUROSCIENCE

The information of neuropsychology came from many disciplines: anatomy, biology, biophysics, ethnology, pharmacology, physiology, physiological psychology and even philosophy. The central focus is to have a better understanding of human behavior based on the function of the human brain.

2.2.1   Basic Neuron Structure and Signal Transfer

The basic structure of the neurons includes the dendrites, cell body and axon. A neuron may have one to twenty dendrites, each of which may have one to many branches. The function of dendrites is to collect information and transfer it to the cell body. The axons from other neurons are connected to the dendrites. The information that appears in the neuron is the action potential on the membranes of the nerve cell. The information from other neurons is collected at dendrites and is processed at the cell body. The cell body integrates the information and then sends the integrated information through axons to the end feet, where it is passed on to a target neuron. Each neuron has a single axon. The information transferred from the axon of one neuron to the dendrite of the other neuron is via chemical synapse where there are synaptic transmitters that synthesize there. When an action potential arrives to the axon terminal, the chemical transmitters are released.

On the membranes of dendrites, there are receptors to receive the chemical transmitters. There are three types of effects that happen on the postsynaptic cell (Kolb, 2001): a) excitatory action; b) inhibitory action and c) initiating other chemical reactions. A schematic diagram of a neuron structure is shown in Figure 2-1. In the figure, "+" symbolizes the excitatory action and "-" symbolizes the inhibitory action.

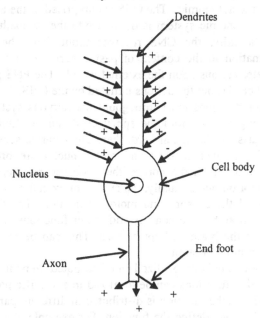

Figure 2-1. Schematic diagram of a neuron structure

There are various types of synapses in the central nervous system. An axon terminal can end on a dendrite, on another axon terminal or on another cell body. It may also end on a blood capillary or end freely in extra-cellular space. Dendrites may also make synaptic connection with each other. These different types of synapses increase the flexibility of behaviors of the nerve system. Even though synapses vary in both structure and location, they all do one of two things: either excite target cells or inhibit them.

2.2.2 The Principles of Brain Functioning

The discussion in Section 2.1 indicated one of the basic principles about how the brain functions: information is processed in a sequence of "in–integration–out." As we know now, most neurons have afferent (incoming) connections with many (tens or even hundreds) other neurons. The information is integrated in the neuron cell body and its efferent (outgoing) connections to many other cells. Basically, the nervous system works through excitation and inhibition.

Anatomically, the nervous system is divided into two main divisions, central (CNS) and peripheral (PNS). The peripheral portion is further divided into somatic and autonomic. The CNS is comprised of the brain and spinal cord, and the autonomic system is divided into the sympathetic and parasympathetic. Basically, the CNS receives input from the senses, processes this information in the context of past experience and inherent proclivities and initiates actions upon the external world. The PNS provides the sensory input and carries out the actions initiated by the CNS.

Sensory and motor divisions exist throughout the nervous system. The peripheral nerves going to and from the spinal cord can be divided into sensory and motor parts. In the spinal cord, there are separate sensory and motor structures. The distinct sensory and motor nuclei are present in different levels of the brain. In the cortex, there are separate regions for sensory input and motor output. Actually, the entire cortex can be viewed as being organized around the sensory and motor distinction. The CNS has multiple levels of function. Similar sensory and motor functions are carried out in various parts of the brain and spinal cord. This can be regarded as another principle of brain functioning.

Functions in the brain are both localized and distributed. In neuroscience, we normally assume that functions can be localized in a specific part of the brain. It is also found that the function is distributed in different parts of the brain. It depends on how we define the function. For example, the function of "language" has many aspects, such as "word" can be spoken, written, signed and even touched. It is not surprise that language function is widely distributed in the cortex.

The brain's circuits are crossed. Each of the brain's halves receives sensory stimulation from the opposite side of the body and controls muscles on the opposite side as well. The brain is both symmetrical and asymmetrical. The left and right hemispheres look like mirror images of each other, but they sometimes handle different tasks. For example, control of mouth movement is only on one side of the brain. Some motor functions are also controlled by one side of the brain to avoid the conflict of having different instructions from different sides of the brain.

Most neuroscientists subscribe to some assumptions: the brain is the source of behavior and there are certain reasons why the physical brain is organized with that of the mind. Regarding the brain functions, neuroscientists assumed that there are many relatively independent modules in the cognitive system. The unit of brain structure and function is the neuron. Each module can function to some extent in isolation from the rest of the processing system. Brain damage will typically impair only some of these modules. Investigation of cognition in brain-damaged patients can tell us much about cognitive processes in normal individuals. Researchers have tried to localize different functions on the brain by studying different functionally impaired patients and by electrophysiological confirmation of the localization. There is a hierarchical organization in brain function. The nervous system has three levels: the spinal cord, the brainstem (which is the central structure of the brain including the hindbrain, midbrain, thalamus and hypothalamus) and the frontal cortex. Each successively higher level that happens in the frontal cortex controls more complex aspects of behavior but does so through the lower levels (such as through the spinal cord).

When it comes to the location of different functions in the brain, there are opposite hypotheses. Neuron hypothesis assumes that the nervous system is composed of discrete, autonomous cells, or units, that can interact but are not necessarily physically connected. Nerve net hypothesis insists that the nervous system is composed of a continuous network of interconnected fibers. These assumptions and hypotheses lead to the research methodology design, results anticipation and theory development in neuroscience. Brain systems are organized both hierarchically and in parallel. It was found that functionally related structures in the brain are not always linked serially. The brain has many serial connections, but many expected connections are not found. This is probably due to certain interconnections that we don't understand clearly.

2.2.3   Methodology of Neuropsychological Studies

One of the traditional methods is to compare the cognitive performance from brain-intact and brain-impaired patients. The information can be used to support or to reject the theories within cognitive psychology. Such theories specify the processes or mechanisms involved in normal cognitive functioning. A typical example is Shallice and Warrington's (1970) investigation on a patient who had suffered damage to a part of the brain specialized for speech perception and production. This patient appeared to have severely impaired short-term memory but essentially intact long-term memory. This investigation supports partly the cognitive theory about the distinction between long-term memory and short-term memory.

In the last few decades, the electronic, digital and computing technologies have achieved fast development, and such development affects the neuroscience study. New measurement technologies appeared and provided the new opportunities and new methods to study the brain function. One of the most interesting factors is that we can study the human mental behavior without opening the brain, or waiting for special damage on the brain to appear. We can carry out neuroscience study on normal human beings. Some commonly used brain function measurement technologies will be briefly introduced here. For detail of their application, one should read some neuropsychology books.

An electroencephalogram (EEG) measures the summarized graded potentials from many thousands of neurons. EEGs reveal some remarkable features of the brain's electrical activity. By interpretation of the EEG records, one can find out that the brain's electrical activity is never silent; it has a large number of patterns, some of which are extremely rhythmical. An EEG's changes are coordinated to the behavior changes.

Among different patterns in an EEG, there is a record called event-related potential (ERP), which has been especially interesting to behavior science. It is a change in the slow-wave activity of the brain in response to sensory stimulus. One problem with ERPs is that they are mixed in with so many other electrical signals in the brain that they are difficult to spot just by visually inspecting an EEG record. One way to detect ERPs is to produce the stimulus repeatedly and average the recorded responses. Averaging tends to cancel out any irregular and unrelated electrical activity, leaving in the EEG records only the potentials that the stimulus generated. This characteristic has limited the ERPs application.

The electrical currents of neurons generate tiny magnetic fields. A special recording device known as a SQUID (superconducting quantum interference device) can record this magnetic field—magnetonencephalogram (MEG). Similar to ERP, the MEG procedure requires that many measurements be taken and averaged.

Positron emission tomography (PET) is a technique to measure the changes in the uptake of compounds such as oxygen or glucose due to the brain metabolism. Magnetic resonance imaging (MRI) is an imaging procedure in which a computer draws a map from the measured changes in the magnetic resonance of atoms in the brain. MRI allows the production of a structural map of the brain without actually opening the skull. For example, when a region of the brain is active, the amount of blood flow to it and oxygen consumption increases. A change in the oxygen content of the blood alters the blood's magnetic properties. This alteration, in turn, affects the MRI signal. Because oxygen consumption varies with behavior, it is possible to map and measure changes that are produced by behavior. The measurement of the magnetic properties of the brain for the distribution of elements, such as oxygen, due to certain functions is called function MRI (fMRI).

The major advantages over PET is that fMRI allows the anatomical structure of each subject's brain to be identified, and brain activity can then be related to a localized anatomical region on the brain image. It has better spatial resolution, but it is very expensive to perform such measurement.

Table 2-1 gives the summary of these electron or magnetic graphics methods. The advantage of these methods is that they provide the opportunities to study the brain functions and to locate the respective function center in the brain without damage to the brain. The disadvantage of these technologies is that the measurement equipment is huge and expensive, and the measurement has to take place in an advanced laboratory. It requires special training to be able to perform, measure and analyze the records.

Table 2-1. The summary of different electron or magnetic graphics methods

Name	Measurement mechanism
Electroencephalogram (EEG)	Measures the summarized graded potentials from many thousands of neurons.
Event-related potential (ERP)	A change in the slow-wave activity of the brain in response to sensory stimulus. It is part of EEG.
Magnetonencephalogram (MEG)	Measures the tiny magnetic fields generated by the electrical currents of neurons.
Positron emission tomography (PET)	Measure the changes in the uptake of compounds, such as oxygen or glucose, due to the brain metabolism.
Magnetic resonance imaging (MRI)	An imaging procedure in which a computer draws a map from the measured changes in the magnetic resonance of atoms in the brain.
function MRI (fMRI)	The measurement of the magnetic properties of the brain for the distribution of elements, such as oxygen, due to certain functions.

Just like many other scientific researches, people employ different research methodologies in neuroscience studies. The methodology that people use, either by studying brain-damaged patients, or by applying different electron or magnetic graphic methods on normal subjects, can influence the research results. Sometimes, by the development of the technologies and applying different research methodologies, some well-established theories may be questioned as well. For example, when studying the onset dynamic of certain activities and allocating it in the brain, the measurement speech of the technology may affect the results (Gage, 1998).

2.3 PERCEPTION

The perception started from receptors, or sensors, that are located on different parts of organs and skin in the human body. They convert the sensory energy (for example, the sound waves or light waves) into neural activity and conduct it through neural relays to the cortex, and the information will be coded in different ways there. The perception is a process of receiving the outside world information by the receptors and transferring the information to the brain. The brain recognizes the meaning of the information.

Different receptors perceive different kinds of energy, for example, visual receptors perceive light waves and auditory receptors perceive sound waves, and turn them into nerve discharges, and they follow certain pathways of nerves to the brain. Neuropsychologists still do not know how human beings or animals recognize that one set of discharges represents pictures and another set is the taste. It is assumed that everything we know comes to us through our senses. But our senses sometimes can deceive us, so some others proposed that we must have some innate knowledge about the world to be able to distinguish between real and imaginary sensations (Kolb, 2000).

2.3.1   The Sensory System

The sensory system starts from the receptor. The receptors are specialized parts of cells. They are different in each sensory system, and they perceive different kind of energy. The receptors act as a filter. They are designed to respond only to a narrow band of energy within the relevant energy spectrum. For example, the human ear can respond only to the sound waves between 20 to 20,000 Hz, and the human nose is not as sensitive as a dog's. Receptors normally receive information from a specific environmental field, such as a visual field. The receptive fields of individual receptors are overlapping. By comparing with neighboring receptors detect, it can also locate the sensory information in space. If one receptor is more activated by a stimulus than another, the object is located more precisely in the receptive field of those receptors. The detection of stimulus is often determined by receptor density and overlap.

Some receptors are easily activated but also stop responding quickly. They are called "rapidly adapting receptors." They react as soon as stimulated energy appears. The reaction can decrease if the stimulated energy does not change its intensity. In other words, they react quickly to the changes of the stimulated energy. For example, our ears can adapt to the noise. When we stay in a noisy environment for a while, we normally do not feel the noise level as high as in the beginning. There are also slow adapting receptors that will provide the information as the stimulated energy is still there. A typical example is the pain receptors. Our pain receptors do not adapt to the stimulation, as pain is a warning signal for the body to indicate certain injuries or damages in the body.

Each sensory system requires three to four neurons, connected in sequence, to get information from the receptor cells to the cortex. There are changes in the code from level to level and it is not a straight-through or point-to-point correspondence. Three important events can occur at the synapses between the neurons in the relay: a) A motor response can be produced, for example, axons from pain receptors stimulate first synapse in the spinal cord, where they can produce a withdrawal reflex. b) The message can be modified. c) Systems can interact with one another.

How does action potential in the neuron cells code the differences in sensations, and how do they code the features of particular sensation? For some sensory modality, we know part of the answer; for most of them, we still don't know. The increase or decrease of the discharge rate of a neuron is correlated to the presence of the stimulus. The amount of change can code the intensity. As in the visual sensory system, redder makes more discharge activity and greener makes less activity. There are different theories for explanations: a) neural areas that process these sensations in the cortex are distinct. b) We learn through experience to distinguish them. c) Each sensory system has a preferred link with certain kinds of movements, which ensures that the systems remain distinct at all levels of neural organization (Kolb, 2000). The distribution of the receptor projecting on the cortex is very complicated. There are at least seven visual systems projecting into different brain regions.

2.3.2   Sensory Modality

Normally, sensory modalities are touch, taste, smell, vision and hearing. There are actually many sub-modalities, or subsystems, within each of these modalities. These sub-modalities are distinctive with respect to their receptors, the size of the fibers that go from the receptors to the brain their connections within the brain, and the actions they produce. Each sub-modality is designed for a very specific function. For example, in the visual system, there are pathways to different locations of the brain for control of different functions as a) daily rhythms of such behaviors as feeding and sleeping in response to day-night cycles; b) changes in pupil size in response to light-intensity changes; c) head orienting, particularly to objects in peripheral visual fields; d) long-term circadian rhythms; e) movement of eyes to compensate for head movements; f) pattern perception, depth perception, color vision and tracking moving objects; and g) voluntary eye movement.

When neuropsychologists try to map the cortex for different functions of the human, it was found that the entire cortex behind the central fissure has some kind of sensory function. A view is emerging that the cortex is fundamentally an organ of sensory perception and related processes. Just as there are many visual systems projecting into different brain regions, there are multiple subsystems projecting to the visual cortex as well. It is demonstrated that some areas in the cortex were identified that had functions in more than one modality, for example, the vision and touch are in the same area. Gesture language and vocal language depend on similar neural systems. These areas, known as polymodal cortex or multimodal cortex, presumably function to combine characteristics of stimuli across different modalities (Kolb, 2000). There are three distinct regions of multimodal cortex. The existence of these three areas implies that there is probably more than one process that requires multimodal information, although it is not known exactly what these processes might be (Kolb, 2000). The knowledge of sensory modality and sub-modalities and the structure of their maps in the brain may lead to us to have a better understanding of how we understand and interact with the real world, and it may also lead us to understanding better about how we define the modality and how does different modalities interact to each other.

At the same time, the integration perception of multisensory input can also be an interesting issue. Viewing an incongruently articulating face categorically alters the auditory speech perception. This is the famous McGurk effect (1976). When it comes to the multisensory-audio/visual-input situation, it was believed that the integration of perception is dominated by one modality. Vision dominates audition is regarded as the most common case. A recent study indicated that either audition or vision domination is a matter of relative issue, not as an all-or-nothing condition (Andersen, 2004).

2.3.3   Cortical Connection and Function

As we discussed above, when the receptors perceive the stimulation, it turns the special stimulated energy (such as light waves, or auditory waves, etc.) into the active potentials and transmits them through different pathways to the cortex. The recognition happens in the cortex. It was also found that there are several locations in the cortex working together for recognition of the stimulation and reaction to it. Different perceptions can also end up on the same cortex location, as the multimodal cortex. Therefore, it is important to study the cortical connection. There are several interesting properties in the cortical connection (Kolb, 2000):

a) There is not a single area in the cortex that could represent entire perceptual or mental states.
b) All cortical areas have internal connections that connect units with similar properties.

c) There is a re-entry of the connection, which means that when a given area A sends information to area B, area B reciprocates and sends a return message to area A.

Now the most agreeable cortical function model is the distributed hierarchical system that is suggested by Felleman and van Essen (Felleman, 1991) as shown in Figure 2-2. There are several levels of processing in the associated area. The areas at each level and across the levels are interconnected with one another.

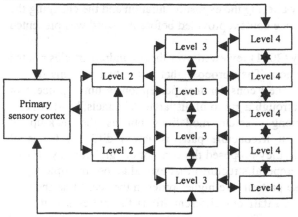

Figure 2-2. Distributed hierarchical mode of central nerve system

2.3.4   Perception and Recognition Theories

There are two different theories regarding perception. a) the *direct perception* (or bottom-up perception) theory: the array of information in our sensory receptors, including the sensory context, is allegedly sufficient to permit the individual to perceive anything. In other words, high-level cognitive processes are not necessary for perception; b) *constructive perception* (top-down perception): the perceiver builds (constructs) cognitive understanding (perception) of a stimulus, using sensory information as the foundation for the structure but also using other sources of information to build the perception (Eysenck, 1995; Sternberg, 1999). In this theory, perception is an active and constructive process. Perception is not directly given by the stimulus input but occurs as the end product of the interactive influences of the presented stimulus and internal hypotheses, expectations and knowledge, as well as motivational and emotional factors. This means that perception is prone to error, since it is affected by hypothesis and expectation (Eysenck, 1995; Sternberg, 1999).

Some studies support the direct perception theory, and some support the constructive perception theory. Nowadays, most psychologists agree that visual perception may be largely determined by bottom-up processes when the viewing conditions are good but may increasingly involve top-down processes as the viewing conditions deteriorate because of very brief presentation time or lack of stimulus clarity.

In word perception, the role of bottom-up processing and top-down processing was varied by altering the exposure duration and the changing the amount of relevant sentence context provided before the word was presented (Eysenck, 1995).

Stoffregen and Bardy (2001) reviewed over three hundred articles related to sensation and perception. They propose that perceptual systems do not function independently. They consider that perception is not a process of picking up information through a group of disparate "channels," and not as a set of interactions among discrete modalities but as the pick-up of information that exists in irreducible patterns across different forms of energy. By this approach, they proposed the concept of "global array" (GA). GA consists of spatial-temporal structures that extend across multiple forms of ambient energy. These patterns are higher-order in the sense that they are super-ordinate to (and qualitatively different from) the patterns that exist within single-energy arrays. Thus, in principle, information in the global array may be detected without prior or concurrent sensitivity to structure in single-energy arrays.

This theory can have broad implications for research on perception and action. In studies of perception, researchers may need to take into account the global information that is always available. They argue that researchers commonly present to subjects only a single form of energy. It is assumed that the application of stimulus energy to only one sense. Methodologically and analytically, it is grounds to ignore the energy available to other senses. However, when an experimenter stimulates a single modality, there is an influence on structure in the global array. If the GA exists, then it is a new challenge for the study of sensation and perception, especially in perceptual-motor learning and development, as well as the challenge of developing new research methodology.

The GA theory has taken the ecological consideration. It questions the validity of most traditional laboratory studies in this area when researchers tried to isolate the stimulation from the environmental context. Ahrens (Stoffregen, 2001) argues that it is impossible to discern the theory of GA. This theory appears to conflate sensation and perception. "They provide no evidence for the synergy between energy arrays they claim is the hallmark of the GA, nor do they provide evidence for novel sensors of higher order patterns; and they discount the behavioral and neurophysiologic evidence that sensory integration occurs within the brain."

The occipital lobe, an anatomical region in the cortex, is involved in the perception of form, movement and color. There are distinguishing locations in the brain that can be identified with different information that is involved in visual object recognition such as visual knowledge, semantic (associative functional) knowledge, and object naming (Eysenck, 1995). At the same time, there are multiple visual systems in the cortex for various aspects of stimulus recognition or for the guidance of movement in space. The multiple visual systems in the cortex are important for visual-spatial recognition. There are different mechanisms behind the facial recognition and word recognition, since facial recognition may take place on the right side of the hemisphere and word recognition may take place on the left side (Kolb, 2000).

One of the major functions of visually guided action is to allow us to interact effectively with our immediate environment. Processing the information about object location and object identification happens on separate parts of brain systems. In the cortex, there are three function zones that can be identified for control of movement. One (anterior zones) is primarily involved in somatosensory functions. The somatosensory system has more than twenty different types of receptor cells covering body tissues, except the brain itself. These include the hair, touch, temperature, nociceptors, joint, and muscle and tendon receptors. One region (superior parietal region) is primarily devoted to the visual guidance of movements of the hands and fingers, limbs, head and eyes. This region can be conceived of as having "spatial" function. We don't know what the nature of this spatial function actually is. The inferior parietal region has a role in processes related to spatial cognition and in what have been described as quasi-spatial processes such as ones used in arithmetic and reading (Kolb, 2000).

2.4   LANGUAGE

Language is one of the special capacities that separate human beings from other animals. The distinction between human language and the "communication" between different animals by sound, such as bird songs, is we use "words" and have the regulation of ordering words in different ways to give the meaning. In language, the fundamental sounds to make words are called *phonemes*. Phonemes are combined to form *morphemes*, which is the smallest meaningful unit of words. A string of words that follows certain grammar is called *syntax*. The stringing together of sentences to form a meaningful narrative is called *discourse*. The collection of all words in a given language makes up the *lexicon*. *Semantics* is the meaning that corresponds to all lexical items and all possible sentences. The vocal intonation that can change the literal meaning of words and sentences is called *prosody*. Language has the ability to represent an extension of the development of multiple sensory channels. We organize sensory inputs by assigning tags to information, which allows us to categorize objects and ultimately concepts. Language involves the motor act to produce *syllables*. A syllable is made up of consonant and vowels. Language requires the ability to impose grammatical rules, which dramatically increases the functional capacity of the system.

2.4.1   Neurology of Language

Neuroscientists can identify four important cortical regions that control language. In the left hemisphere, there are the Broca's area and Wernicke's area. Broca's area is believed to be the set of speech production and Wernicke's area for language comprehension. When we hear a word that matches one of those sound images stored in Wernicke's area, we recognize the word, while to speak words, Broca's area will play the important role, since the motor program to produce each word is stored in this area. If people suffer Wernicke's aphasia, they can speak fluently, but their language is confused and makes little sense (Kolb, 2000). There are another two regions of language use that are located on both sides of the brain: the supplementary speech area and the facial regions of the motor and somatosensory cortex.

How does the brain categorize sensory input and processing language? It was believed that there are three types of systems involved in this process. It was found that there are neural systems to categorize nonlanguage representations of sensory input. In these systems, it not only allows simple sensory stimuli to be combined but also provides a way of organizing events and relationships. In the left cerebral hemisphere, there are smaller regions in neural system that represents phonemes, phoneme combinations and syntactic rules for combining words. There is a third neural system that interacts with the category system and speech system to form the concept (Kolb, 2000).

The neurological models of language include nine facts (Kolb, 2000):

a) Capacity to categorize and to attach phonemic tags to the categories.
b) Visual and auditory language input has different processes.
c) Ability to produce syllables and quickly shift from one to another is important to produce language.

d) Categorizing verbs from movement is the key point for generating grammar in language.
e) Language is a relatively automatic process because it comes from repetition of well-learned responses or the mimicking of input.
f) Memory is required for language process.
g) Both left and right hemispheres are involved in different processes of language.
h) Language representation changes with experience.
i) There are individual differences on the cortical representation of language.

The language process includes the perception of speech or reading the text, word recognition and language comprehension. In the following sessions, we will have a brief introduction for each part of the process.

2.4.2 Speech Perception

Language can be perceived by listening to speech or reading text. Physiologically, speech perception follows the basic mechanism of auditory perception. The sound stimulates the auditory system with its air pressure. The auditory system then transfers changes in air pressure into what we perceive as sound. Sound has three fundamental physical qualities: frequency, amplitude and complexity. Perceptually, these qualities translate into pitch, loudness and timbre. From acoustic perspective, all human language has the same basic structure, despite differences in speech sounds and grammar.

Speech perception and reading text differ in a number of ways (Eysenck, 1995). One is received from auditory channels, and the other is visual. The differences of language perception via speech or reading text are summarized in Table 2-2.

Table 2-2. The differences between speech perception and reading text

Reading the text	Listening to speech
Each word can be seen as a whole	Spoken word is spread out in time
Text is less ambiguous	More ambiguous and unclear signal
The text can be continuously available	Higher memory demand—the information is transient, a word is heard and then it ends.
There is not so much redundant information behind the word itself	Contains numerous hints, such as pitch, intonation, stress and timing, to sentence structure and meaning.
Text is accessible only in the focus visual area	Can take input from any direction

Normally, listening to speech is more difficult than reading a text, as listening to speech demands more from memory, and the words that have already been spoken are no longer accessible. Speech normally contains numerous hints (known as prosodic cues) to sentence structure and the intended meaning in the form of variations in the speaker's pitch, intonation, stress and timing, which makes speech easier to understand. Young children often have good understanding of spoken language but struggle to read even simple stories.

There are several fundamental ways that make the difference between speech input and other auditory input. First, the sounds of speech come largely from three restricted range of frequencies, called *formants*. *Vowel* sounds are in a constant frequency band and *consonants* show rapid changes in frequency. Second, the same speech sounds vary from one context in which they are heard to another, while they are all perceived as being the same. For example, the English letter "d" is pronounced differently in words "deep," "deck" and "duke" when we measure its sound spectrogram. The auditory system must have a mechanism for categorizing sounds as being the same. This mechanism should be affected by experience. So a word's spectrogram is dependent on the words that precede and follow it. Third, speech sounds change very rapidly in relation to each other, and the sequential order of the sounds is critical to understanding.

Listeners use more than acoustic information when interpreting a spoken message during human to human communication. The history behind the conversation, the context during the conversation and the familiarity with the speaker and the subject are all aids to the perception. When available, visual cues, such as lip reading also assist. Lip reading provides the main cue for deaf people it also provides important cues for normal conversation to solve the confusions in the acoustic information. Because of these reasons, it adds certain perception difficulties for people to have telephone conversations and more difficult to listen to synthetic speech, even when the sound is very close to natural human speech.

The functional neuroanatomy of speech perception has been difficult to characterize, due to the fact that the neural systems supporting speech perception vary as a function of the task. Hickok and Poeppel (2000), after reviewing many research articles, proposed that auditory related cortical fields in the posterior half of the superior temporal lobe, bilaterally, constitute the primary substrate for constructing sound-based representations of speech. There are at least two distinct pathways that are involved in task-dependent speech perception, one is a ventral pathway and the other is dorsal pathway, as shown in Table 2-3.

Reading requires several different perceptual and other cognitive processes, as well as a good knowledge of language and grammar. Reading is basically a visual input in which the eye movement and word identification are the important parts. Both bottom-up and top-down processes operate within all three levels: feature, letter and word levels of recognition.

Table 2-3. The pathways for speech perception

	Ventral pathway	**Dorsal pathway**
Cortex location	Vicinity of the temporal-parietal-occipital junction in both hemispheres, but the computational mechanisms involved is probably different.	Inferior parietal and frontal system.
Function	Interfacing sound-based representations of speech with widely distributed conceptual representations.	Receptive phonetic processing such as performance on syllable discrimination and identification tasks. It involves coding speech in an articulatory context.
Tasks	Requires access to the mental lexicon*.	Requires explicit access to certain sub-lexical speech segments.
Serving	Perception of the semantic content of sensory information.	Interface system between sensory and motor processes. It mediates between auditory and articulatory representations of speech.

*The entries in the mental lexicon include phonemic, semantic and perhaps morph-syntactic information in speech.

2.4.3 Word Recognition

There are several different theories regarding the word recognition. Literatures support the suitability of different theories that apply to different situations. All of the theories have strengths and weaknesses. They can be considered from different angles to understand the process. There has been relatively little research on auditory word recognition and comprehension in the brain-damaged patient.

a) *Phonetic refinement theory*, or bottom-up theory (Pisoni, 1985): When the auditory signal is received on the receptors, it shifts to higher-level processing. The words are identified on the basis of successively paring down the possibilities for matches between each of the phonemes and the words we already know from memory.

b) *Phonemic-restoration effect,* or top-down theory (Warren, 1970): The speech we perceive may differ from the speech sounds that actually reach our ears because cognitive and contextual factors influence our perception of the sensed signal.

c) *Categorical perception* (Sternberg, 1999): Even though the speech sounds we actually hear comprise a continuum of variation in sound waves, we perceive discontinuous categories of speech sounds.

d) *Motor theory* (Lieberman, 1967): The motor signal produced by articulation is more informative than the speech signal, and it is our reliance on the motor signal that allows spoken word recognition to be reasonably accurate.

e) *TRACE model* (McClelland, 1986): It is assumed that bottom-up and top-down processing interact during speech perception. Bottom-up activation proceeds upwards from the feature level to the phoneme level and on to the word level, whereas top-down activation proceeds in the opposite direction from the word level to the phoneme level and on to the feature level.

All of these theories differ in different degrees, yet they all have some common assumptions. First of all, the auditory sensor input activates several candidate words in the early process of word recognition. The activation levels of candidate words are graded rather than being either very high or very low. Several parallel processes may happen, rather than occurring serially. Most of the theories propose that both bottom-up and top-down processes combine in some fashion to produce word recognition.

2.4.4   Language Comprehension

In any language, there is a possibility to generate an infinite number of sentences, but these sentences are nevertheless organized to systematically follow certain roles. These roles are regarded as grammar (or syntax). There are three main levels of analysis in the comprehension of sentences: syntactical (grammatical) structure, literal meaning and interpretation of intended meaning. Comprehension would be impossible without access to stored knowledge. The relationship between syntactic and semantic analysis is not clear. It is possible that syntactic analysis influences semantic analysis. Another possibility is that semantic analysis occurs prior to syntactic analysis. It is also possible that both analyses are carried out independently of each other.

The inference drawing process plays a crucial role. Many researchers are trying to identify more precisely which inferences are normally drawn. Different theories regarding language comprehension are about when and why the inferences are made. These theories are applicable in different situations. There is not an entirely satisfactory research method to study this issue. A few dominant theories are introduced here (Eysenck, 1995):

a) *Constructionist approach*: Numerous elaborative inferences are typically drawn while reading a text.
b) *Minimalist hypothesis*: Inferences are either automatic or strategic (goal-directed). In automatic processes, two kinds of inferences are constructed—concurrently process the parts of a text that are locally coherent representations and the information that is quick and easily available.
c) *Search-after-meaning theory*: Readers engage in a search after meaning based on their assumption of goals, coherence and explanation between the texts.

The minimalist hypothesis is preferable in many different aspects. There is a big gap between the minimalist hypothesis and the search-after meaning theory.

By studying language disorder symptoms, one can distinguish many different language functions. These disorders are: a) comprehension disorder such as poor auditory comprehension and poor visual comprehension; b) language production disorder such as poor articulation; word-finding deficit; unintended words or phrases; loss of grammar and syntax; inability to repeat aurally presented material; low verbal fluency; inability to write; and loss of tone in voice. Different language functions are distributed in a large region of the cortex. Language is organized like other cerebral functions in a series of parallel hierarchical channels. The evolution of language may not represent the development of a single ability but rather the parallel development of several processes such as the ability to categorize and the ability to use gestures for communication.

2.5 LEARNING AND MEMORY

Learning is a process that results in a relatively permanent change in behavior. *Memory* is the ability to recall or recognize previous experience; certain neural structures and circuits are associated with different types of learning and memory. Learning and memory involve a series of stages. When the learning material is presented, the learning process "encodes" the material (encoding stage). Some of the encoded information is stored within the memory system (storage stage). The stored information from the memory system can be retrieved through recovering or extracting.

Theories of memory generally consider both the structure of the memory system and the processes operating within the structure. *Structure* refers to the way in which the memory system is organized, and *process* refers to the activities occurring within the memory system.

2.5.1   Learning Theory

There are multiple forms of learning. The primary distinction can be made between Pavlovian conditioning, in which some environmental stimulus (such as a tone) is paired with a reward; and operant conditioning, in which a response (such as pushing a button) is paired with a reward (Kolb, 2001). Normally a learning theory contains five components (Coe, 1996): a) *Experiences*: learning is the result of experiences. The sensation-perception continuum forms the basis of the experience. b) *Schemata*: A schema is a mental framework or model that we use to understand and interact with the world. Experiences use sensation-perception to either create a new schema or modify an existing one. The easiest way to leverage a user's schemata is to use examples that center on schemata he already has in memory. Simplicity is the way to help users create schema. c) *Habits*: Connections between symbols and their corresponding actions are called habits. Habits have strength and family (a set of related habits). d) *Reinforcement*: It is the process of using events or behaviors to produce learning. There are both positive and negative reinforcement. To be effective, reinforcement must be systematic. Continuous reinforcement is reinforcing a behavior each time it occurs, and this is the quickest route to learning. Intermittent reinforcement is sometimes reinforcing a behavior, sometimes not. Vicarious reinforcement is learning to perform those actions we see others rewarded for and to eschew those actions we see others punished for. e) *Interference*: Old habit families interfere with new learning.

2.5.2   Memory

The study results from neuroscience have strong reasons to imply that every part of the nervous system is connected to learning, even the spinal cord. This would mean that the areas that process auditory information house auditory memory, areas that process visual information house visual memory, areas of the brain involved in producing movement house motor memories and so forth. It would also mean that memory could be further subdivided according to the specialized regions within each which houses the major functional modalities. Neuropsychologists believe that there are many kinds of memories, each of which has its brain location. Different types of memories can be found in the literature (Kolb, 2000):

a) *Short-term memory*: Memory for things that are retained only for a short period of time.
b) *Long-term memory*: Memory for things that are remembered for a long period of time.
c) *Implicit memory*: An unconscious, not intentional form of memory.
d) *Explicit memory*: Memory which involves conscious recollection of previous experiences.
e) *Reference memory*: Memory which refers to knowledge of the rules of a task.
f) *Working memory*: Memory which refers to events that can happen in a trial.

The distinction of the above types of memories is more or less from a functional perspective. There is no specific brain location that holds one specific type of memory. Among them, the long-term memory and the working memory have been especially interesting.

Short-term memory and long-term memory are parallel mechanisms in which material is processed separately and simultaneously (Kolb, 2000). Short-term memory is the memory that we use to hold digits, words, names or other items in memory for a brief period of time. It is sometimes called "working memory." Actually, the working memory and short-term memory is not exactly the same thing. Repeating a list of digits may be useful for putting the list in long-term memory, but it is not the short-term memory trace that is being placed in long-term memory. It is likely, however, that items can be pulled from long-term memory and held for some use in short-term memory. Short-term memory can be doubly dissociated from long-term memory as it can be impaired itself. It can have neural structures. There are probably a number of kinds of short-term memory, each with different neural correlates.

The processes of explicit and implicit memories are housed in different neural structures with different functions. Implicit information is encoded in very much the same way as it is perceived. This is data-driven or "bottom-up" processing. Explicit memory depends upon conceptually driven or "top-down" processing, in which the subject reorganizes the data.

Based on animal and human studies, Petride and Mishkin (1994) have proposed models for explicit and implicit memory. They demonstrated different areas in the cortex that are involved in explicit memory such as object memory, spatial memory, emotional memory, etc. At the same time, they found a brain circuit for implicit memory.

2.5.3   The Mechanism of Memory

Hebb's (1949) *cell assembly theory* has been almost the only theory from the neuron level to explain the mechanism of memory. He proposed that each psychologically important event—whether a sensation, percept, memory, thought or emotion—can be conceived as the flow of activity in a given neuronal loop. The synapses in a particular path become functionally connected to form a cell assembly. It is assumed that if two neurons (A and B) are excited together, they become linked functionally. In Hebb's view (1949), one cell could become more capable of firing another because synaptic knobs grew or became more functional, increasing the area of contact between the afferent axon and the efferent cell body and dendrites. Some studies have proved that there are qualitative changes in the synapses. These include changes in the size of various synaptic components, in the number of vesicles, in the size of postsynaptic thickenings, and in the size of the dendrites spines (Kolb, 2000).

Cell assembly represents the idea that networks of neurons (cell assemblies) could represent objects or ideas, and it is the interplay between those networks that results in complex mental activity (Kolb, 2001).

At least three problems arise (Kolb, 2000): a) A sensory experience could not change every neuron in the relevant systems, otherwise all subsequent experiences would be changed. b) If sensory experiences change sensory systems, thus permitting memories of the events, how do we remember ideas or thought? c) If experiences result in widespread changes in synapses, how do we find specific memories? Even though people are not satisfied with Hebb's theory, there is no better theory and these questions are waiting for answers.

2.5.4   Working Memory

Baddeley and Hitch (1974) suggested replacing short-term memory with working memory. This suggestion has been extensively accepted in cognitive science. The working memory system consists of three components: a) A modality-free *central executive resembling attention*. It has limited capacity and is used when dealing with most cognitive demanding tasks. It is like an attention system; b) an *articulatory or phonological loop* which holds information in a phonological (i.e., speech-based) form. Articulatory loop is determined by temporal duration in the same way as a tape loop. An experimental study showed that the subject's ability to reproduce a sequence of words was better with short words than with long words (Eysenck, 1995). A phonological loop consists of: a) A passive phonological store that is directly concerned with speech perception;

b) An articulatory process that is linked to speech production and gives access to the phonological store. c) A visuo-spatial *sketchpad* that is specialized for spatial and/or visual coding. An auditory presentation of words permits direct access to the phonological store regardless of whether the articulatory control process is being used. In contrast, visual presentation of words only permits indirect access to the phonological store through subvocal articulation (Eysenck, 1995).

The central executive is the most important component in the working memory. There has been relatively little clarification of the role played by the central executive. The capacity is very difficult to measure (Eysenck, 1995). A major function of the phonological loop is to facilitate the reading of difficult material, making it easier for the reader to retain information about the order of words in text during the process of comprehension (Eysenck, 1995). The characteristics of the visuo-spatial sketchpad are less clear than those of articulatory loop. Baddeley (1974) defined it as "a system especially well adapted to the storage of spatial information, much as a pad of paper might be used by someone trying, for example, to work out a geometric puzzle."

There are several advantages with working memory model (Eysenck, 1995):

a) It is concerned with both active processing and transient storage of information.
b) It is better placed to provide an explanation of some findings from neuropsychology with brain-damaged patients. If brain damage affects only one of the three components of working memory, the selective deficits on short-term memory tasks would be expected.
c) It incorporates verbal rehearsal as an optional process that occurs within only one component.

2.5.5 Forgetting

Everybody has the experience that after a certain period of time forgets a lot of things learned, done and talked about. What makes us forget about things? What is the mechanism behind it? Why are there certain things we wish to remember, such as the knowledge we learn the school, but keep forgetting, while some bad experiences that we wish to forget, we remember for the rest of our life? The scientists are not able to answer all these questions yet. There are not studies from neuroscience to demonstrate why forgetting happens.

There are three theories about forgetting; one is interference theory, one is decay theory (Sternberg, 1999) and the third is *functional decay theory*.

Interference theory assumes that our subsequent ability to remember what we are currently learning might be disrupted or interfered with by what we have previously learned (proactive interference) or by what we learn in the future (retroactive interference). *Decay theory* asserts that information is forgotten because of the gradual disappearance, rather than displacement, of the memory trace. This theory is very difficult to test! Whereas interference theory views one piece of information as blocking another, decay theory views the original piece of information as gradually disappearing unless something is done to keep it intact. Evidence exists for both theories, at least in short-term memory. The *functional decay theory* has proposed that forgetting the current task is functional because it reduces interference—if the current task has already decayed by the time the next task comes along, it will cause less interference once the next task is current. Altmann and Gray's (2000) experiment indicates that forgetting helps by preventing stale information from interfering with fresh information. Forgetting is difficult when the world changes faster than memory can decay. Cognition responds to this limitation by starting to forget an item as soon as it is encoded.

2.6   CONCLUSION

Neuropsychology is the science of studying the relationship between brain function and behavior. Within the limited space, we can only introduce some basic concepts about how the brain is constructed and functions in general. We also tried to introduce the recently development in the neuroscience regarding different cognitive behavior such as language process, learning and memory.

The brain consists of a very large number of simple processors and neurons, which are densely interconnected into a complex network. These large numbers of neurons operate simultaneously and co-operatively to process information. Furthermore, neurons appear to communicate numerical values of information from input to output.

The knowledge from neuroscience may tackle some cognitive theories and information processing theories, as they normally consider information is coded as symbolic messages. In the speech and language process studies, research has concentrated on modeling human behavior. The data is taken mainly from cognitive psychology, or linguistics, rather than neuroscience. Even parallel distributed processing, or neural network, or neuro-computing, normally bear no relation to brain architecture and lack biological realism (Christiansen, 1999).

There should be historical reasons behind these phenomena. One of the main reasons is because we still lack enough knowledge about how the brain works. The knowledge gained from traditional methodologies of studying brain-damaged patients is far from enough to support the computational design and applications. The development of the new technologies, especially the electron or magnetic graphics technologies, provides a new possibility for neuroscience to study the functional mechanism of the brain. The results from neuroscience studies based on new methodologies may provide the opportunities for the computational models searching for biological realism. The requirement from computational modeling may also set considerable influence on neuroscience development, especially on language process and memory research.

Chapter 3

ATTENTION, WORKLOAD AND STRESS

3.1   INTRODUCTION

Attention, multiple task performance and workload are three important factors when we design any system for occupational use or when the users have to take care of a few different things at the same time such as the car driver is driving, while wanting to change the radio channels or change the music CD. These three factors are connected to each other and have been interesting to cognitive scientists for a long time for the speech interactive system when it is implemented into the working environment, together with other systems. These three factors should play important roles for the interaction design.

In the later half of last century, there was a huge amount of research published on these aspects. Unfortunately, there is relatively little research on attention, multiple tasks performance and workload when speech technology is applied. Some studies has proved that the recognition accuracy of ASR decreased tremendously in stress (Chen, 2001; 2003). Many people experienced the big differences of the recognition accuracy of ASR in the laboratory and in the application settings. People used to blame ASR technology not being good enough. Actually, the causes are far more complicated than just the technology itself. The changes of attention, multiple-task performance, different workload and stress can contribute factors to the causes.

In this chapter, I would like to discuss the research results on attention, workload and stress and how affect human speech and performance.

3.2   ATTENTION

3.2.1   Neuroscience Perspective

From a cognitive perspective, "attention is the means by which we actively process a limited amount of information from the enormous amount of information available through our senses, our stored memories and our other cognitive processes" (Sternberg, 1999). Neuroscience does not know yet what attention is. It is a "hypothetical process that either allows a selective awareness of a part or aspect of the sensory environments or allows selective responsiveness to one class of stimuli" (Kolb, 2000). This concept implies that somehow we focus a "mental spotlight" on certain sensory inputs, motor programs, memories or internal representations. This spotlight might be unconscious in that we are not aware for the process, or it might be conscious, such as when we scan our memory of some experience. We called this "mental spotlight" the "attention."

Research has proved that attention is the function of some regions of the brain. Neurons in the pulvinar (a region of the thalamus) respond more vigorously to stimuli that are targets of behavior than they do to the same stimuli that are not targets of behavior (Petersen, 1987). The same study also showed that knowledge of the task demands could somehow alter the activity of neurons in the visual system. This is essence of a top-down process of attention.

A human being has limited capacity for mental activity, and this limited capacity should be allocated among concurrent activities. A laboratory study showed that perception systems do not always work at peak efficiency. This may be because we can only process so much information at once. If a task is easy, little attention capacity is used; the operator can do a few things at a time, driving and talking, for example. But when one of the tasks requires high mental effort, more mental capacity must be shifted to the task and other performance is briefly interrupted. For example when it is a complex traffic situation, people have to stop talking. In other words, if we are overloaded, there is a "bottleneck" in processing. Some process in the central nerve system must be active to shift attention to different tasks in response to changes in task demands. Neuroscience does not know how this "spotlight" or the attention shift happens in the brain.

People can selectively pay attention to certain things or mental process and shift their attention in different situations. Attention is sub-divided into focused attention and divided attention. In the following text, we will have some discussion on them.

3.2.2 Focusing Attention

We are able to focus, look at, and consistently prioritize almost any feature that distinguishes one string of stimuli from another. Although some features of the stimulus such as its spatial location and physical characteristics appear to be easier to focus upon and follow, voluntary selection is highly flexible and strategic. Many different models were developed to reflect different strategies of focused attention (Kahneman, 1973).

- **Focused visual attention**

Under the general mode, attention is allocated evenly, in parallel, across the whole visual field. Under the specific mode, attention is focused on one display location, facilitating processing there, while inhibiting it at all others. Physical properties of the stimuli can catch visual attention, but the mind can also guide it, the later one regarded as the top-down selective-attention processes controlling the "mind's eye" (Posner, 1980).

The focused visual property is related to its physical structure. The eye has a focused visual region—fovea, which is about 2 degrees of visual angle. Focused visual attention happens inside this fovea visual region. Two strategic modes of focusing in the visual field are found (Castiliano, 1990; Egeth, 1977; Ericksen, 1986; Jonides, 1983): *Pursuit* movements occur when the eye follows a target moving across the visual field; *Saccadic* movements are discrete, jerky movements that jump from one stationary point in the visual field to the other. There was external research in this area in the '70s and '80s (Wickens, 2000).

For the display design, people applied different cues to direct the visual search. Some of the commonly used intelligent cues can be a flashlight, highlighting the keywords or using different colors. The reaction time between the cues appears and the target is detected is called "stimulus-onset asynchrony" (SOA). For more effective target detecting, the SOA should be between 200 ms (millisecond) to 400 ms. This data indicated that, for example, is, to design the visual display; the information on the screen should not change faster than 200 ms!

- **Selective auditory attention-bottleneck theories**

Similar distinctions were reported by Kahneman (Kahneman, 1973 a; 1973b) for the auditory selective attention. There are three bottleneck theories that tried to explain how the selective auditory attention happens. All three theories agreed that human sensors can receive much more information than this information being processed in the brain. The differences between these theories are where the bottleneck happens.

Broadbent's filter theory (Broadbent, 1958) assumed that two stimuli or messages presented at the same time gain access in parallel to the sensory buffer. One of the inputs is then allowed through a filter on the basis of its physical characteristics, with the other input remaining in the buffer for later processing. This filter is necessary in order to prevent overload of the limited-capacity mechanism beyond the filter. In other words, we filter information right after it is registered at the sensory level. This theory assumes that the unattended message is always rejected at an early stage of processing, but this is not always correct.

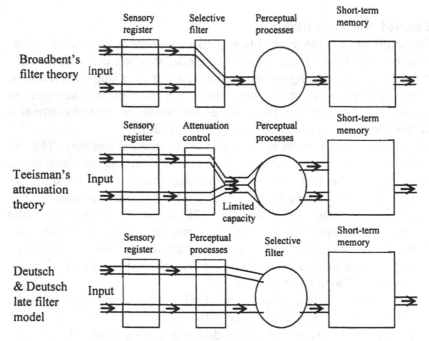

Figure 3-1. Different bottleneck theories

Treisman's attenuation model (Treisman, 1964) claimed that the location of the bottleneck was more flexible. The selective attention involves a different kind of filtering mechanism: a) *preattentively analysis* of the physical properties of a stimulus such as loudness, pitch and so on. This preattentive process is conducted in parallel on all incoming sensory stimuli; b) *pattern analysis* of the stimuli—if they show the target pattern, we pass the signal on to the next stage. If it does not show the pattern, we pass on only a weakened version of the stimulus. We focus attention on the stimuli that make it to this stage.

Deutsch and Deutsch's late filter model (1967) was simply to move the location of the signal-blocking filter to follow, rather than precede, at least some of the perceptual processing needed for recognition of meaning in the stimuli. Figure 3-1 shows the comparison of the three bottleneck models. Actually, these bottleneck theories are not only applied to selective auditory attention, but also explain the human information process limitation.

An unattended channel of auditory input remains in pre-attentive short-term auditory store for three to six seconds. The contents of this store can be "examined" if a conscious switch of the attention is made. Thus, if your attention wanders while someone is talking to you, it is possible to switch back and "hear" the last few words the person spoke, even if you were not paying attention to him or her when they were uttered. Even in the absence of a conscious attention switch, information in unattended channels may make contact with long-term memory. When the unattended material is sufficiently pertinent, it will often become the focus of attention. Typically, if someone calls your name in an unexpected moment, it can still catch your attention and cause you to search for the source of the signal.

3.2.3 Dividing Attention

Dividing attention comes primarily from experiments of dual-task performance (Gopher, 1986). Subjects have to cope with the demands of concurrently presented tasks; either two time-shared tasks have to be treated as equally important, or one is designated as primary and the other secondary. A secondary task should be attended to only insofar as performance of the primary task is fully achieved. Divided attention and dual-task performance will be discussed in more detail in section 3.

During multiple tasks performance, people normally will develop a strategy in order to better cope with task demands and performance objectives within the boundaries of their processing and response limitations. This strategy influences the performer's mode of response to the requirements of the task (Logan, 1985).

Auditory and visual attentions influence each other. For example, if the auditory target is attended in one side of the spatial location, not only auditory discrimination but also visual discrimination becomes better for targets on the indicated side. Conversely, when visual targets were very likely on one side, auditory discrimination improved there as did visual discrimination (Eimer, 2000). Neurons in the posterior parietal cortex and ventral premotor cortex have been found that respond to both tactile and visual stimulation from similar locations in external space. It seems plausible that such neurons may be involved in establishing the cross-modal links in spatial attention modulations of visual and audio modalities (Eimer, 2000).

3.3   MULTIPLE-TASK PERFORMANCE

A lot of speech interactive systems will be used in the place where the users are in busy situation with both hands and eyes such as in-vehicle computer systems and in most military applications. This is a typical situation when more than one task needs to be performed at the same time. How does the user timeshare, or how does he or she cope with the situation when he or she must perform two or more activities in a short period of time? How can a system be designed to improve timesharing tasks performance? Before studying the efficiency and performance of different speech interactive systems, we need to understand the theories and factors related to multiple-task performance. The dual-task performance is also commonly used as a research methodology to study the human workload, which we will discuss later in this chapter.

3.3.1   The Resource Concept

The resource concept that Wickens proposed (1991) assumes that the human operator has a limited capacity for processing resources that may be allocated to task performance. Two tasks demand more resources than one; therefore, timesharing can lead to a situation in which one or both have fewer resources than required, and, hence, the performance of one or both may deteriorate.

What are the "resources"? According to Wickens, they are a convergence of multiple phenomena that have a number of different manifestations, some intuitive, some empirical, and some based on plausible models (1991). The resources cannot be directly measured, as each in isolation may be explained by a different concept. Resources can also be the mental effort that is invested in order to improve performance. The relationship between effort (resources invested) and performance is known as a performance-resources function (PRF). All tasks may generate the same level of performance with full effort allocated to the task (resources investment), but some tasks may require fewer resources to achieve maximum performance. Besides, people may apply different strategies to achieve the same task performance. Typically, some invest a full effort to achieve very high performance and some apply a "quick and dirty" solution to achieve "reasonable" performance. When it comes to multiple task performance, the strategies people apply to distribute the resources across different modalities and tasks can be very different. There is no single objective measurement that one can use to measure the resources. The effort can be measured by physiological and psychological measurements, which can be similar as the workload measurement that we will discuss later in this chapter.

3.3.2   The Efficiency of Multiple Task Performance

To make the multiple-task performance available in continuous, graded quantity, there are three characteristics of resources that are relevant to performance within the dual-task domain: their scarcity, their allocation and their relation to task difficulty (Kahneman, 1973). To have a better timeshare performance, a good scheduling of time and efficient switching between activities are important ingredients to success. Wickens' multiple resources theory (2000) has often been used to explain the timesharing task design and studies.

There are three factors that will influence the effectiveness of multiple-task performance (Wickens, 1991):

a) Confusion of task elements due to the similarity. The similarity can be in different aspects, such as similar voice, similar color, similar shape, similar contents, etc. For example, if one is trying to listen to two speakers at once, whose voices are of similar quality, and who are discussing a similar topic, it is quite likely that one will mistakenly attribute the words of one to the message of the other (Navon, 1984).

b) Cooperation between task processes as the high similarity of processing routines. They can yield cooperation or even an integration of the two task elements into one.

c) Competition for task resources when people have to switch his/her attention and effort between the tasks. For example, when both tasks require intensive visual searching the competition of these two tasks will affect the performance.

For the task performance, two processing stages can be recognized; one is the perceptual and one is the response. The perceptual-cognitive activities are display reading, information monitoring, voice comprehension, mental rotation, situation assessment, diagnosis or calculation. The response processes include control manipulation, switch activation, typing on the keyboard or even voice input. Tasks with demands in the former category may be timeshared efficiently with those having similar demands in the latter category. Processing spatial and analog information is assumed to use different resources from those involved with processing verbal and linguistic information. This dichotomy is applicable to perception (speech and print versus graphics, motion and pictures), central processing (spatial working memory versus memory for linguistic information), and response processes (speech output versus spatially guided manual responses). The third dichotomy contrasts perceptual modalities of visual versus auditory input (e. g., reading uses separate resources from listening to speech). This is the famous multiresources theory developed by Wickens (Wickens, 2000). A better timesharing efficiency of these three dichotomies is shown in Table 3-1.

Table 3-1. Modalities giving better timesharing efficiency among perception and responses

Perceptual modalities	Perceptual/cognitive	Response
Verbal and linguistic	Print reading	Speech
	Voice understanding	
	Rehearsal	
	Mental Arithmetic	
	Logical reasoning	
Spatial and analogical	Velocity flow field	Manual control
	Spatial relations	Keyboard press
	Mental rotation	
	Image	
	Transformations	

Wickens' multi-resources theory in some way provides some cognitive theory for how to select the modality for specific information input and output. Unfortunately, there are very few multimodal interface designers who have looked at this theory for their design. At the same time, there are some arguments on this theory, especially when it comes to the critical situation when focused attention has to be applied. The typical example is when one is driving a car; one can listen to music or talk to people at the same time. But when it comes to complicated traffic situation, one may stop talking or listening and concentrate on the driving.

Besides the bottleneck theories (Figure 3-1), the capacity sharing model and cross-task model give different explanations (Pashler, 1994). People share processing capacity (or mental resources) among tasks. More than one task is performed at any given moment; thus, there is less capacity for each individual task, and performance can be impaired. The psychological refractory period (PRP) can be identified (Pashler, 1994). PRP is the time needed when people respond to each of two successive stimuli effects; the response to the second stimulus that often becomes slower when the interval between the stimuli is reduced. A PRP effect can be found even when pairs of tasks use very diverse kinds of responses, indicating a stubborn bottleneck encompassing the process of choosing actions and probably memory retrieval generally, together with certain other cognitive operations. Other limitations associated with task preparation, sensory-perceptual processes and timing can generate additional and distinct forms of interference. Interference might be critically dependent not on what sort of operation is to be carried out but on the content of the information actually processed: What sensory inputs are present, what responses are being produced and what thoughts the person is having.

3.3.3 Individual Differences

Individual differences in timesharing multiple-task performance have been studied for about hundred years. There is a lack of a methodology for examining individual differences in multiple-task performance, and there is no evidence for general timesharing ability (Brookings, 1991).

Individual attention in dual-task performance is varied in different age groups. The picture with regard to aging, divided attention and dual-task performance is a complicated one, but it may be safe to say that in all but the simplest tasks, older adults perform worse under dual-task conditions than young adults do (McDowd, 1991). When older adults attempt to conduct two activities or perform two motor tasks simultaneously, they tend to engage a strategy resembling serial processing, often at the expense of one of the tasks, whereas the younger adults perform smoothly in a fashion that more closely represents parallel processing.

Experimental design requires a careful control for potentially confounding and intervening variables. These variables can be subject sampling bias (e.g., health, physical fitness, previous experience, educational background, intellectual capacity), non-standard task characteristics (e.g., disproportionate or age-sensitive intra-task loadings on processing resources or stages, conflicts in stimulus or response modalities, lack of specified workload intensities in terms of difficulty and complexity) and inter-study differences in experimental procedures for data collection and analysis.

3.4 STRESS AND WORKLOAD

It is known that the automatic speech recognition accuracy decreases dramatically when it is implemented in the real-time application environment. Two of the main factors that cause the decrease are the environmental factors and the workload induced stress and human speech changes.

In the workplace, stress can be induced by temperature, noise, vibration, G-force, high workload, sleep deprivation, frustration over contradictory information and emotions such as fear, pain, psychological tension, etc. These conditions are known to affect the physical and cognitive production of human speech and communication. The change in speech production is likely to be disastrous not only to human understanding of coded speech, but also to the performance of communication equipment. Before we discuss about how stress can affect speech, we should have a correct understanding of the relationship between workload and stress.

3.4.1   Stress

Stress in the physiological state is concerned with preparing the organism for action. In the psychological literature "stress" is considered as the non-specific response of the body to any demand made upon it, while "stressors" are those events that occur in an individual's environment that make the demands (Seyle, 1983).

For the study of speech under stress, a couple of definitions of stress were suggested (Murray, 1996) regarding its psycho-physiological aspects and its effects on speech. One of them is definition five "stress is observable variability in certain speech features due to a response to stressors" (Murray, 1996). By this definition, stress is the responses to the stressors, the change of the responses is observable and it can be reflected in certain speech features. The implication of being "under stress" is that some form of pressure is applied to the speaker, resulting in a perturbation of the speech production process and, hence, of the acoustic signal. According to such definition, a few things need to be identified: a) *Stressors*: What are the stressors? What are their characteristics? How are they measured qualitatively and quantitatively? b) *Responses*: How is the speech feature changed corresponding to the stressors? How are changes measured? What is the mechanism behind the changes? In the rest of this chapter, we will try to find out some of the answers, or try to find the way to get the answers.

3.4.2   Workload

Workload, as the word states, is the load induced by work. Workload is psychosomatic responses appearing as a result of factors associated with accomplishing one's duties and an imbalance between work conditions, work environment and the adaptive capacity of the people who work there. Workload expresses the degree of qualitative and quantitative load induced by work and is closely related to stress and fatigue.

Workload studies address the following aspects (Wilson, 1995):

a) Subjective aspect: symptoms of self-conscious fatigue and other psychosomatically related complaint symptoms and physical conditions.
b) Objective aspect: operational behavior such as subsidiary behavior, performance and human error.
c) Psychophysiological function aspect: Maladjustment of psychophysiological functions such as the lowering of cerebration.

d) Other transitions in fatigue phenomena in the workplace such as work history, labor conditions, adaptation symptoms, and proclivity to contract disease, physical strength, physical constitution, age and gender.

3.5 THE RELATIONSHIP BETWEEN STRESS AND WORKLOAD

Stress and workload are two concepts that are used to describe similar phenomena and are therefore often confused. It is often assumed that a high (subjective) workload results in stress responses such as psychosomatic and psychological complaints.

However, the relationship between workload, especially the mental workload, and stress is more complicated. On one hand, people are able to work intensively under a high level of workload without experiencing stress. On the other hand, stress responses may occur when the workload is too low, as is the case in monotony and isolation.

The characteristics of the biobehavioral states of mental workload and stress are summarized by Gaillars and Wientjes (1994) as shown in Table 3-2. The distinction is based on experimental results and theoretical viewpoints coming from quite different research areas.

Table 3-2. *Summary of the characteristics of two biobehavioral states*

	Mental load	Stress
Mechanism	Effort	Emotion
Brain center	Cortex	Limbic system
Response	Functional	Over-reactivity
After-effects	Recovery, fatigue	Accumulation, disturbance, disease
Control	Sense of control	Loss of control
Mood	Positive	Negative
Coping style	Task-oriented, active, engaging	Self-protection, defensive, palliative
Outcome	Predictable, positive	Uncertain, negative

3.5.1 Stressors Classification

Three classes of stressors according to their characteristics have been identified, namely:

(a) Physical stressors, such as extremes in temperature, noise, vibration, G-force, etc.

(b) Cognitive stressors, which can be applied to the "intellectual" functions, such as excessive workload.

(c) Affective stressors, which are predominantly subjective in nature, their effects only being predicted by reference to the life history and personality of the person under stress.

For speech interaction system design, short-term physical and cognitive stressors are of most interest in human factors applications (Hapeshi, 1988). The long-term affective stressors are important only in that they are likely to influence the way in which an individual is affected by short-term stressors.

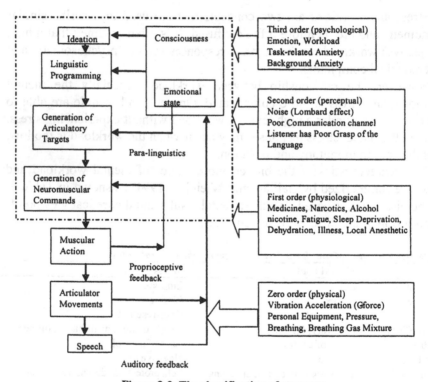

Figure 3-2. The classification of stressors

Murray, et al., (1996) proposed the classification of stressors based on the idea that a given stressor can act primarily on a particular stage of the speech production process (Figure 3-2). Those stressors that can make physical changes to the speech products apparatus itself are classified as the "zero-order" stressors. The "first-order" stressors result in physiological changes to the speech production apparatus, altering the transfer of neuromuscular commands into movement of the articulators. The "second-order" stressors are those that affect the conversion of the linguistic programmer into neuromuscular commands. The "third-order" stressors have their effects at the highest levels of the speech production system.

There are a few weak points by such classification, for example, environmental stressors are in the zero-order (noise can be in the second-order), and those factors causing physiological changes are in the first-order stressors. At the same time, workload is classified as the highest order (the third order). As we discussed in the earlier section, workload has its objective and subjective aspects. The workload can cause both physiological and psychological changes. It is also clear that the physical factors in the working place will directly affect to the perception of workload. The self-awareness makes it possible that almost all stressors may have a third-order effect.

3.5.2 Stress: Its Cognitive Effects

Stress, as a cognitive state, is part of a wider process reflecting the person's perception of and adaptation to the demands of the (work) environment (Cox, 1995). This approach emphasizes the person's cognitive appraisal of the situation, and treats the whole process of perceiving and reacting to stressful situations within a problem-solving context. The effects of stress to individual person are very different. Cognitive appraisal appears to take account of individuals' perceptions of the *demands* on them, the *constraints* under which they have to cope, the *support* they receive from others in coping and their *personal characteristics* and coping *resources*— their knowledge, attitudes, behavioral style and cognitive skill.

Demands are cognitive, emotional, behavioral or physiological in nature and requests for action or adjustment. They usually have a time base. The absolute level of demand would not appear to be the important factor in determining the experience of stress. More important is any discrepancy, which exists between the perceived level of demand and individuals' perceptions of his or her ability to meet that demand. Demand also has something to do with personality. Some people may set higher demands on themselves than the task required, and some may set less demand. The relationship between the discrepancy and the intensity of the stress experience may not be linear (Cox, 1995) when considering the situation of when demands are larger than abilities or demands are smaller than abilities.

Constraints operate as restrictions or limitation on free action or thought, reflecting a loss or lack of discretion and control over actions. These may be imposed externally. For example, constraints may be imposed by the requirements of specific jobs or by the rules of the organization. They may also reflect the beliefs and values of the individual.

3.5.3   Stress: Its Physiological Effects

It is commonly agreed that the hypophysiotrophic area of the hypothalamus is the area in the brain that receives the information of stress, but the mechanism is not known. It was found that qualitatively different stimuli of equal stressor potency (as judged by their ability to elicit the triad or adrenocorticotrophic hormone and corticoid production) do not necessarily cause the same syndrome in different individuals, and even the same degree of stress induced by the same stimulus may provoke different lesions in different individuals (Hamilton, 1979).

There are three stages in which the body reacts to stressors (Hamilton, 1979):

a) *Alarm reaction*: This occurs upon sudden exposure to noxious stimuli to which the organism is not adapted.
b) *Resistance*: This is marked by full adaptation to the stressor during which symptoms improve or disappear. There is, however, a concurrent decrease in resistance to other stimuli.
c) *Exhaustion*: Since adaptability is finite, exhaustion inexorably follows if the stressor is sufficiently severe and applied for a prolonged period of time.

Fortunately, most of the physical or mental exertions, infections and other stressors that act upon us produce only changes corresponding to the first and second stages.

3.5.4   Stress: Its Effects on Speech

Any form of physical exertion will have a principal effect of altering the breathing rate, though individuals automatically compensate to a certain extent (Lieberman, 1988). The pattern of respiration in different stress states could be related to the speech produced: increase the intake of breathes, or the breathing rate quickens. The changes in breathing rate will influence glottal pressure and, hence, affect the fundamental frequency. Furthermore, the increasing respiration rate will result in a shorter duration of speech between breaths (Baber, 1996). Physiologically, stress can result in the reduction of saliva in the mouth, leading to a sensation of dryness, which could influence the speech produced (Williams, 1972). A stressful situation can influence speech by altering the mean (Coleman, 1979; Fonagy, 1978; Niwa, 1971) and the range of the fundamental frequency (Utsuki, 1976). Stress can lead to higher frequency energy in the speech signal (Roessler, 1976).

Hecker, et al., (1968) conducted experimental studies on the effect of stress on speech. Stress was induced by requiring the subjects to carry out a specific task; such as reading and adding meter displays. The results of their experiment revealed that task-induced stress could produce a number of characteristic changes in the speech signal. Most of these changes were attributable to modifications in the amplitude, frequency and detailed waveform of the glottal pulses. Interestingly, the manifestations of stress varied considerably from subject to subject, although the test phrases of most subjects exhibited some consistent effects. Thus, while the acoustic characteristics associated with a given effect of stress for one individual may be quite different from those for another, these effects occur in a consistent manner for each individual and, hence, could be used to predict whether the utterances were produced under stress or not. In Howells' (1982) study, verbal protocols were collected from running commentaries made by helicopter pilots during flight and waveform analysis carried out on them. Statistically significant associations were found between marked changes in the speech waveform and certain combinations of pilot activity associated with the commentary. The results from this study suggest that speech spectra are closely related to task loading.

Baber and Mellor (1996) have suggested that, if the primary consequences of stress on speech relate to shifts in fundamental frequency and speech rate, then it is possible to envisage adaptive recognition algorithms that are capable of dealing with these effects. Realistically, it is very difficult to find a signal acoustic correlate between stress and speech. Thus, it may be difficult to provide compensation in ASR devices.

3.5.5   Fatigue

There appears to be no work on the specific effects of fatigue on speech or speech recognition accuracy, probably because fatigue could be considered as a side-effect of stress. There are two types of fatigue regarding speech (Hapeshi, 1988): vocal fatigue, following long-continued use of the voice or from shouting in a noisy environment, and mental fatigue from working many hours, not necessarily using the voice to any great degree. In hostile environments, fatigue can refer to a behavioral manifestation of stress. However, fatigue may also occur in environments not normally extreme enough to be regarded as stressful.

It is difficult to locate literature on the effects on ASR operation of fatigue arising from extensive physical workload. Zarembo (1986) reported that the use of ASR on the floor of the New York Stock Exchange was subject to error due to users' voices becoming hoarse by the end of the day; this could be attributed both to physical and mental stress induced fatigue.

There are several studies on recognition performance over prolonged periods of ASR use (Pallett, 1985). Frankish and Noyes (1990) reported that a prolonged use of ASR results in shifts in recognition accuracy. Their data indicated that the shifts pertain to overall recognition accuracy rather than to variability, implying a significant change in speech style. Frankish and Noyes (1990) hypothesized that users of ASR go through a brief warm-up period each time they use a device, and speech settles into an appropriate style after a few minutes. They proposed that the changes in recognition accuracy observed could be dealt with by simply retraining the device, either through automatic updating of templates or by the user's re-enrolling specific vocabulary items.

Whatever causes fatigue in the user of an ASR, the problems are the same as for stress; namely, how is the speech signal modified by fatigue and if speech recognition is affected, how can the system be made more robust? This question will be discussed later in this chapter.

3.6   THE MEASUREMENT OF STRESS AND WORKLOAD

Here we will discuss some principles of the measurement of stress and workload, psychologically and physically. In a global sense, there is a lot of research and literature regarding stress and workload measurement. The results can be applicable to the study of speech under stress.

3.6.1   The Measurement of Stress

The measurement of stress would be required from, at least, three different domains (Cox, 1995):

a) The objective and subjective antecedents of the stress experience, which include the audit of the work environment (both its physical and its psychosocial aspects)
b) Self-report on the stress state, can be a survey of workers' perceptions of and reactions to work.
c) The various changes in behavior, physiology and health status, which may be correlated with the antecedents.

In most of published literature, the most common way of measuring the stress is self-report only. It is insufficient, if there is not an established relationship with other data. It lacks the reliability and validity of such a measurement.

Cox and Griffiths (1995) have provided some measurement methods for stress. The Stress Arousal Checklist (SACL) uses self-reported mood for tapping the individual's experience of stress. It presents the respondent with thirty relatively common moods-describing adjectives and asks to what extent they describe their current feelings. The model of mood that underpins the checklist is two-dimensional: one dimension appears to relate to feelings of unpleasantness/pleasantness or hedonic tone (stress) and the other to wakefulness/drowsiness or vigor (arousal). Other methods, such as the measurement of health and, the general well-being questionnaire (Cox, 1995) are used to evaluate the long-term effects of exposure to stress in general. These methods were not interesting or noticed by speech technology engineers and have never been used in any kind of speech studies related to stress.

3.6.2   Workload Assessment

It often happens that people use the assessment of workload to indicate the level of stress. The measurement methods for stress and workload overlap in a big scope. When we discuss stress measurement, we cannot avoid discussing the methods for workload assessments.

In a work environment, the differences of an operator's characteristics can very much influence information processing of the individuals. These differences arise from a combination of past experience, skill, emotional stress, motivation and the estimation of risk and cost in a task. The influence of these individual differences directly affects the experience of workload and stress.

Mental workload is mostly defined as the ratio between task demands and the capacity of the operator. Mental workload is high when the difference between task demands and the capacity is small. To evaluate mental workload, both task demands and an operator's capacity have to be assessed. The analysis of the task demands provides insight into the cause of high workload, and the measurement of capacity provides information about the mental effort that has to be invested in the task.

Normally there are three different ways to approach the workload assessment (Meshkati, 1995): performance measurement, subjective measurement and psychophysiological measurement. Any single measurement may not provide enough information understand the full scale of human skill and abilities required to operate and interact with these complex systems. While the full suite of methods may not be required or feasible in every instance of testing and evaluation, one should consider using as complete a set of methods as possible and reasonable.

In a complex task environment, it is difficult to determine task demands. In an attempt to cope with the task demands, operators adapt by changing their strategies, delaying low priority tasks and lowering performance criteria. So workload cannot be regarded as a static concept. It is not possible to assess the workload by a single measurement, from single aspects to evaluate it or from one moment in the whole time period.

3.6.3   Performance Measurement

Primary task performance measurement is probably the most obvious measurement method. If we wanted to know how driving is affected by traffic conditions, fatigue or lane width, we should measure the driving performance as a criterion. There are several approaches to the measurement of performance. The major challenge is to select the method that is most sensitive to the changes of the performance due to different levels of mental workload. Here I would like to include a brief overview of these approaches.

a. *Analytical approach*: An analytical approach looks in detail at the actual performance of the task to be assessed, examining not only overall achievement, but also the way in which it is attained. The advantage of this method is that the various decisions and other processes that make up performance are considered in the context in which they normally occur, so that the full complexities of any interaction between different elements in the task can be observed. For any one of the performances, several scores need to be taken. The component parts of a complex cycle of operations should be measured separately. Errors may be recorded as well as time taken, and different types of errors need to be distinguished. The reader should review Welford (1978) and Chiles and Alluisi (1979) for details.

b. *Synthetic methods*: These methods start with task analysis of the system, in which the proposed operating profile is broken down into segments or phases that are relatively homogeneous with respect to the way the system is expected to operate. There are many different ways of performing the task analysis (Wilson, 1995). For each phase of the mental workload, the specific performance demands placed on the operator are then identified through task-analytic procedures. Performance times and operator reliabilities are assigned to the individual tasks and sub-tasks on the basis of available or derived data (Meshkati, 1995).

c. *Multiple measurement of primary task performance*: These techniques might be considered useful for workload assessment when individual measures of primary task performance do not exhibit adequate sensitivity to operator workload because of operational adaptation due to perceptual style and strategy.

A secondary task is a task that the operator is asked to do in addition to his or her primary task. If he or she is able to perform well on the secondary task, this is taken to indicate that the primary task is relatively easy. If it is the other way around, then this is taken to indicate that the primary task is more demanding. Some examples of secondary tasks are arithmetic addition, repetitive tapping, choice reaction, critical tracking tasks and cross-coupled dual tasks. Ogden, et al., (1978) reviewed 144 experimental studies which used secondary task techniques to measure, describe or characterize operator workload. The extra load imposed by the secondary tasks as a factor might produce a change of strategy in dealing with the primary task and consequently distort any assessment of the load imposed by the primary task alone (Meshkati, 1995). Individual differences can be large regarding the secondary task performance. Some investigators have addressed additional tasks that may increase arousal and motivation in secondary task performance.

In a complex task environment, performance along with measure often cannot index workload, particularly for each subtask. There are also no general agreed upon procedures to combine scores on different aspects of the task into one score that reflects total performance. Furthermore, it is often difficult to know which task is critical at a particular moment in time. Another complicating factor is that operators will try to keep their performance at an acceptable level. Operators adapt to increasing task demands by exerting additional effort to maintain a constant level of performance. The level of performance, therefore, only provides valuable information when techniques are used to index the invested effort. Under high workload, operators have to invest more mental effort in order to maintain an adequate level of performance (Gaillard, 1994).

Behavior data are suboptimal for the assessment of human information processing and mental workload. These data, as well as subjective measurement data which we will discuss in the following section, are usually only available at the end of a chain or cluster of processes and, hence, contain no information about the timing of single cognitive processes.

3.6.4 Subjective Rating Measures

These include direct or indirect queries of the individual for his or her opinion of the workload involved in the task. The most commonly used methods are self-rating scales or interviews/questionnaires. Several subjective evaluation methods have been developed and used in different kinds of research work such as subjective workload assessment techniques (SWAT) (Reid, 1988), NASA Task Load Index (NASA-TLX) (Hart, 1988), the Bedford workload scale, the Coop-Harper scale, the McDonnell rating scale and the NASA Bipolar rating scale (ANSI, BSR/AIAA, 1992) (Meshkati, 1995).

SWAT (Reid, 1988) uses a three-dimensional workload construct: time load, mental effort load and psychological stress load. It is intended to be a global measure of workload that is applicable in a large range of situations. NASA-TLX (Hart, 1988) used six-dimensional rating scales such as mental demand, physical demand, temporal demand, performance, effort and frustration level. Researchers then weighted each dimension according to the specification of the task in the calculation of the total score (Hart, 1988).

Compared with SWAT, NASA-TLX is more sensitive to the changes of the workload; therefore, it has become the most commonly used method to have the subjective evaluation of mental workload. The experience by the author after applying these methods to several research works are that one shall explain the meaning of the questions carefully to the test person in his or her natural language and provide hint about how to answer each question. Otherwise, different test persons can have very different understanding of the questions, and it can influence the results.

The subjective assessment of workload can be affected by the context of the task performance (Colle, 1998). It means if the subjects knew the performance intensity changes between one test to another, they tend to rate their score coordinate to the intensity of the performance. The effects of context sensitivity can be reduced by presenting the full range of task difficulty on practice trials with instructions to use the full range for ratings (Colle, 1998). Another way is to manipulate the intensity without labeling or making the difficulty levels known to the participants. Therefore, participants could not easily remember a name or a highly identifiable characteristic and remember the ratings previously given to that condition.

Rating scales generally provide a good indication of the total workload. In many instances, however, operators may not have sufficient time to fill out a rating scale while they are working. So, this method cannot be sensitive to measuring the performance when the demands for the performance are changing over time. Moreover, operators appear not to be able to discriminate between the demands of the task and the effort invested in task performance. In other words, on the basis of rating scales it is not clear whether an operator works hard or thinks that he has to work hard.

3.7   PSYCHOPHYSIOLOGICAL MEASURES

Various methods are used to investigate psychophysiological functions in order to evaluate the degree of workload, stress and even fatigue in the workplace. Some of the methods are possible to non-intrusively and continuously monitor the state of the human operator. Tasks do not have to be interrupted in order to obtain the effects. Operators do not have to suspend task performance or break their train of thought to provide an immediate perspective on their interaction with the system that they are operating. For the most part, psychophysiological measures do not require modification of the operated system, as can be the case with performance measures that require data from the system. The continuous nature of these measures provides several advantages to testing and evaluation.

Compared with subjective and performance measures, psychophysiological measures require knowledge and experience for proper use. However, this is the case with all of the methods used in test and evaluation. While the skills might be more specialized, the principles of physiological data collection and analysis are not difficult to learn and apply. There are several good handbooks, guides and journal articles on the proper collection and analysis procedures for these data. Here I will give some brief introduction to different methods and readers should look for original articles or books for details. Wilson[2] (2002) has given a review of the application of different kind of psychophysiological measures in different areas.

Several studies have used physiological measures to classify operator state related to mental workload. Most of these studies employed EEG data. These studies used either simple, single task paradigms (Gevins, 1998; Gevins, 1999; Nikolaev, 1998; Wilson, 1995) or relatively few peripheral physiological variables with complex task performance (Wilson, 1991). They reported overall successful task classification in the high 80% to 90 % correct range. These excellent results classifying high mental workload or altered operator state are very encouraging. They suggest that these methods could be used to provide accurate and reliable operator state assessment to an adaptive aiding system.

Unfortunately, psychophysiological measurement methods are seldom be used in speech studies. Most research on speech under stress take performance measurement and some of them applied subjective evaluation very few studies applied HR measurement. I will discuss it in more details in the later sections in this chapter.

---

[2] HFA workshop material, Swedish Center for Human Factors in Aviation, IKP, Linköping University, SE-581 83, Linköping, Sweden

3.7.1 Psychological Function Test

The critical flicker frequency (CFF) test is one of the methods. The flicker of the light can be sensed when the frequency of the flashing of the light per unit of time is low. The frequency at the boundary between flicker and fusion may be low when a fatigue phenomenon arising from the central nervous system appears. A drop in the CFF value reflects a drop in the sensory perception function that is attributable to a decrease in the level of alertness (Kumashiro, 1995). This measurement is often used for studying a driver's drowsiness and driving behavior.

A reaction time test is commonly used as well. It is to measure the information processing capability. A blocking test is used to measure the performance under uncontrolled conditions. The concentration maintenance function (TAF) method has a standardized performance process (Kumashiro, 1995) to test the state of concentration in continuous target aiming. It is described as a continuous curve. These curves are used to judge fatigue and in particular evaluate psychological stress in places prone to fatigue.

3.7.2 Physiological Function Test

As a human being is a biological entity, any physical and mental activities can be reflected by physiological changes. Workload and stress can also be measured physiologically. Individuals who are subjected to some degree of mental workload commonly exhibit changes in a variety of physiological functions. There is a need to find the optimum approach to determining the psychophysiological function. It requires multiple techniques to ascertain physiological phenomena with multiple channel recorders. Since physiological variables are also influenced by other factors, such as environmental condition and physical activity, an estimation of mental workload should be based on several measures. The advantage of physiological measures is that they are unobtrusive and objective and may provide continuous information about mental effort. Here we will discuss several commonly used methods in more detail. These methods are:

a) Eye blinks
b) Body fluids
c) Skin responses
d) Respiratory and Cardiac function
e) Brain electrical activity

- **Pupil diameter and eye blinks**

Eye blinks can be derived from an electrooculogram (EOG) measured with electrodes above and next to one of the eyes. The algorithm for the detection of blinks and calculation of the duration deviates from a conventional algorithm in which a blink is defined by the decrease in amplitude and the duration by the time between the halfway amplitude of the blink onset and offset (Goldstein, 1992; Wilson, 1987). Some of the studies found that (Veltman, 1996) eye blinks are sensitive to visual load but not to cognitive load.

- **Body fluids**

Hormones produced in the body in response to environmental changes can also be measured. These include cortisone and epinephrine levels in the blood, urine and saliva. Changes in hormone levels can take 5 to 30 minutes to become evident in body fluids following significant environmental events. This is a handicap in some situations where the task is associated with rapidly changing events. However, in other situations where overall effects on operators over longer periods of time are concerned, hormonal measures have value.

- **Respiration and cardiac function measurement**

When subjects invest high mental effort to cope with high workload, it can result in a decrease in parasympathetic and an increase in sympathetic activity (Gawron, 1989; Mulder, 1987), which results in peripheral reaction in heart rate, respiration and blood pressure.

Respiratory function tests include respiration rate, lung capacity and maximum air ventilation. Cardiac function measurement includes several parameters such as inter-beat-interval (IBI), heart rate (HR), heart rate variability (HRV) and blood pressure. Heart rate and its derivatives are currently the most practical physiological methods of assessing imposed mental workload. HR and HRV are assumed to be influenced by both sympathetic and parasympathetic activity of the autonomic nervous system. An increase in workload results in an increase of HR, whereas HRV decreases. There are many different ways to calculate the HRV such as standard deviation of IBIs, mean-squared sequential differences (MSSD), spectral analysis of IBIs and filters. HRV can be divided into different frequency bands, for example, between 0.02 to 0.06 Hz is called low or temperature band, which can be easily affected by environmental temperature. Between 0.07 to 0.14 Hz is middle or blood pressure band and 0.15 to 0.5 Hz is high or RSA (respiratory sinus arrhythmia) band (Veltman, 1996).

Wierwille (1979) and Meshkati (1983) have given a detailed review on the experimental studies of the connection between HRV and mental workload. An index for HRV can be obtained by spectral analysis. Measures of HRV have been assessed through the use of three major calculation approaches (Meshkati, 1988): a) scoring of the heart rate data or some derivative (e.g., standard deviation of the R-R interval); b) through the use of spectral analysis of the heart rate signal; and c) through some combination of the first two methods. There are two advantages to such a measure. First, the absolute advantage refers to the sensitivity of the measure to change in mental workload as demonstrated in the previously cited investigations. The second advantage is its practical utility and relative simplicity in both administration and subsequent interpretation when compared with alternative physiological techniques. However, it should be acknowledged that since HRV is a particularly sensitive physiological function, it is vulnerable to potential contamination from influences of both stress and the ambient environment.

In several laboratory studies, HRV has been found to be a sensitive index for mental effort (Mulder, 1987) in field study, HRV has been found to be less sensitive (Veltman, 1996). The insensitivity of HRV appears to be caused by the intrusion of respiratory activity. Respiration causes a change in HR; during inspiration the HR increases, whereas the HR decreases during expiration. These respirations-related HR changes are called respiratory sinus arrhythmia (RSA). When HRV is used as a measure of mental effort, it is important to measure respiration, too.

Blood pressure (BP) is controlled by many different short- and long-term mechanisms. Changes in the mean level of systolic and diastolic BP are mainly caused by sympathetic activity. An increase in sympathetic activity causes an increase in BP. BP is not influenced by respiration (Veltman, 1996).

- **Brain electrical activity**

In many situations, there is a need to focus on fluctuation in the levels of cerebral cortex activity. An EEG provides information concerning the electrical activity of the CNS. The principles and different measurements of EEG are described in Chapter 2. Rokicki (1995), based on 14 years of observation, reviewed the rationale for using psychophysiological metrics, particularly an EEG, during the test and evaluation process, and discussed the requirements for its use and acceptance by decision-makers and subjects.

Event related potentials (ERPs) derived from an EEG (see Table 2-1 for summary) represent the brain activity associated with the brain's processing of discrete stimuli and/or information. ERPs are thought to represent the several stages of information processing involved in cognitive activity related to particular tasks. Other measures, including positron emission tomography (PET), functional magnetic resonance imaging (fMRI) and the magnetioencephalogram (MEG), provide fantastic insights into CNS functioning. However, they are presently limited to laboratory environments. They will no doubt provide impetus to field applications, but in their current state their use is not at all feasible for test and evaluation applications with most systems. The restraints involve the recording equipment, which requires subjects to remain immobile in the recording apparatus and does not permit ambulation. Also, high levels of shielding from external magnetic and electrical fields are required, which limit real world application at present (Wilson, 2002).

Wilson used a trained artificial neural network (ANN) to recognize the physiological data from the experiment (Wilson, 2002). In his study, the input to the ANN consisted of spectral EEG features including delta (1-3 Hz), theta (4-7 Hz), alpha (8-13 Hz) and beta (14-30 Hz) bands. Other features included ECG interbeat intervals, blink rate and blink closure duration and respiration rate. The results demonstrate that an ANN using central and peripheral nervous system features can be trained to very accurately determine, online, the functional state of an operator. This is especially significant in light of the complex, multiple tasks that were performed by the subjects. High levels of accuracy are possible using ANNs. No doubt using both central and peripheral nervous system features enhanced the performance of the ANN. Additional accuracy may be achieved by including performance features when possible (Wilson, 1999).

Instead of the statistical method, Moon, et al., (Moon, 2002) employed a smoothing algorithm for the power of alpha and beta band frequencies of EEG curve and the variation of pulse width obtained from ECG curve to generate their corresponding trend curves. Using the values taken from the trend curve, they (Moon, 2002) defined the mental workload as a function of the three variables. They found that all three variables, α-power, β-power and the variation of pulse width increase during the period when the mental workload is expected to increase. They also found that the results of using nonlinear fuzzy rules reflect the mental workload changes more clearly than the results of using linear functions.

The ERP usually consists of several components that are assumed to be real-time correlates of specific cognitive processes. The relatively early components mainly reflect automatic processes such as stimulus detection, while the later components mainly reflect controlled processes. The latency of a component reflects the timing of a process, the component amplitude reflects the strength of the process and the component topography (the amplitude distribution across the scalp) reflects the participation of different brain areas in the process (Hohnsbein, 1998). ERP is a good method to study the information about single cognitive processes and mechanisms and their dynamics with high resolution in time. ERP can be used, for example, as a tool to assess mental effort, to specify bottlenecks and mental overload or to elucidate the reasons for performance differences between conditions and between subjects. The study (Hohnsbein, 1998) showed that the ERP analysis could reveal specific reasons for the observed performance differences. This also makes it possible to choose suitable strategies to enhance the performance of subjects, (e.g., preparation and motor control training procedures).

However, the ERP approach has some problems. First, the affiliated ERP components may overlap in time since in most tasks the cognitive processes run partly parallel (particularly under time pressure conditions). Second, repeated stimulation and similar psychological conditions across trials are necessary (Hohnsbein, 1998). Indeed, the crucial task in ERP research is to establish relationships between processes and components.

Working memory refers to the limited capacity for holding information in mind for several seconds in the context of cognitive activity (see Chapter 2). Overload of working memory has long been recognized as an important source of performance errors during human-computer interaction. An effective means of monitoring the working memory demands of computer-based tasks would therefore be a useful tool for evaluating alternative interface designs.

The requirement for the method for monitoring working memory load must not interfere with operator performance, must be employable across many contexts and must have reasonably good time resolution. Furthermore, such a method must be robust enough to be reliably measured under relatively unstructured task conditions yet sensitive enough to consistently vary with some dimension of interest. Some physiological measurements, especially EEG, appear to meet these requirements. Gevin, et al., (1998) uses EEG to assess working memory load during computer use, and the EEG spectral feature was analyzed with ANN pattern recognition. The test data segments from high and low load levels were discriminated with better than 95% accuracy. More than 80% of test data segments associated with a moderate load could be discriminated from high- or low-load data segments.

3.8   ENVIRONMENTAL STRESS

It is generally known that adverse environments have a negative effect on the employment of speech technology. It is still difficult to define precisely what constitutes an adverse environment on speech technology, especially the automatic speech recognition (ASR) system (Baber, 1996). Since many speech technology systems will be operated in unfavorable conditions, such as in workshops, public places, traffic controls, vehicles or cockpits, the negative effects should be understood and predicted. All kinds of environments have the potential to create negative effects on human-ASR interactions. For many speech technology applications, ASR is the gate to the speech systems.

One of the main problems of using speech interaction is the speech recognition error caused by misinterpretation or user's error of speech (Chen, 2003; 2001). A user normally needs to find the different strategies to correct the mistakes made by the ASR. It irritates the user and, hence, increases the workload and stress. The studies regarding different environmental factors to ASR will be briefly discussed here. The effects of noise, vibration and G-force to human behavior are of interest for the military application.

3.8.1 Acceleration

Although acceleration can happen in all moving vehicles, the level of acceleration that can affect the speech products of humans only happens in high-performance jet aircraft and space rockets. There the occupants are subject to very high levels of acceleration. General human performance decrements under high acceleration have been observed in a number of studies (Baber, 1996). The effects of acceleration on performance can be summarized as below:

a) At about 3 g, pilots have experienced difficulties in visual perception, particularly in tasks requiring fast and accurate reading/monitoring of displays.
b) In excess of 5 g, serious problems seem to arise. The capacity of doing tasks related to manual control-particularly fine motor activity and reaction time performance decrease. Subjective workload ratings increase from around this level.
c) In excess of 6 g, there are disruptions to memory, particularly to immediate recall, and the ability to retain and follow procedures.

Acceleration increases in forces applied to the speech production apparatus and changes in breathing rates. Displacement of the vocal tract as a result of increased acceleration leads to marked shifts in the format frequencies of speech (Moore, 1986). In excess of 6 g, speech articulators in the vocal tract can be displaced, though intelligible speech can be produced up to 9 g.

Bond, et al., (1987) compared speech produced at normal and high acceleration levels with individuals wearing oxygen masks. They found, although there was no change in amplitude as a consequence of increasing acceleration, there was an increase in fundamental frequency and a compacting of the vowel space between F1 and F2 (the frequencies at which a vocal tract resonates, particularly during the production of vowels). South (1999) measured the first and second formant frequencies of nine vowels from one speaker recorded under high levels of acceleration with and without positive pressure breathing. Under acceleration alone, or with positive pressure breath, the changes in F_1 and F_2 are different but showed reasonable consistency.

One way to overcome the problems of high-G causing changes in the person's speech and thus adversely affecting the performance of the speech recognizer may be to adapt the recognizer's word models to the current environment using information on G-level and breathing gas pressure supplied by the aircraft's systems. Baber, et al, (Baber, 1996) suggested that future research could, for instance, define approximate shifts in a person's speech relative to changes in G and then develop an algorithm to track the shifts in an individual's speech in order to compensate. Jones (Jones, 1992) has pointed out that, although G-forces were found to have significant effect on speech, the level at which recognition was seriously impaired (in the region of 7 g) is above the level at which manual input would be affected. In general, flight trials with trained aircrew show that physical stressors are less likely to affect recognition accuracy than cognitive and emotional stress, and we have very little knowledge on this part of effects from high-G to speech products.

3.8.2   Vibration

Vibration can occur in a wide range of domains such as aircraft, road, and marine transport. Vibration becomes a nuisance as it has a detrimental effect on human performance and on human-human/human-machine communications. Human visual performance and motor performance are known to deteriorate with the increasing of vibration (Lewis, 1978a; 1978b).

When considering the potential effects of vibration on speech production, variations in airflow through the larynx will probably lead to some degree of "warbling" in speech sounds, and some frequencies of vibration will induce breathing irregularities, which in turn will affect speech sounds (Griffin, 1990). Moore and Bond (1987) suggested that under low-altitude, high-speed flight, vibration would be sufficient to buffet the body and lead to changes in speech pattern; that is, the fundamental frequency increases and the space between F1 and F2 becomes more compact. These effects, in addition to the modulation or "shakiness" imposed on the voice, are due to the whole body vibration. The vibration level and which part of the body is exposed to the vibration source are the key factors for the effects to the speech products.

Dennison (1985) conducted a study and exposed the subjects to six types of simulated helicopter vibration after enrolling them on a speech recognizer. Tests on an ASR device with a vocabulary of fifty items revealed no difference in recognition accuracy between words spoken under vibration stress and words spoken with no vibration. This was the case even though all templates had been trained with no vibration present. It appears that physical stress induced by vibration alone is not sufficient to affect ASR recognition accuracy. One reason for this is that subjects were exposed to vibration that simulated only routine manual performance; therefore, psychological stress remained low (Malkin, 1986).

3.8.3 Noise

Noise is viewed as "an unwanted sound" (Haslegrave, 1991). Such a definition makes this subject very complicated. Environmental noise is a big problem for ASR applications for a number of reasons:

a) Variations in noise levels and the frequency of the noise may create possible problems for ASR. The ASR device may misrecognize noise as an utterance, or else an appropriate utterance by the user is rejected due to noise corrupting the input signal (these difficulties may be particularly serious if the background noise is itself speech).
b) Noise appears to affect the level of processing of information. Auditory feedback for some ASR applications will be difficult or impossible in a high noise environment.
c) Levels of 80, 90 and 100 dB lead to an increase in the amplitude, duration and vocal pitch of speech relative to quiet and also observed changes in format frequency (Van Summers, 1988).
d) People in a noisy environment tend to concentrate on surface aspects of information rather than on deeper aspects (Hockey, 1979; Smith, 1991). Noise can be stressful or fatiguing for the user particularly if it is persistent and loud (changing voice patterns in users suffering from stress or fatigue may cause errors). In noisy environments, people tend to increase the volume of their speech in order to hear what they are saying. This is known as the "Lombard effect" (Lane, 1970).
e) Individual differences and preferences will play a major role in tolerance to various noise levels. Noise has a tendency to bias response in favor of dominant responses.

There are a number of solutions to background noise. A relatively simple hardware solution is to use uni-directional noise-canceling microphones. This will eliminate or reduce the extraneous noise entering the ASR device and help prevent misrecognition. Alternatively, noise-cancellation can be carried out during the processing of the input signal, prior to applying the speech recognition algorithm. Essentially, this involves "subtracting" the noise spectrum (recorded from an independent microphone) from the spectrum containing speech and noise. Another solution to background noise that is intermittent is to "train" a template for the noise, so that if the noise is recognized by the device, no response is made.

In the past ten years, the effects of noise on ASR operation have been extensively studied. Electrical engineers and scientists have put a lot of effort on increasing the recognition accuracy by developing different computing models based on acoustic signal analysis of the noise and speech in the noise environment. Some commercially available ASR devices are equipped to cope with high ambient noise levels (Swail, 1997). Solutions focus on the development of sophisticated noise canceling and adaptation algorithms, which minimize the effects of noise on ASR performance. Compared to the human behavior change due to exposure to the noise, the speech intelligibility affected by noise seems easier to handle.

Noise can affect the intelligibility of speech, and it seems to affect the female voice more than the male voice (Nixon, 1998) at high noise levels such as 115 dB. The intelligibility degradation due to the noise can be neutralized by use of an available, improved noise-canceling microphone or by the application of current active noise reduction technology to the personal communication equipment. When both male and female voices are applied to the ASR system in the same noise environment, the recognition accuracy is the same (Nixon, 1998).

Noise can affect the users in different way. It may cause stress and fatigue. This effect can be avoided or reduced by using ear protection or headphones or by reducing the length of time to which users are exposed to the noise. Although users will be protected when noise levels are very high, wearing hearing protection does not eliminate the problem of noise in ASR applications since even moderate noise can affect the user's behavior and cause recognition errors.

Auditory feedback may also be masked by ambient noise. These problems can be overcome by different solutions. Headphones or visual feedback could be used instead. If, for any reason, headphones are impractical or not effective, feedback could be in the form of cutaneous stimulation, perhaps in the form of a pad of vibrating "needle" (Hapeshi, 1988).

3.8.4 Auditory Distraction

Auditory distraction is caused mainly by those "irrelevant sounds" rather than environmental noise. It can be another's speech, music, which is not in the best interest for communication. Banbury (Banbury, 2001) has given a good review of the properties of the irrelevant sounds that cause disruption when performing complex mental tasks:

a) Acoustic change is the main causal factor of the disruption. Thus, repeated sounds, tones or utterances are not disruptive.
b) Acoustic changes may be manifested by changes in pitch, timbre or tempo but not sound level.
c) Nonspeech sounds, such as tones, can be as disruptive to performance as speech when equated in terms of their acoustic variation.
d) Irrelevant sounds do not have to be presented at the same time as the material they corrupt. They produce an effect during stimuli presentation as well as during retention intervals.
e) The disruption is enduring. The weight of evidence suggests that habituation does not occur. If habituation does occur, relatively short periods of quiet have been observed to cause rapid dis-habituation to the irrelevant sound.
f) Memory is very susceptible to interference by irrelevant sound, specifically the cognitive functions pertaining to the maintenance of order in short-term memory.
g) Other tasks, such as proofreading and comprehension, may also be disrupted by the semantic properties of the irrelevant sound.
h) The effect of irrelevant sound is not one of masking, nor is it attributable to the similarity of the sound to the target items. Rather, it appears to be caused by a conflict between two concurrent processes of seriation: one stemming from the rehearsal of material in memory; the other arising from the obligatory processing of auditory information.

For speech interaction design, the disruption from irrelevant sound is a big problem for the system performance.

3.9 WORKLOAD AND THE PRODUCTION OF SPEECH

Up to now, we still lack the fundamental understanding of how human speech degrades under severe environmental and stress conditions. Workload leading to changes in speech can be considered from two perspectives. A mechanical perspective would consider that workload can induce measurable physical changes to the speech production apparatus, e.g., muscle tension could increase (resulting in changes to speech level and fundamental frequency), breathing rate could quicken (leading to increased speech rate, reduced inter-vowel spacing and increased glottal pulse rate). A psychological perspective would consider that the effects of workload would vary, according to the possible competition between task and speech, and that the competition could be mediated by strategies (Baber, 1996a; 1996c). If there is no signal acoustic correlate between stress and speech, then it may be difficult to provide compensation in ASR devices.

Many workload situations have a characteristic of increase in time-stress, such as finishing a couple of tasks or a series of tasks within the limited time. The performance decrement was related to both a reduction in recognition accuracy and an increase in the number of errors in the vocabulary used to perform the task. It is interesting to note that not all speakers responded to the time-stress in a similar fashion (Baber, 1996a; 1996c). Baber's study of reading car number plates, using the ICAO alphabet under slow and fast conditions (Baber, 1995; 1996c), raised the question of how people deal with stressors, e.g., the fact that some speakers show an increase in speech level and others show a decrease suggests that changes in speech level are not simply physiological reactions to a stressor. There is evidence that speakers can adopt strategies for dealing with the time-stress other than speaking quickly (Baber, 1996a; 1995).

A few questions are asked: Is it possible to compensate for the production of speech under load? Is it possible to produce algorithms that track the graceful degradation of utterance consistency in speakers under workload? What level of training is sufficient to reduce the problems of speech consistency? In order to determine the likely effects of workload on the use of speech recognition systems, it is necessary to develop an idea of what activities are being performed when a person speaks to a device. A simple model is shown in Figure 3-3. The model is developed based on Baber's, et al., (Baber, 1996) proposal.

Based on a speaker's knowledge of the language, the meaning of the message to be communicated is generated. At this level, the context of the conversation, the goals of the speaker, the basic intention to utter a message and the information from a database will affect the process. At the next level, the syntax of the utterance and the vocabulary to build the utterance will be processed. Finally, an articulatory program needs to be constructed and run to produce the utterance. For the utterance to be recognized by the ASR system, the utterance and the acoustical signal from this utterance has to be matched between the human speech and the system.

The cognitive effects of the stressors may, at higher levels, influence the selection of vocabulary and the syntax before the utterance is generated. Another effect can be no response to the requirement of generating the utterance or delay of the response. It may also, in the lower level, cause the vocal changes of the speech production. Many studies are focused on the analysis of the vocal changes due to stress and try to improve the recognition accuracy by building up different algorithms. This will be discussed extensively in the following section. There is very little study about stress and language behavior at the high level of the model (Figure 3-3).

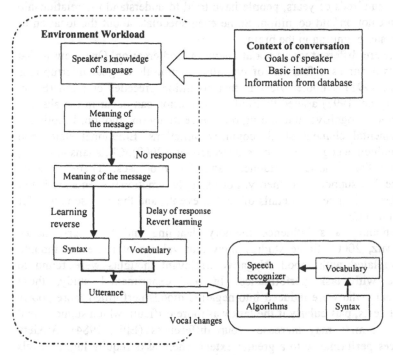

Figure 3-3. The process of how an utterance is developed

3.10   ANALYSIS OF SPEECH UNDER STRESS

An understanding of speech changes under stress is necessary for employing speech recognition technology to the stressful environment (Brenner, 1994; Vloeberghs, 2000). Here we are interested mostly in those effects from short-term physical and cognitive stressors. Many studies related to analyze the speech voice are related to emotion and sometimes it is difficult to distinguish between emotion and stress. People sometimes use the word "emotional stress." Emotion and speech is a hot research topic recently. Speech carries rich information of a speaker's emotional states, but emotion and speech are not our interests in this book. However it can be very useful for us to have some understanding of the relationship between emotion, cognition and stress.

3.10.1   Emotion and Cognition

Over hundreds of years, people have tried to understand the relationship between emotion and cognition. Some even speculate about the location of emotion and cognition in the brain.

There are different views about cognition and emotion. One view is that emotions occur as a function of cognitive appraisals or goal interruptions (Lazarus, 1991). Another view is that cognition precedes emotion (Stein, 1992; Zajonc, 1997) argued that although emotion can occur in the absence of cognition, cognitive functioning relies essentially on emotion. Emotion is a fundamental element of all cognitive functions. Emotional states and events influence cognitive processes (Isaacowitz, 2000). Christianson (1992) suggested that emotion influences attention processes. There is some evidence that supports this theory. For example, emotional events are better remembered than central details of neutral events, and the reverse is true for peripheral details.

Emotional states influence memory, learning and task performance (Isaacowitz, 2000). Some studies show that valence of irrelevant thought was correlated with mood. Negative irrelevant thoughts were found to interfere with task performance. In one experimental study, those participants who were induced into negative moods performed more poorly on recall tests, particularly if the task was more difficult, which suggests that greater demands may increase group differences (Ellis, 1984). Anxiety influences performance to a greater extent when tasks require higher levels of cognitive organization (Eysenck, 1982). It was also found that depressive symptoms were related to poorer cognitive performance on tasks involving both memory and psychomotor speed (La Rue, 1995).

Bower (Bower, 1981) proposed an association network model of spreading activation. He used nodes to represent concepts. Activation of one node automatically activates related nodes. Theoretically, different emotions are represented by specific nodes in associative networks, each linked to related conceptual modes. These concept nodes can be automatic responsive, cognitive appraisals and memories for events that occurred when emotion was experienced. So activation of a specific emotion mode is distributed, thereby activating relevant physiology and memory of related events. When people learn information in a particular mood state, they disproportionately encode and remember mood-congruent information. If people are in a happy mood, they are more likely to generate positive characteristics.

It is clear at this point that emotion affects task performance, and hence induces different experience, of stress. That is probably why, especially in speech study, it is hard to distinguish between effects from stress and from different emotional states.

3.10.2 Speech Measures Indicating Workload

The ability to detect stress in speech has many applications in voice communication algorithms, military voice applications and law enforcement. The special advantage with speech measure is unobtrusive and does not require attaching any equipment to the person being tested. The measurement can be taken on radio-transmitted speech or recorded speech. There are studies analyzing the formant location and formant bandwidth in vocal spectrum. They show significant changes under various types of stress conditions (Bond, 1989; Steeneken, 1999).

Different stressors affect the nervous system and, in turn, the physiological consequences of the nervous system's response define the changes in voice characteristics that carried out the stress information (Scherer, 1981). It was evident that the acoustic correlates of stress in the human voice are subject to large individual differences and most researchers indicated that there are no "reliable and valid acoustic indicators of psychological stress" (Scherer, 1986).

Anyhow, several aspects of speech have appeared to respond to workload demands. But their robustness and sensitivity for the speech changes and workload are still being questioned (Johnson, 1990; Ruiz, 1990; Scherer, 1981; Williams, 1981). The candidates' measurements of speech normally were speaking fundamental frequency or pitch, speaking rate and duration, and vocal intensity or loudness.

- **Speaking fundamental frequency (pitch)**

Pitch has been widely used to evaluate stress. Under stress, the pitch of the voice may increase. Fundamental frequency (F_0) reflects the physical tension of the laryngeal muscles and, to a lesser degree, subglottal pressure related to stiffness of the vocal folds. It is among the most frequently recognized speech indices of stress (Ruiz, 1990; Scherer, 1981; Williams, 1981). There are several ways of considering pitch in a speech: subjective assessment of pitch contours, statistical analysis of pitch mean, variance and distribution (Hansen, 1996). Mean pitch values may be used as significant indicators for speech in soft, fast, clear, Lombard, questioning, angry or loud styles when compared to neutral conditions.

- **Speaking rate and duration**

Speaking rate is a common measure of psychological state (Brenner, 1991; Johnson, 1990; Ruiz, 1990), and its predicted increase would be related to a general speeding up of cognitive and motor processes, which often appears under stress. Speaking rate was measured in syllables per second.

Duration analysis was conducted across whole words, and individual speech classes (vowel, consonant, semivowel and diphthong) (Hansen, 1996). Individual phoneme class duration under many conditions is significantly different for all styles (Steeneken, 1999). It is another way of measuring the speaking rate. Some of the study results are concluded here (Vloeberghs, 2000):

a) In slow, clear, angry, Lombard and loud speech conditions, mean word duration will change.
b) In other styles of speech, the slow and fast mean word duration will change.
c) In styles of speech except slow, the clear mean consonant duration will differ.
d) Duration variance increased for all domains (word, vowel, consonant, semivowel and diphthong) under slow stress.
e) Under fast stress condition, duration variance will decrease.
f) For angry speech, duration variance will increase.
g) Clear consonant duration variance was significantly different from all styles.

- **Vocal intensity (loudness)**

The expected increases in the intensity of the voice would likely then reflect increased thoracic air pressure which often occurs under stress (Ruiz, 1990; Scherer, 1981). In general, the average intensity is observed, which increases in noise (Lombard reflex), with anger or some types of high workload. It was also found that mainly vowels and semi-vowels show a significant increase in intensity while consonants did not (Steeneken, 1999). The analysis of intensity was conducted on whole word intensity and speech phoneme-class intensity (vowel, semi-vowel, diphthong, consonant) (Hansen, 1996). Statistical tests were performed on mean, variance and distribution across the database.

Jitter, a subtle aspect of audible speech, is a measure of the minute changes in the period of successive fundamental frequency cycles.

Lieberman (1961) proposed that jitter decreases in response to psychological stress.

3.10.3 Acoustic Analysis of Speech

A variety of calibrated data was collected by a NATO project (Vloeberghs, 2000) covering a moderate range of stress conditions. Parameters indicating a change in speech characteristics as a function of the stress condition (e.g., pitch, intensity, duration, spectral envelope) were applied on several samples of stressed speech. The effect on speech obtained for perceptual (noise) and some physical stressors is evident. More difficult to determine is the effect on speech obtained for psychological and physiological stressors (Steeneken, 1999).

In an extreme psychological stress, by asking the subjects to count the numbers in a very limited time period while doing a tracking task, Brenner (Brenner, 1994) reported that pitch, intensity and speaking rate responded to the changing workload demands. His study did not find quantitative relationship between the changes and the workload. The degree of the changes was relatively small, even though significant differences were appeared. Between the easy and the difficult tasks, average sound pressure level increased by 1 dB, average frequency by 2 Hz and average speaking rate by 4%. At a much lower level of stress, the same increases in these three aspects of speech have been reported in response to workload changes in cockpit-like applications in the laboratory (Griffin, 1987). Protopapas and Lieberman (1997) reported that the range of F_0 does not inherently contribute to the perception of emotional stress, whereas maximum F_0 constitutes the primary indicator. It should be noted that while many studies have reported similar results, some experiments have found no change or even a decrease in these three speech measures (Johnson, 1990; Streeter, 1983).

Many studies suggest that the increase in loudness, pitch and speaking rate may reflect some valid underlying physiological process that characterizes human speech response over a wide range of stress situations. Another thing that one should pay attention to is that heart rate and vocal intensity are closely associated with respiration, the experimental settings and other factors that may influence the results.

Ruiz, et al., (1996) indicated in their study that the parameters show various kinds of behaviors in reaction to various kinds of situations of stress. They suggested that it is important to explore the physical determinants of the reported variations. They could enlighten the development of new, compound indices, built up in a multivariate framework and able to take into account the various aspects or stress-induced variations of the speech signal.

Some consistent patterns can be identified in the voice of pilots under stress. In a study reported by Howells (1982) verbal protocols were collected from running commentaries made by helicopter pilots during flight and waveform analysis was carried out on these. Statistically significant associations were found between marked changes in the speech waveform and certain combinations of pilot activity associated with the commentary. The results from this study suggest that speech spectra are closely related to task loading and could be used to measure workload on the flight deck.

3.10.4   Improving ASR Performance

Today, commercial-based speech recognition systems can achieve more than 95% recognition rates for large vocabularies in restricted paradigms. However, their performance degrades greatly in stressful situations. It is suggested that algorithms that are capable of detecting and classifying stress could be beneficial in improving automatic recognition system performance under stressful conditions. Furthermore, there are other applications for stress detection and classification. For example, a stress detector could be used to detect the physical and/or mental state of the worker and that detection could put special procedures in place such as the rerouting of communications, the redirection of action or the initiation of an emergency plan (Vloeberghs, 2000).

Most of the stress studies have shown that speech variability due to stress is a challenging research problem and that traditional techniques generally fall far short of improving the robustness of speech processing performance under stress. One approach that can improve the robustness of speech processing (e.g., recognition) algorithms against stress is to formulate an objective classification of speaker stress based upon the acoustic speech signal (Womack, 1996; Zhou, 1999). It is suggested that the resulting stress score could be integrated into robust speech processing algorithms to improve robustness in adverse conditions (Womack, 1996). Another approach to improve recognition robustness under adverse conditions is re-training reference models (i.e., train-test in matched conditions). This method sometimes can be difficult.

Zhou, et al., (1999) reviewed some recent methods for stress classification for speech recognition systems, as well as Vloeberghs, et al., (2000). It can be summarized as:

a) Traditional methods analyze the changes in fundamental frequency, pitch and spectral, microprosodic variation index, spectral-based indicators from a cumulative histogram of sound level and from statistical analyses of formant frequencies, and distances of formants from the center of the first three formants. Evaluation results showed that a microprosodic variation index is effective in detecting mild stress while the fundamental frequency itself is more efficient for severe stress.

b) Neural Network Stress Classification: The evaluation proved that stress classification can help in improving speech recognition performance.

c) Watelet-Based stress classification: Some are better than others.

d) Linear speech feature.

e) Nonlinear TEO (Teager energy operator) features.

Fernandez and Picard (2003) uses features derived from multi-resolution analysis of speech and the TEO for classification of driver's speech under stressed conditions and can achieve better speech recognition performance (80% versus 51%).

Still, all of these methods have certain problems, and stress affecting the speech recognition system performance still remains as one of the unsolved research problems.

3.11   RESEARCH PROBLEMS

There are two research streams in this area. One is on artificial laboratory works to simulate different kinds of stress, where the experiments are conducted in the simulators with different degrees of fidelity to the real events. Another one is relying on real events. Apart from the methodological variability, differences in the concept of stress appear obviously. It seems difficult to agree that the stress introduced in the laboratory setting should provoke the same kind of stress in the real workstation. Moreover, most of the laboratory works focus on a single stressful situation. Studies on real situations, normally from a specific event in the real life with uncontrolled factors, affect the results. The analysis methodology is therefore often specifically to the study. Thus, it is quite difficult to check for the validity of laboratory-induced stresses and to correlate findings from other experiments. Research dealing with both real and artificial stress and using unified methodologies, therefore, appears highly commendable, in order to collect comparable results.

Ruiz, et al., (Ruiz, 1996) compared speech data from real situations and laboratory stress and found that artificial stressing conditions are able to lead the subject to produce vocal changes similar to those obtained in natural situations. At the same time, they pointed out that trained subjects and non-trained subjects, people in real life stress and in laboratory stress, were likely to respond to stress in different way. Investigation would pay more attention to local than to global changes.

Previous research directed at the problem of analysis of speech under stress has generally been limited in scope, often suffering from one to five problems. These include: a) limited speaker populations, b) sparse vocabularies, c) reporting only qualitative results with little statistical confirmation, d) limited numbers and types of speech parameters considered, and e) analysis based on simulated or actual conditions with little confirmation between the two.

Normally, to be able to quantitatively measure the workload and stress, three types of measurements are taken simultaneously, as indicated in previous sections: performance measurement, subjective measurement and psychophysiological measurement. During speech data collection under stress, there is often a lack of behavior data correlated to the speech data. Most of all, people seldom took psychophysiological measurements during the data collection. A few of them took some subjective measurements. This probably in another way contributes to the factor that it is difficult to compare different studies, both laboratory experiment and real situations.

Here we made analysis of ten literatures from publications studying speech under stress (Baber, 1999; 1991; Fernandez, 2003; Rahurkar, 2002; Ruiz, 1996; South, 1999; Stedmon, 1999; Vloeberghs, 2000). NATO's report on "The impact of speech under 'stress' on military speech technology" (Vloeberghs, 2000) included eight stress speech databases. After taking a detailed look at these studies we can find that people ignore or did not measure the stress level properly. Most of the studies only take the measurement on the speech, some studies measured the performance and some applied subjective evaluation scales to indicate that the simulated stress has effects. As there are so many different types of stressors and the cognitive effects are so different, even during the exposure to the stress, moment to moment can be very different, and the results of these studies are faulty, in context-depends conclusion. Most of them have very little external validity, and there are large differences from study to study.

To solve these problems, it is necessary to find a comparable measurement method to measure the stress level and to be able to compare between each other. I would like to propose a research agenda. It shall be taken in three steps. First, establish the relationship between the operator's stress and its psychophysiological measurements both in the field and in the laboratory. Second, establish the correlation between psychophysiological measurements and the acoustic feature of the operators both in the field and in the laboratory. Third, if the above two relationships can be established, then it should be possible to simulate the stress based on different performance context in real life from a laboratory setting. If the psychophysiological measurements in the laboratory simulation are comparable with the field study, it is reasonable to believe that the speech records in the laboratory are representative of its respective field experience. It will provide the opportunities to compare the studies between different research findings. Figure 3-4 shows the connections.

The experiences from the research work of performance under high workload and stress, as we have discussed in previous chapters, have pointed out that psychophysiological measurement, such as an ECG and EEG, can provide the detailed information about the effect of the stress in a time series without interruption to the performance and speech. The psychophysiological measurement can also indicate the stress effects independent from the experimental settings or the specific events when data is collected. It makes the cross-comparison possible. If one can connect the vocal analysis from stress with the respected psychophysiological measurement data, then the external validity of the study will increase.

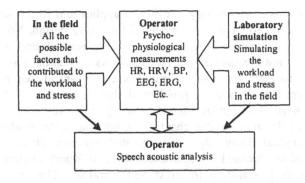

Figure 3-4. The measurement structure of speech under stress

The present statistical modeling paradigm of ASR is also called "knowledge-ignorant models" because it has nothing to do with linguistic knowledge and has more or less reached its top level. These systems are often overly restrictive, requiring their users to follow a very strict set of protocols to effectively utilize spoken language applications (Lee, 2004). The dramatic declines of the ASR system accuracy in adverse conditions affect the usability of ASR system. However, as stress is closely related to the human emotional states and the research on speech and emotion still have long a way to go, the problems with speech interaction system application in stress situations will still contain a lot of problems.

Lee (2004) proposed that the past successes of the prevailing knowledge-ignorant modeling approach could still be further extended if knowledge sources available in the large body of literature in speech and language science can be objectively evaluated and properly integrated into modeling.

The future development will be knowledge-supplemental modeling of the ASR, which means to integrate multiple knowledge sources, to overcome the acoustic-mismatching problems and "ill-formed" speech utterances, which contain out-of-task, out-of-grammar and out-of-vocabulary speech segments. The new approach will give a better understanding of human speech, not just acoustic signals. It provides an instructive collection of diagnostic information, potentially beneficial for improving our understanding of speech, as well as enhancing speech recognition accuracy. This approach probably is one of the reasonable solutions for the speech interaction system design for the application in a stress environment.

3.12   CONCLUSION

To apply the speech technology in different applications in the working environment, one has to have an overview of the entire design. Normally, speech technology is integrated into a system, and the user will deal with the entire system and not the speech interface itself. As a potential positive of using speech technology can be to let the user be eye-and-hand-free from the devices, the users always face the problem of sharing attention with other performances. A good design of attention-sharing is the key point for the success of the application. In this chapter, we introduced the fundamental psycho physiological knowledge and updated theories about human attention, divided and focused attention and multiple task performance.

Human speech is sensitive to stress and workload. The recognition accuracy of ASR decreases dramatically when the user is tired or stressed. Speech under stress is always a challenge issue for ASR application. There is a long history of research work on this topic, but the outcome is not as positive as it should be. One of the problems is that people working in this field lack a basic understanding of workload and stress and their psycho physiological mechanism. Psycho physiological measurements of the stress and workload are important for such kind of study, otherwise, it is not possible to compare of each speech records. Different measurement methods are introduced in this chapter. A review of the research work on speech and stress is also presented in this chapter.

Chapter 4

DESIGN ANALYSIS

4.1   INTRODUCTION

In Chapters 2 and 3, we discussed some fundamental knowledge of human cognition and behavior from neurological and psychological perspectives. When we have the basic knowledge about human cognition, we can prepare for the speech interaction design. There are a few books that talk about the technologies of speech interaction design, especially for spoken dialogue system design (Balentine, 2001; Bernsen, 1998; Cohen, 2004; Gibbon, 2000; 1997; Kotelly, 2003). In most such dialogue systems, users are interacting with the system via telephone, for example, a user calls the call center and asks for different information or books tickets, etc. Such a dialogue system has little connection with other systems. The dialogue design itself is the key technology to having the system functional. As speech technology develops, people are interested in exploring the advantages of speech interaction with complicated systems. Compared to the traditional manual input via keyboard or different control panel and visual output interface, speech provides the user a possibility of hands and eyes free from the control devices. It provides new opportunities for the operators to perform certain tasks that are otherwise almost impossible or poorly done, a typical example being documentation of medical treatment in emergency situation (Gröschel, 2004). The speech interaction design for in-car computer systems and many military applications are such cases. The characteristics of such applications are that speech interaction is just a part of the complicated systems, the user needing to handle not only the speech interaction, but also many other things that probably have higher priorities than the speech itself, such as driving the car and taking care of the patient.

To be able to design speech interaction in a complicated system, before a design activity is carried out, we need to do certain analysis of the design. To design a speech interaction system in real world application, the design analysis shall answer the following questions: Where do we need speech interaction? How would the speech interaction system affect the entire performance of the big system? What can be the impact of the speech interaction system toward human cognitive behavior, working life and daily life? What would the interaction look? We need to have a deep understanding of human work.

There are different methods to carry out such analysis. With different theories to guide the analysis, the outcome can be very different. This chapter will give a brief discussion of different cognitive theories for the design work analysis. It includes ecological cognition theory, distributed cognition concept, cognitive system engineering (CSE) theory and information process (IP) theory.

Information process theory has been dominating in the cognitive engineering research and design work since the '50s' and still has a very strong impact on the engineering world. Most of the research works related to the human behavior in the industry including empirical experimentation works, task analysis and design in the working place, are based on the idea of information process.

The ecological theory has been developed since the '60s. If we said IP is a symbolic processing theory, then ecological theory is trying to give the meaning back to the human behavior. The knowledge in an IP system is based on engineering principles and methods that are shared with other engineering disciplines, together with a specialist knowledge base that describes what we know about human capacities and limitations, and then an ecological approach tries to deal with the constraints imposed by the technology, human capabilities and social/organizational constraints. Ecological interface design theory does not give any answers to the detail of the interaction design while, an IP approach provides the most necessary knowledge for detail of design. These two theories can compromise each other. Before a new design is carries out, the work analysis should be performed from ecological perspective. When it comes to the detail of the design, however, the IP approach to the task analysis should be applied. When it comes to the detail of the human-computer interaction, then the GOMS analysis method should be the choice.

4.2 INFORMATION PROCESSING THEORY

In the '70s and early of '80s, the information processing (IP) paradigm was considered to be the appropriate way to study the human cognition process. Human cognition can be understood by comparison with the functioning of computers. It has the basic characteristics of symbolic processing. People's intentional beings are interacting with the external world through certain symbolic processing. The aim of cognitive psychological research is to specify the symbolic processes and representations that underlie performance on all cognitive tasks. The symbol system depends on a neurological substrate but is not wholly constrained by it. At the same time, the limitation of the cognitive process can be reflected by its structure, resource and reaction time.

The most representative work in information process approach is Wicken's information process model (2000), as shown in Figure 4-1.

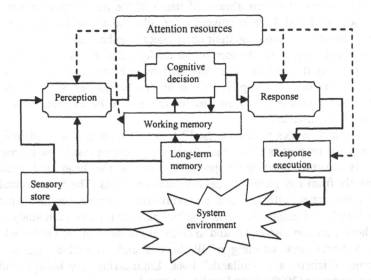

Figure 4-1: Wickens' model of human information processing stages.

This information process is presented as a series of stages. Information and events in the environment gain their accesses to the brain through sensory systems, such as visual and auditory receptors. All sensory systems have an associated short-term *sensory store* that resides within the brain. Raw sensory data in the sensory store must be interpreted, or given meaning, through *perception*. The perception processing normally is automatic and rapid (requiring some attention), and is driven both by sensory input and by input from long-term memory about what events are expect. The amount of information in the sensory store is much larger than what has been perceived. In Chapter 2, we discussed some mechanisms and theories about perception in neuroscience perspective. The bottleneck theories (in Chapter 3) explained why only some information in sensory store is perceived.

Cognitive operations are processes of rehearsal, reasoning or image transformation. These processes are carried out by using *working memory*. They generally require greater time, mental effort or attention. The working memory is a vulnerable, temporary store of activated information, and it is driven by attention, which is resource-limited. Sometimes material that is rehearsed in working memory can get access to *long-term memory*, which is less vulnerable and, hence, more permanent. This cognitive process triggers an action—the selection of a *response*. The *response execution* requires the coordination of the muscles for controlled motion to assure that the chosen goal is correctly obtained. The mental operations require the mental resources. *Attention* can be selectively allocated to channels of sensory materials to process. It affects almost all stages of the information process. In both Chapters 2 and 3 we discussed externally about working memory, long-term memory, attention and their neurological mechanisms.

The feedback loop at the bottom of the model indicates that actions are directly sensed by the human, or if those actions influence the system with which the human is interacting, it will be observable sooner or later. The feedback process makes the "flow" of information to be initiated at any point, and it is continuous.

Many research works following the IP framework are trying to identify the fundamental properties of internal cognition processes inside the human brain. It is assumed that these properties are in the head and exist independently from any particular environment and tasks. The fundamental limitations associated with attention and workload became a central theme for laboratory-based empirical work. The key part for the attention study is to know how attention distributes and to discover the locus of the bottleneck. To have a better understanding of IP theories and respective cognitive engineering, Wichens and Hollands' book Engineering psychology and human performance (2000) is the best book to read.

4.3   THE ECOLOGICAL PERSPECTIVE

In IP framework, there is no place where humans set up their goals and what the information and the environment mean to them, and there is not an explanation about how goals and meaning affect people's attention and information processing. The term *meaning* is synonymous with "interpretation." The meaning of something is the interpretation of a particular agent.

Ecological theory took a totally different approach to understanding human behavior. The key issue in ecological theory is that human behavior is goal directed, not just stimulus or feedback-driven. People endow their experiences with meaning. People can see the world around them in terms of relevant cues and expectancies in service of larger goals. People can manage entire situations, anticipate the changes in the world and assess situations in terms of possible solutions, not as causes for trouble experienced.

In ecological perspective, "meaning" is independent from any observer or interpretation. "Meaning" is synonymous to significance. If a pattern is significant with respect to domain functions, the expert can perceive it as significant. With an ecological approach, meaning is a raw material—it is constructed and it is a product. It exists independent of any processing. What the experts do is to discover it and it should be correspondent to certain function goals of the domain work (Flach, 2000). Discovery is a dynamic interaction between an agent and environment. The correspondence is a state of process. High correspondence results in a well-balanced process or a tight coordination between agent and environment.

For the ecological approach, the functional goals of a work domain and the understanding of the structure of the environment were central to building theories of cognition. The structure of the environment is a critical aspect of the "entire system." If the structures of the laboratory task environment are not representative of a work situation, then any generalization from the laboratory study is suspect (Flach, 1998). Meaning here has a basis in the situation. Objects, events or situations are meaningful, relative to the situation.

In the ecological approach, the stimulation energy is not the main concern, rather the stimulus information, not information in the technical sense of communication theory but in the colloquial sense of specificity. The key difference between IP framework and ecological perspective is the shift from stimulus energy to stimulus information. It was argued that the stimulus energy provides activation of the sense organ, but it is stimulation information or structure that provides meaning and is pertinent to perception (Flach, 1998; Garner, 1974; Gibson, 1953). For example, the same object will give different visual sensations, depending on the position of the observer, the position of the light source and the positions of other objects. But still, humans can perceive it as the same object.

The meaning from stimulated information has objective and subjective properties (Flach, 1998). The objective properties refer to the concrete significance for a particular actor. The subjective properties refer to the recognition or interpretation by human beings. This interpretation depends on the specific person's personal aspects, his knowledge and experience in the specific area. The coordination between the objective and subjective properties of meaning requires a structured medium, or "feedback."

Figure 4-2 illustrates an ecological perspective on meaning. Meaning is viewed in terms of three classes: constraints on goal (values); on action and on information. Each of these constraints is a joint function of the human actor and the environment (Flach, 2001). The objectives of ecological study are these constraints and relations among these constraints.

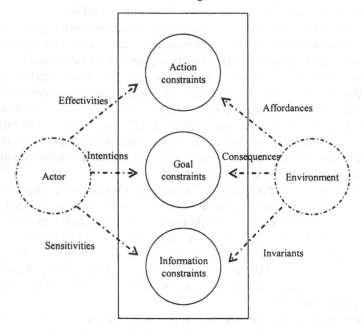

Figure 4-2. The ecological perspective on meaning

In the ecological approach, the fundamental aspect is to identify the objective property of meaning within work situations. A general framework for identifying the "dimensions" of a situation is critical (Flach, 1998). The dimensions include the "consistencies" and "habits" that are unique to a particular work domain. The second aspect is to evaluate the possible strategies for control and the relative costs and benefits of these strategies in the context of specific work ecology. The nature of the structure of the medium and the information value or degree of specificity within the medium (or interface) is a fundamental concern. The third concern is how the abilities and preferences of individuals and the structure and culture of organizations shape the choice and application of the different strategies that are available within an ecology (Flach, 1998). For the interface designer, the central question from this approach is: how to create structure in the medium (interface) to support appropriate control strategies.

4.3.1 Ecological View of the Interface Design

The ecological approach to the interface design has been developed from designing complex sociotechnical systems. It is based on a sociotechnical system viewpoint. It takes the technical/engineering system as the design center. As the approach difference, the research methods also differ between the ecological approach and cognitive psychological approach. Table 4-1 shows the basic differences between the two approaches.

Table 4-1. The principle difference between cognitive psychology approach and ecological approach

Cognitive psychology approach	Ecological approach
The coupling between perception and action happens in the head	The coupling between perception and action through situation (outside the head)
Discover the neural mechanism or knowledge/schema	Structure the links that support the expertise in the situation
Analysis goal is to discover the cues that experts use to interpret a situation	Analysis goal is to discover the constraints on action, the consequences of actions and the information that specifies the constraints and consequences.
Observation of individual's capacity, and awareness	Observation goal is to understand the situation
Laboratory work: based on a parsing of awareness; insights about the structure of cognitive mechanism; meaning associated with particular situation was considered a confound	Capture meaningful chunks of the situation
Simulation: It is used as the tool for laboratory work	Goals, displays, context, resources or plant dynamics can be manipulated at multiple levels that reflect the multiple levels of meaning in a situation

The interface design from an ecological perspective will give support to the workers in adapting to, and coping with, events that are either familiar or unfamiliar to them and that have or have not been anticipated by system designers. The ecological interface design differs from traditional interface design as the idea is to present internal functional and intentional relationships of the work, rather than by giving advice with respect to preconceived ideas by the designer about the proper way of approaching a task (Rasmussen, 1995). Such a design will provide the workers with more flexibility in coping with the dynamic environment and developing their skill in the work.

The operators' decision and performance are influenced not only by technology itself, but also by the organization, management infrastructure and the environment context. The understanding of human capabilities and limitations and identification of the functionality that is required to accomplish intellectual tasks are important in the design. In this view, the cognitive constraints are work demands that originate with the human cognitive system, while environmental constraints are work demands that originate with the context in which workers are situated.

Vicente (1999) argued that the traditional approach to the design (from task analysis to prototype building and testing it in a scenario) suffers from two limitations: strong device-dependence and incompleteness. The knowledge gains from testing prototypes are limited by the number and range of tasks that workers are asked to perform during testing. Prototyping and usability testing could re-design an existing system, but couldn't suggest wholly new directions.

Cognitive work analysis (CWA) is the first and most important step for the design analysis. Vicente has systematically devoted the cognitive work analysis idea and methods in his book (1999). Figure 4-3 indicates the working model of cognitive work analysis.

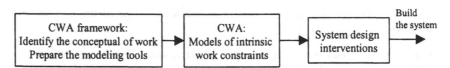

Figure 4-3. Ecological approach to work analysis

The ecological approach to the sociotechnical system design came from human error analysis of a catastrophic accident, such as nuclear power plant accident analysis. Vicente in his book also tried to apply the idea to the design of human-computer interaction systems, as he especially pointed out that "ecological perspective is not specific to the nuclear industry. It generalizes meaningfully to any application domain where there is an external physical or social reality—outside of the person and the computer—that imposes dynamic, goal-relevant constraints on action (1999)."

The ecological approach has not been popularly accepted. The successful application cases are not so many; one of them is the DURESS case (Vicente, 1995). The ecological approach concentrates on the semantics of the work and did not provide the designer any handy design tools. As Figure 4-3 shows, it is an open loop. Besides, a deep analysis and understanding of the work domain requires skilled specialists to perform and analyze and it is quite time-consuming. Most of the engineers do not have the special training to perform such an analysis. Still, the approach itself is interesting.

4.4 DISTRIBUTED COGNITION

To understand the complex and temporally rich interplay of body, brain and world requires some new concepts, tools and methods—ones that are suited to the study of emergent, decentralized and self-organizing phenomena (Clark). The distributed cognition can be one such concept. The distributed cognition does not look for an individual cognitive process that happens in the brain, but the cognitive process of the interaction between people and with resources and materials in the environment (Hollan, 2000). This theory argues that both the cognitive process itself and the knowledge used and generated are often distributed across multiple people, tools and representations. For example, when a research project is being carried out, the knowledge for the research work may be distributed among different colleagues and in different literatures, books and even the research equipment and computers that collect and store the information. In distributed cognition, resources for cognition include the internal representation of knowledge in the human brain and external representations. This external representation includes potentially anything that supports the cognitive activity. It could be a physical layout of objects, notes in a book, diagrams, text or graphics in computers, symbols, etc.

The cognition is distributed, which means the cognitive process might also be distributed. Three different kinds of distribution emerge from the cognitive process (Benyon, 2005): a) across different members of a social group; b) coordination between internal and external structures; and c) a time sequence followed in such as way that the products of earlier events can transform the nature of later events. In other words, distributed cognition is looking for the interaction among many brains, between humans and the environment, such as the process inside the computer, or the machines where a user interacts. It also concerns the range of mechanisms that may be assumed to participate in cognitive processes. A cognitive process is delimited by the functional relationship among the elements that participate in it. The spatial collocation of the elements becomes less important. In distributed cognition, one expert finds a system that can dynamically configure itself to bring subsystems into coordination to accomplish various functions (Hollan, 2000).

The distributed cognitive theory can be quite useful for analyzing the communication and cooperation among a group of peoples within a complicated social-technical system (Preece, 2002). For example, it is a good method to study the way people work together to solve problems, the distributed problem solving that takes place and how knowledge is shared and accessed. The various communicative pathways have been taking place as a collaborative activity progresses. In the problem-solving process, the role of verbal and non-verbal behavior can be interesting, what is said, what is implied by glances, winks, etc., and what is not said and the various coordinating mechanisms that are used, e.g., rules, procedures.

4.5   COGNITIVE SYSTEM ENGINEERING

Hollnagel, et al., has formulated the cognitive system engineering (CSE) (1983) concept. CSE did not focus on human cognition as an internal function or as a mental process, rather on human activity or "cognition at work." Its view changes from the interaction to the cooperation between humans and machines. It views the human-machine ensemble as a single joint system (Hollnagel, 2001). Figure 4-4 shows the basic cyclical model of Hollnagel's CSE. Traditional cognitive engineering studies focused on describing the human as a system component and were satisfied with representing the work environment by the inputs and outputs. The users are treated as single individuals. Their actions are just response to the input stimulation, not the anticipation of the information. Influence from situation or context is indirect through input channels to the users. Cognition at work may involve several people distributed in space or time, which makes cooperation and coordination at least as important as human information processes. The cyclical model has overcome the weak points by considering the users are parts of a whole system. The actions become continuous anticipation and action. The influence from environmental situation, or the context, is not just regarded as input information but also affects the user's way of working—specifically how events are evaluated and how actions are chosen.

The cyclical model aims to describe the necessary steps in controlled performance, regardless of whether the control is carried out by an artifact, a human being, a joint cognitive system or an organization (Hollnagel, 2001). An effective control requires the operator, or the controlling system, to be able to interpret the events and find and choose effective action alternatives. In this cyclical model, a time frame is introduced: event evaluation time (T_E), action selection time (T_S) and the time available for the next event to happen (T_A). If ($T_E + T_S$) exceeds T_A, the operator will lag behind the process and may gradually lose control. If $(T_E + T_S)$ is less than T_A, the operator will be able to refine the current understanding of the situation. Besides the three types of times, the time needed to carry out a chosen action (T_P) will also be considered, it normally is context-and operator-experience dependent, and sometimes can be predicted exactly.

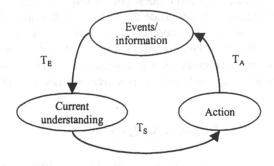

Figure 4-4. The basic cyclical model of CSE.

T_E = *event evaluation time*
T_S = *action selection time*
T_A = *time available for next event to happen*
If $(T_E + T_S) > T_A$: operator may gradually lose control
If $(T_E + T_S) < T_A$: operator may refine the current understanding of the situation

Time needed to evaluate events, (T_E) can be reduced in several ways. Prime among these is the design of information presentation. A number of high-level design principles have been put forward with varying degrees of success, for example as adaptive displays (Furukawa, 1999), multimedia interfaces (Bergan, 1992; Guha, 1995) and ecological interface design (Schiff, 1995). The adage has been to provide the operator with "the right information, in the right format, at the right time." This, however, has in most cases been simpler to say than to do, and the attempts to define commonly accepted standards have so far met with limited success.

The time needed to select an action, (T_S), can also be improved in several ways. Prime among these is the use of procedures and extensive training. Another approach is to improve the human-machine interface. The time needed to carry out a chosen action, (T_P), can be reduced, for instance, by using automation or by amplifying human performance capabilities in other ways.

CSE has tried to shift the focus from the traditional internal function of either a human or machine to the external function of the joint cognitive system (JCS). In JCS, the boundaries are defined by the purpose of the analysis, not its structure or composition (Hollnagel, 2003). For example, when the driver is driving a car, the driver and the car can be considered as one JCS. The boundary is established by two criteria (Hollnagel, 2003), whether the functioning of a constituent is important for the JCS and whether it is possible for the JCS effectively to control a constituent in the sense of being able to manipulate it so that a specific outcome results.

In the view of CSE, cognition does not belong to human property anymore; it is a continuous activity inside the JCS, by both humans and even some artifacts. Cognition is distributed across multiple natural and artificial cognitive systems, rather than being confined to a single individual. Sets of active cognitive systems are embedded in a social environment or context that constrains their activities and provides resources (Hollnagel, 2003).

4.6   WORK AND TASK ANALYSIS

The point of doing any kind of work and task analysis is to derive implications for a systems design. The purpose of performing a work analysis is to have deep insight and knowledge of work domain. It focuses on the semantic knowledge of work. It helps the designer understand the domain of the work and understand what the constraints are from the work, from the environment, from the technology and from the operators. This analysis provides the task allocation that is suitable for speech interaction and the demands for such interaction. It may also provide the knowledge on the changes of the performance of the whole system when the new interaction is introduced. This knowledge may indicate the necessity of other design changes in the system to reach a better performance. Work analysis is guided by ecological theory, distributed cognition and CSE theories. Task analysis has broad usage. It studies the detail about how the task is performed. There are many different ways to do the task analysis and get the analysis data. For different usages and different type of systems, the analysis methods can be different. Due to the limited space of this book, I will briefly discuss the basic principles of how to perform the task analysis. As a classic method, there are many books and articles that discuss how to perform the task (Diaper, 2002; Diaper, 1990; Hackos, 1998; Kirwan, 1992; Schraagen, 2000; Sheridan, 1997; Stammers, 1990). The information processing theory guides the task analysis. If any reader has very limited time and would like to find a suitable method to perform the task analysis, Neville Stanton and Mark Young's book *A guide to methodology in ergonomics—design for human use* is a good book to read (1999).

4.6.1   Task Analysis

Task analysis is a study of what an operators (or team of operator) is required to do in terms of actions and/or cognitive processes to achieve a system goal, or to examine the tasks that must be performed by the users when they interact with the systems (Kirwan, 1992). There are many different task analysis models found in the literature. Many task analysis methods were described by Luczak (1988). Through this paper one can find most of the important references that related to the practical issue and methods published before 1988. Some important publications after 1988 can be found in other books (Diaper, 1989; Kirwan, 1994; Kirwan, 1992).

In general, to perform a task analysis, the first step is to define the concept and the scope of the task. Normally, three interactive components of tasks must be considered (Stammers, 1990):

a) Task requirements: refers to the objective and the goals that the task performers are intending to achieve.
b) Task environment: refers to factors in the work situation that limit and determine how an individual can perform.
c) Task behavior: refers to the actual actions that are carried out by the user within the constraints of the task environment in order to fulfill the task requirement.

There are many task analysis methods available in the literature. Which method is chosen depends on the purpose and the characteristics of the tasks and the analysis. Some methods serve for general understanding of the job, others serve directly for the interface design.

One often used task analysis method is called Hierarchical Task Analysis (HTA). Patrick, Spurgeon and Shepherd's handbook (1986) gave the details about how to use this method. HTA broke down the task into task components at a number of levels of description with more detail of the task being revealed at each level. The components include goals, operations and plans.

HTA is easy to implement once the initial concepts have been identified. It mainly provides descriptive information from direct observation, so it normally does not provide design solutions and it does not handle cognitive components of the tasks.

There are many task analysis methods needed to decompose the big task or the main goal into smaller tasks or sub-goals. During the task analysis, one of the difficulties is to allocate the task. Task analysis and task allocation are highly interrelated. If the task analysis is to answer "what" the task is doing, then the task allocation is to answer "how the task is to be done."

While there are many accepted techniques for doing task analysis, there is no commonly accepted means to perform task allocation. Sheridan (Sheridan, 1997) pointed out three reasons for the difficulties associated with allocating the task:

a) There are many different ways to break the task into pieces, depending on how you looked at it. Those pieces are normally interacting with each other in many different ways.
b) The interaction between human and computer can have infinitive ways and the task allocation between human and computers can also have infinitive possibilities.
c) It is difficult to quantify the criteria for judging the suitability of various human-machine interactions.

4.6.2   Cognitive Task Analysis

Cognitive task analysis (CTA) is the extension of traditional task analysis techniques to yield information about the knowledge, thought processes and goal structures that underlie observable task performance (Chipman, 2000). Different people have different definitions to confine the term exclusively to the methods that focus on the cognitive aspects of the task. A NATO technical report (RTO-TR-24) has given an extensive review of the CTA studies. Most of the CTA research and methodologies were developed in the '80s and early '90s. GOMS (goals, operators, methods selection roles) is one of the commonly used methods specially developed for the human-computer interaction studies.

Chipman, et al., (2000) pointed out that artificially separating cognitive functions from the observable behavior of task performance may not provide a better understanding of the job. Furthermore, given the purpose and constraints of particular projects, several cognitive task analysis approaches should be appreciated. One single approach will not fit all circumstances.

The process to perform the CTA is strongly dependent on the purposes of the analysis. In principle, a few key phases can be categorized as:

a) Identify which task(s) need to carry out the CTA. These tasks shall be important, typically and frequently appearing in the interactions between humans and the systems. The methods one can use normally are observation, interview and questionnaires to the users, or preferably to the experts.
b) Identify knowledge representations that are needed in the tasks.
c) Use the proper technology to perform the knowledge-elicitation from the experts.

4.6.3 GOMS—A Cognitive Model

GOMS was first developed by Card, et al., (1983). The primary application of GOMS is the design of HCI. GOMS has four basic elements: goals, operators, methods for achieving the goals and the selection rules for choosing among competing methods for goals. GOMS focuses on the cognitive processes required to achieve a goal using a particular device. GOMS analysis is a description, or model, of the knowledge that an answer must have in order to carry out tasks on a device or system. It is a representation of the "how to do it" knowledge required by a system in order to get the intended tasks accomplished. Therefore, the model developed from GOMS is used to outline the cognitive performance of a person by decomposing a problem into hierarchical sub-goals and goal stacks. The outcome of the GOMS analysis provides engineering models of human performance. The engineering models criteria distinguish them from psychological-oriented cognitive models by the ability to make a priori predictions, the ability to be learned and used by practitioners as well as researchers, coverage of relevant tasks and approximation.

The aims of GOMS analysis are to describe tasks in terms of the following:

- **Goals**—what are people trying to do using the systems? Usually it has a hierarchical structure. Accomplishing a goal may require accomplishing one or more sub-goals. A goal description is an action-object pair. For example: delete word, create file, etc.
- **Operators**—the action that the system allows people to make. An operator is something that is executed. Here the operator does not mean the user or a human, rather it means the actions that the user executes or actions the software allows the user to take. There are two types of operators, the external and mental. The external operators represent the observable actions through which the user exchanges information with the system. The mental operators represent the internal actions performed by the user, which are hard to observe.
- **Methods**—sequences of subtasks and operators. It can be well-learned sequences of steps that accomplish a goal, or exact sequences of operators that may be performed to achieve a goal. If a user can perform a task, it means the user has learned a method for the task and the method is a routine cognitive skill. No planning is required to determine physical actions.
- **Selection rules**—the rules that people use to choose between methods of achieving the same subtask (if there are options). If there is more than one method to accomplish a goal, a selection rule will route control to

the method needed to accomplish the goal and personal rules that users follow to decide what method to use

GOMS techniques are a big family. Most of them are based on a simple cognitive architecture known as the Model Human Processor (MHP), similar to the information processing theory as described earlier in this chapter. GOMS provides a framework for modeling aspects of human performance and cognition. It provides a rich set of techniques for evaluating human performance on any system where people interact with computers. GOMS analysis can provide much insight into a system's usability such as task execution time, task learning time, operator sequencing, functional coverage, functional consistency and aspects of error tolerance. Some type of GOMS analysis can be conducted at almost any stage of system development, from design and allocation of function to prototype design, detailed design and training and documentation for operation and maintenance. Such analysis is possible for both new designs and redesign of existing systems.

John and Kieras (John, 1996; John, 1996; Kieras, 1997) described the current family of GOMS models and the associated techniques for predicting usability and list many successful applications of GOMS to practical design problems. The tutorial materials for how to use the GOMS can be found in different articles (Card, 1983; John, 1995; Kieras, 1997; 1988).

4.6.4   Cognitive Work Analysis

Cognitive work analysis is the ecological approach to the study of human behavior in complicated systems. Human behavior is goal-orientated and everything happening in the working environment has its meaning to the operator. The meaning and the goal are closely related to its specific context where the operator performs. With cognitive work analysis approach, none of the detail information that normal task analysis methods provided is of interest. The goal of the cognitive work analysis is to discover the constraints on action the consequences of actions, and the information that specifies the constraints and consequences. It only provides the guidance about the goal state and the constraints on action. It will not provide the information about how the task should be accomplished (Vicente, 1999). Different task analysis methods may provide this information. Figure 4-5 shows the concept model of cognitive work analysis.

Design of the interface includes defining the contents of the information, formulating the structure of information presentation and selecting the right form in which to present the information. The domain of complex work and the limitation of the human operator dominate the design activities. The purpose of work domain analysis is to find a domain representation formalism that can describe the domain complexity. The analysis of the human operator is to understand the mechanisms that people have for dealing with the complexity of the work by communicating with the system.

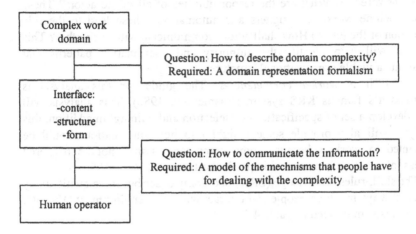

Figure 4-5. The concept model of ecological work analysis

There are five phases of cognitive work analysis and each of them will lead to respective system design interventions (Vicente, 1999): First is the *work domain*. Here, the analysis tries to answer three questions: a) What information should be measured? The answer to this question shall leads to the sensor design of the system. b) What information should be derived? The answer to this question leads to the design of the model. c) How should information be organized? This analysis leads to data collection and building up the database that keeps track of the relationship between variables, providing a coherent, integrated and global representation of the information constrained therein.

The second is the *control task*. Here there are two main questions: a) What goals must be pursued, and what constraints are on those goals? The answer will guide the decision whether the system shall be procedures or automation. b) What information and relations are relevant for particular classes of situation? This analysis can be used to design context-sensitive interface mechanisms that present workers with the right information at the right time.

The third one is *strategies*. The analysis here deals with how the work shall be done. The analysis shall answer two questions: a) what frames of reference is useful? The answer is used to decide what type of human-computer dialogue modes shall be designed. b) What control mechanisms are useful? This analysis will help to specify the process flow for each dialogue mode.

The fourth is *social-organizational*. Here, two important questions need to be answered: a) What are the responsibilities of all of the actors? These actors include workers, designers and automation. These lead to the role allocation of the job. b) How shall actors communicate with each other? This analysis will help identify the authority and coordination patterns that constitute a viable organizational structure.

The fifth is *worker competencies*. The guide for this analysis is Rasmassen's famous KRS system (Rasmussen, 1986). This analysis will help develop a set of specifications of selection and training. In addition, this analysis will also provide some insight into how information should be presented to workers because some competencies may not be triggered unless information is presented in particular forms (Vicente, 1992).

The skill, rules, knowledge (SRK) taxonomy describes three qualitatively different ways in which people can interact with their environment (Vicente, 2002). It is summarized in Table 4-2.

Table 4-2. The interaction quality of the SRK and its design principle

	Interaction with the environment	Design principles
Knowledge-based behavior	Involves serial, analytical problem solving based on a symbolic mental model	The interface should represent the work domain in the form of an abstraction hierarchy to serve as an externalized mental model for problem solving
Rule-based behavior	Involves associating a familiar perceptual cue in the world with an action or intent, without any intervening cognitive processing	There should be a consistent one-to-one mapping between the work domain constraints and the perceptual information in the interface
Skill-based behavior	Involves parallel, automated, direct behaviors	Workers should be able to act directly on the interface

Figure 4-6 shows the principles of design related to the SRK analysis. The skill-level-based behavior is easy to perform, so it is regarded as the low level behavior. Knowledge-based behavior is most difficult to perform, so it is regarded as high level behavior. If the design can help the operator to perform the task in lower level behavior, one should not make the design unnecessarily complicated so the operator has to perform it in higher-level demands.

There are three prospective design principles that are often applied different types of control/interaction systems (Torenvliet, 2000):

a) To support skill-based behavior, operators should be able to act directly on the interface
b) To support role-based behavior, the interface should maintain a consistent one-to one mapping between the work domain constraints and the perceptual cues provided in the interface
c) To support knowledge-based behavior, the interface should represent the work domain in the form of an abstraction hierarchy (Rasmussen, 1985)

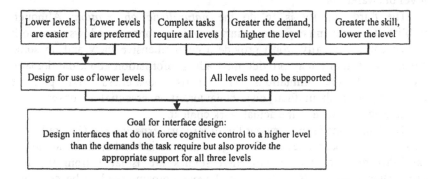

Figure 4-6. The application of SRK analysis to the interface design

For control design, there are three dimensions that a work analysis should address (Flach, 2000):

a) What are the constraints on action? What can be done?
b) What are the consequences of action? What are the functional constraints, or costs and benefits, associated with action?
c) What are the constraints on feedback (information)? How are the constraints on, and consequences of, actions specified in the perceptual array?

To be able to carry out the ecological task analysis, there are five normal activities involved (Flach, 2000). The first step is to do the *table-top analysis*: a literature survey. The kinds of information that might be found would include a) constraints on action; b) functional constraints; and c) constraints on information. Be careful to not completely trust conventional wisdom—there are many examples where conventional wisdom does not generalize to specific contexts.

The second activity is to carry out the *knowledge elicitation* from experts: Many aspects of practice may be highly dependent on context, so it is difficult to abstract a general principle of the sort that can easily be communicated in a textbook. The goal here is to get a comprehensive, authoritative and/or normative view of the situation. There are a number of methods for knowledge elicitation such as concept maps, planning goal graphs, knowledge audits, critical incident reports, etc, that have been developed. Applying multiple methods to multiple experts may help to differentiate those aspects that are significant with respect to the work domain from those aspects that are significant to a particular perspective or level of awareness.

The third activity is to do the *naturalistic observation*: Although published reports and expert accounts provide important windows on a work domain, they typically won't provide a comprehensive view and, in some cases, they actually may provide misleading information about the domain. Because of the limitations of experts' awareness and possible misconceptions about their own behavior, it is essential to observe the situated performance in the actual work context.

The fourth one is to carry out the *laboratory experiments*: Generally, the ecological approach does not favor general experimental studies, due to the fact that most of the experimental studies are isolated from the real application context. Since the natural work domains tend to be complex, some events might be very important to make a complete understanding of the domain. There can be advantages to more controlled observations where the phenomenon can be parsed into more manageable chunks or where the same event can be observed repeatedly.

The fifth activity is to make the *synthetic task environments*: The availability of high fidelity simulations and virtual reality systems creates a great opportunity for an ecological approach to cognitive task analysis. These synthetic environments allow the semantics of complex environments to be brought into the laboratory. These simulations allow constraints at various levels of abstraction to be controlled and manipulated in attempts to identify performance invariants. The synthetic task environment also offers a safe place for the analyst to experience the domain constraints without endangering patients or passengers. However, there is a chicken/egg problem associated with synthetic task environments. What constitutes a high fidelity simulation? From the above four task analysis processes, one can give the definition of the "high fidelity" simulation. The synthetic task environment can then be used to test a hypothesis about what high fidelity means.

4.7   INTERACTION DESIGN PROCESS

4.7.1   Human-System Interface

There are four dimensions of an interface (Torenvliet, 2000): a) The *content of the interface*: What kind of information communication should take place between the users and the system that the interface will handle? b) The *interface form*: In which form, the information should be presented? An interface that directs an operator's attention to critical information should foster functional adaptation. c) *Type of training*: What kind of training is the user required to have for a good performance on such interface? d) *Pre-existing competencies*: What is the competence the potential users may have?

Additional unnecessary complexity caused by implementation 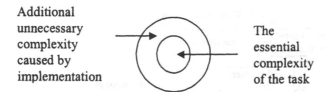 The essential complexity of the task

Figure 4-7. *Two aspects of most existing interfaces*

Normally, most of the available human-system interface has two parts, one is the essential complexity of the task and the other is the additional unnecessary complexity caused by implementation as showed in Figure 4-7. The interface design should try to abolish the unnecessary complex part. A good interface is "transparent," which means the user sees only the task.

It is difficult in current practices in many complex systems, military or otherwise, to achieve a man-system interface that is based on human characteristics alone. This is partly because a lot of designs are still technical driven and the designers have very limited knowledge about the human cognition, and partly because the effects from the human factors issue were not very clear in the design context during the design process, due to the limitation of knowledge and ethical or safety issues. In future systems, where technological development allows—such as in Virtual Environment Interfaces—the limitations of physical factors such as space, cost and safety are liable to be less restrictive. The interface can be more user-friendly and more flexible. Such interface should fulfill the criteria of ease of use, high efficiency, supportiveness and acceptability. The common characteristics of any usability interface should be that (Hackos, 1998):

a) They reflect the workflow familiar or comfortable to the user.
b) They support the user's learning styles.
c) They are compatible to the user's working environment.
d) They encompass a design concept (a metaphor or idiom) familiar to the users.
e) They have a consistency of presentation (layout, icons, and interaction) that makes them appear reliable and easy to learn.
f) They use language and illustrations familiar to the users or easy to learn.

4.7.2   Design Process

Normally, an interaction design basically follows the same process as the software development process, which can be described as a waterfall model (Dix, 2003). An example of a waterfall model is shown in Figure 4-8.

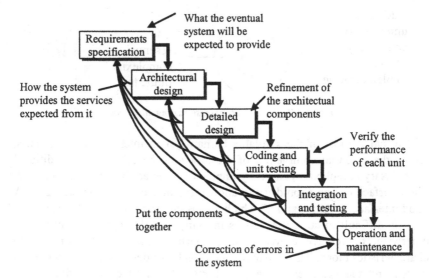

Figure 4-8. The activities in a waterfall model of software design process.

This waterfall model shows the iterative process in the design. A design starts from specifying the requirement for the design. A requirement study should be carried out step by step until finally the system can be integrated and operated. To increase the usability of the system, the iterative process should be carried out. When any step in the design process has finished, not only the technical, but also the usability evaluations, be carried out. The results of the evaluation may cause some improvement of the same step, an improvement of earlier steps or even a new design of the earlier step.

In the first step, *requirements specification* involves eliciting information from the customer about the work environment, or domain, in which the final product will function. The designer and customer try to capture what the system is expected to provide, which can be expressed in natural language or more precise languages. Groups of people who negotiate the features of the intended system with the designer may never be actual users of the system. During this process, a typical ecological work analysis should be carried out. The results from the work analysis may help designers have a deeper understanding of the design requirement and how the design would affect the operation of the system, including the operators.

The second step is *architectural design*. It is a high-level description of how the system will provide the services that are required from major components of the system and how the needs are interrelated to satisfy both functional and non-functional requirements. It provides a decomposition of the system description which allows for isolated development of separate components that will later be integrated. Here, a typical hierarchical task analysis should be carried out. The results from the task decomposition may help the designer construct the design.

The third step is the *detailed design*. It is a refinement of architectural components and interrelations identifying modules to be implemented separately. The refinement is governed by the non-functional requirements. Typically, there will be more than one possible refinement of the architectural component that will satisfy the behavioral constraints. Choosing the best refinement is often a matter of trying to satisfy as many of the non-functional requirements of the system as possible. Here, typically the GOMS analysis can be carried out.

It is only after acceptance of the integrated system that the product is finally released to the customer and the final system certified according to requirements imposed by some outside authority. After product release, all work on the system is considered under the category of maintenance. The majority of the lifetime of a product is spent in the maintenance activity. It includes correcting the errors in the system that are discovered after release and the revision of the system services to satisfy requirements that were not realized during previous development.

4.7.3 Interaction Design

What is interaction design? The interactive design is the design of spaces for human communication and interaction. Rakers (2001) said, "Interaction design is about designing structure by juggling with purposes and functionality, and we arrive at purposes and their associated functionality by studying the user's roles, goals, responsibilities and tasks." To be able to design an interactive interface, the designers should a) have good knowledge and understanding of the target domain's work pattern at the workflow, task-flow and event-flow levels. To have certain domain knowledge is an important factor for the interaction design. The design analysis, especially the task analysis discussed in Chapter 4, served this purpose. b) Have a good understanding of the user's knowledge of the (professional) world, the task-situation and the device. Applying the cognitive work analysis method, the designers can have a deep knowledge of the constraints from work, from environment and from the users. c) Translate this into a matching and meaningful "conceptual model" that together appears as a whole, as the UI structure, interaction and visual behavior to the user (Rakers, 2001). Here it requires the combination of cognitive information processing theory and ecological interface design theory in practice. To be able to do so, it is important to understand how users act and react to events, how they communicate and interact together and how to design different kinds of interactive media in an effective and aesthetically pleasing way. So it is also a multidisciplinary issue.

The process of the interaction design can be to *first*, identify the design requirements. These include specification of user types, roles, goals and responsibility; domain tasks description and requirement; working environment description and requirements; and user interface design requirements (Rakers, 2001). The needs and requirements shall be established into the design language. *Second*, develop a couple of alternative designs that meet those requirements, and select the most suitable ones among the design. *Third*, build interactive versions of the designs so that they can be communicated and assessed. *Fourth*, evaluate what is being built throughout the process (Jennifer, 2002). Here, the domain task is what people really want or need to do, and it is solution-independent.

Before any detailed interface design can commence, it is necessary to identify an appropriate interaction style or styles. The choice of interaction style will be determined by knowledge of the potential users and tasks. For human and computer dialogue systems, there are five aspects in the interactive design that will determine the style of the interactions (Smith, 1997):

a) *Initiation*: Where and in what degree the initiation of the dialogue rests, in computer or in human user.
b) *Flexibility*: The number of ways a user can perform given functions.
c) *Option complexity*: The number of different options available to the user at any given point in the dialogue.
d) *Power*: The amount of work accomplished by the system in response to a single user command.

e) *Information load*: A measure of the degree to which the interaction absorbs the memory and reasoning power of the user.

Rakers (2001) through his many years of design/practice experience tells us that the overall interaction design process does not exist. It may depend on the specifics of the design task, the business objectives, the requirement of the customers or even the designer's role in the design process.

4.8   SCENARIO-BASED DESIGN

Scenarios have been used in software engineering and interactive system design for many years. Scenarios are stories about the interactions between people, activities, contexts and technology. Scenarios offer an effective way of exploring and representing activities, enabling the designers to generate ideas, consider solutions and communicate with others. Carroll (2000) illustrates the usefulness of scenarios in design. He argues that scenarios are effective at dealing with five key problems of design:

a) The external factors that contain design such as time constraints, lack of resources, having to fit in with existing designs and so on. Scenarios promote an activity-oriented view of designed artifacts, allowing life-cycle participation by users.
b) Design moves have many effects and create many possibilities. Scenarios can be written at multiple levels, from many perspectives and for many purposes. For example, a single design decision can have an impact in many areas and these needs to be explored and evaluated.
c) Scenarios can help generalize the scientific knowledge in design.
d) Vivid descriptions of end-user experiences evoke reflection about design issues.
e) Scenarios concretely fix an interpretation and a solution but are open-ended and easily revised.

Scenarios can be used throughout the design process. By creating user stories, the designers can have a better understanding of what people do and what they want. A conceptual scenario can be used to generate ideas and specify requirements for the design. A concrete scenario can be used for prototyping ideas and evaluating it. A use case can be used for documentation and implementation.

As a scenario is so useful in interaction design, it does not mean that there is no negative effect. Diaper (2002) pointed out that scenarios should be written very carefully. There may be important and missing functionality in the scenario. Different people will almost inevitably envision a scenario in different ways. But how do they do this and the assumptions they make when doing it, however, are not usually open to introspection. In the worse case, scenario creators and users are generally unaware of the assumptions on which their scenario envisions are based. The task analysis should be performed for the necessary data collection.

4.9   DISCUSSION

Before any design is carried out, whether it is a new design of the entire system, or part of the system, or to improve the existing system, the design analysis is needed. For speech interaction system design, the design analysis should answer the following questions: Where do we need speech interaction? How would the speech interaction system affect the entire performance of the big system? What can be the impact of the speech interaction system towards human cognitive behavior, working life and daily life? what would the interaction look like?

From different theoretical approaches to the analysis, the outcome can be very different. Here we discussed the information processing theory and ecological design theory. In many literatures, these two theories stand as contradictory to each other. Those who develop the ecological theories normally like to compare their approach with information processing theory and emphasize the differences between these two theories (Flach, 2000; Flach, 2000; Flach, 1998; Vicente, 1999). Table 4-1 indicated part of the argumentation.

In my personal opinion, these two theories compromise each other. An ecological approach from a top-down analysis of the work has a deep understanding of the goal of human work and the constraints from environment, from technology and from organization. The results of the analysis provide the requirements for the interface design. Computer systems are designed to help people carry out the work. The work is the multiple combinations of different tasks. The work and the tasks are, therefore, of central interest to application system developers. The ecological approach did not serve the detail knowledge for the interface design such as how to design the color, the size of the symbols and text on the screen, how to select the right modality for interaction, etc. Cognitive engineering of the information processing theory has rich knowledge for the detail of the interface design. But IP theory lacks the understanding of the goals that lead human attention to perception, interpretation and action. Human attention distribution and shift cannot only be driven by symbolic stimulation and not only by the brightness of the light or the loudness of the sound but by the meaning this light and sound represents. But the symbolic system design in the interface will affect the efficiency of the performance. If the sentence on the computer screen presents the right information on right timing, but hardly can be seen by the operator, the interface would still be regarded as a bad design. Figure 4-9 shows the framework of the design analysis by applying both ecological theory and information processing theory.

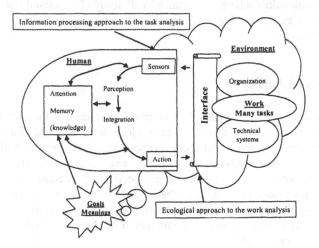

Figure 4-9. The framework of design analysis

Scenario is a commonly used tool for interaction design. The setting of the scenario in detail will affect the quality of the detail. Work and task analysis help collect the necessary data for envisioning the scenario.

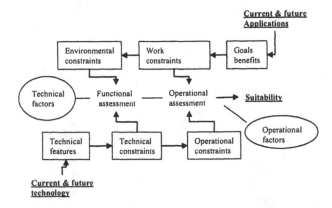

Figure 4-10. Selecting the suitable technology for the design

Another important function of the work and task analysis is to guide the selection of the right technology. Figure 4-10 shows the application of the technology selection.

The suitable technology should be selected for the design shall base on two different assessments. One is the technical assessment that is carried out from the comparison of the evaluation of the technical constraints with the environmental constraints where the technical will apply. The second one is the operational assessment that will consider the work constraints and operational constraints.

4.10   CONCLUSION

In this chapter, we discussed different interaction design theories in general. Information process theory takes a close look into the human information perception from different sensors in the human body, interpretation of the information in the brain and executing the decision by different muscle actions. The ecological perspective of interface design ignores the detail of close interaction on the human-system interface; rather, it takes the whole job into consideration. As these two theories consider the design activities at different levels, a combination of the two theories can lead to a better design of a job.

Chapter 5

USABILITY DESIGN AND EVALUATION

5.1   INTRODUCTION

In Chapters 2 and 3, we discussed mainly the neuropsychological and cognitive theories that are related to the interface design. In Chapter 4, we discussed how to analysis the work and tasks. Now we are ready to design the interaction. In this chapter, we will discuss different design theories and design approaches for interactive interfaces. What is interactive interface design? What does usability mean for the modern user interface? How can we fulfill the usability requirement in design? How can we evaluate the design? In this chapter, we will illustrate different design theories for usable interactive interface design.

5.2   DIFFERENT DESIGN APPROACHES

Throughout human technical development history, there are different design approaches that have been dominating the design activities. From the interface design perspective, the different approaches mainly differ from how to consider the user and the usability toward the technology available in the design. We can divide these approaches into five categories (Flach, 1998).

The *technical centered approach* has taken the technological capabilities into main consideration. This approach emphasizes the capacity of the technology. The design is more or less driven by the technology, and it focuses on the internal architecture and components of the technology in design. The main characteristic of such a design is that there is a display for every sensor and a control axis for every control surface. It has a very limited focus on a user's experience, validation and measurement.

The *control-centered approach* views humans as controllers, or supervisors, of the technology. This approach emphasizes the stability constraints arising from the coupling between the human controllers and the machine dynamics (Flach, 1998; Jensen, 1981).

The *user-centered approach* has taken the cognitive limitations and expectations of human operators into consideration. The purpose of this design process is to increase the usability of the product and its competition capacity in the market. There are different ways and design methods for this approach. Normally, the potential users are involved in the entire design process from the beginning. This approach emphasizes the limitations of the human information processing system. This approach is more or less driven by the user. It specializes user experiences. It works well for improving the existing products, but it shows disadvantages when innovative design is required.

The user-centered design (UCD) process is developed to insure the usability requirements of the user. Practically speaking, there are very few organizations managing to implement the ideas of user-centered design. What are the problems associated with user-centered design? Smith (1997) pointed to four possible issues: user issues; organizational commitment; developer skills; and resource constraints. The user issues include the limitation of user experience, knowledge, expectations, contribution, agreements and diversity. These are the factors make the user involve difficulties.

On the other hand, the *usage-centered approach* focuses on the interaction between humans and work. It emphasizes the usage of the interfaces. It focuses on human behavior as goal-orientated and as event that happens in the living and working environment has meaning to the user. Ecological interface design theory is one of the theories that take the usage-centered approach. The notion of meaning, constraints and goals are closely related to specific context task that the user performs. Constraints from the work domain are the fundamental, and the constraints on human information processing, on technology or on the control loops become secondary (Flach, 1998). In this approach, the potential users are not necessary to involve in the design process. Some user tests should be carried out in the design process. The purpose to carry out such a test is not for the purpose of getting a user's opinion of the design but to understand better the interaction between the user and the products, thus to take the most benefit from the developed technology and increase the usability.

Both user-centered design and usage-centered design has strongly considered the usability of the products. But with the usage-centered approach, usability is not the highest goal of the design. After some year of practice, the advantages and disadvantages of the user-centered approach has been clearly shown to the people, especially when it comes to the innovative design, in which the potential users can hard to define in the early design stages. Even when the users can be defined, the users may not have any idea about how the technical can be used. It is not appropriate to involve a user in the design process. The usage- centered approach is more and more accepted in the practical design.

Compared with the user-centered approach, the theories and practical study methodologies are less well developed. The ecological interface design theory is one of the approaches to the use-centered design, and it works better for the large system design. Even though, so far there are not so many successful application cases that have been well documented with ecological interface design approach in application. We will discuss this issue later.

5.3 THE CONCEPT OF USABILITY

5.3.1 Definitions

In most of the situations, when we talk about usability, we use to refer to the ISO DIS 9241-11 (Guidance of Usability 1998) as shown below. ISO is the International Standards Organization. This definition is used in subsequent related ergonomic standards[3]:

"Usability is the extent to which a product can be used by specified users to achieve specified goals with effectiveness, efficiency and satisfaction in a specified context of use."

ISO 9241-11 explains how to identify the information necessary to take into account when specifying or evaluating in terms of measures of user performance and satisfaction. Here, the efficiency means the amount of effort required to accomplish a goal. The less effort required, the higher the efficiency reached. The effectiveness can be measured in terms of the extent to which a goal or a task is achieved. Satisfaction indicates the level of comfort the users feel when using a product and how acceptable the product

---

[3] http://www.usabilitynet.org/tools/r_international.htm

is to users as a means of achieving their goals. The simple summary is shown in Figure 5-1.

ISO 9241:

    Effectiveness (% of goal achieved)

+   Efficiency (time to complete a task,
            or the error rate,
            or the amount of effort)

+   Satisfaction (subjective rating scale)

=   Usability

Figure 5-1. International standard of usability

For a machine, the effectiveness is easy to understand. For example, if the machine operator's goal were to produce 100 components per day, then if he or she produces only 80 components, the effectiveness level is 80%. When it related to the speech interactive information system, the effectiveness can be understood as the acceptable performance that should be achieved by a defined proportion of the user population, over a specified range of tasks and in a specified range of environments as the system was designed for. If, for example, the dialogue system is designed for booking the train tickets for all available trains in Europe, and some traffic information could not be retrieved from the database, the effectiveness was decreased. The efficiency might be measured in terms of the time taken to complete one task, or the errors the operator made during the performance. For an automatic speech recognition system, typically the recognition error rate has reflected the efficiency of the system. The satisfaction is a subjective aspect; it is difficult to measure and often is strongly correlated with effectiveness and efficiency.

Actually, there are a few different ISO/IEC (International Electrotechnical Commission) that relate to the definition of usability (Bevan, 2001). ISO/IEC FDIS 9126-1 (Software engineering—product quality—part 1: Quality model 2000) has been more narrowly associated with user interface design. It defines the usability as:

"Usability: the capacity of the software product to be understood, learned, and attractive to the user, when used under a specified condition."

ISO/IEC 9126-1 describes the six categories (functionality, reliability, usability, efficiency, maintainability and portability) of software quality that are relevant during product development. By this activity, the software can meet the user's need that is regarded as quality in use, which is similar to the usability definition in ISO DIS 9241-11. Each of the six categories can be considered from different aspects. They are described below:

a) *Functionality*: accuracy, suitability, interoperability and security.
b) *Reliability*: maturity, fault tolerance, recoverability and availability.
c) *Usability*: understandability, learnability, operability and attractiveness.
d) *Efficiency*: time behavior, resource, utilization.
e) *Maintainability*: analyzability, changeability, stability, testability.
f) *Portability*: adaptability, installability, co-existence, and replaceability.

These two standards take different approaches to usability. The ISO/IEC 9126-1 definition of usability is concerned with attributes of the product that make it understandable, learnable, easy to use and attractive. The ISO 9241-11 definition of usability is very broad, implicitly including not only utility, but also computer efficiency and reliability. In the interaction design process, one should combine these two standards (Bevan, 2001).

5.3.2 Usability Design

Usability is indeed a fuzzy concept. It can only be meaningful within a specific context. One particular system placed in one context will probably display different usability characteristics when placed in a second context (Smith, 1997). Usability is not simply a property of a product in isolation, but rather dependent on who is using that product, the goal that they are trying to achieve and the environment in which the product is being used. Usability is a property of the interaction between a product, a user and the task, or set of tasks. To conceptualize usability in the design process, Don Norman (1988) and Ravden and Johnson (1989) have pointed out some principles in the design:

a) *Visibility*: Information presented should be clear, well organized, unambiguous and easy to read.
b) *Feedback*: Users should be given clear, informative feedback on where they are in the system, what actions they have taken, whether these actions have been successful and what actions should be taken next.
c) *Consistency and compatibility*: The way the system looks and works should be consistent at all times and compatible with user expectations.
d) *Explicitness*: The way the system works and is structured should be clear to the user, so the user will easily know how to use it. It should show the relationship between actions and their effects.

e) *Flexibility and constraints*: The structure and the information presentation should be sufficiently flexible in terms of what the user can do to suit different user needs and allow them to feel in control of the system. At the same time, the system should also restrict certain kind of user interaction that can take place at a given moment.

f) *Error prevention and correction*: The possibility of user error should be minimized, automatically detected and easy to handle for those that do occur.

g) *User guidance and support*: Easy-to-read and understand, relevant guidance and support should be provided to help the user understand the system.

In practice, these principles need to be interpreted in the design context. After over fifteen years practice, people have developed many different principles regarding how to support the usability design. Dix (2003) summarized these principles in three main categories as shown in Table 5-1.

Table 5-1. Summary of design principles to support the usability

Learnability	Flexibility	Robustness
Predictability (or operation visibility)	Dialogue initiative (or system/user pre-emptiveness)	Observability (or browsability, static/dynamic defaults, reachability, persistence, operation visibility)
Synthesizability (or immediate/eventual honesty)	Multi-threading (or concurrent vs. interleaving multi-modality)	
Familiarity (or guessability, affordance)	Task migratability	Recoverability (or forward/backward recovery, commensurate effort)
Generalizability	Substitutivity (or representation multiplicity, equal opportunity)	
Consistency		Responsiveness (or stability)
	Customizability (or adaptivity, adaptability)	Task conformance (or task completeness, task adequacy)

The usability design principles proposed by Don Norman (1988), Ravden and Johnson (1989), and by Dix (2003) are quite similar. They differ from different detail levels. The design principles and usability evaluations are closely related to each other. The *heuristic evaluation* developed by Jakob Nielsen (1994), which we will discuss in detail in a later section, has also covered most aspects related to the usability design principles. The six categories in ISO/IEC 9126-1 standards related to software quality took the consideration of earlier studies in usability and different design principles and evaluation aspects.

Learnability is essential for any interactive design, and it is one of the important factors that affect the users' perception of satisfaction. Understanding of the users' learning curve help the interface designer understands the learning process. User learning curves are a way to plot and track the stages of users' learning. They help you determine what type of information users need at each stage of learning, break the information down into components and subcomponents and understand what assumptions are safe to make. Coe has combined different user learning curves into one as shown below (1996).

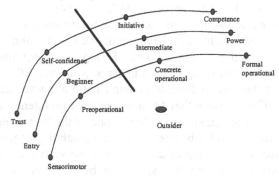

Figure 5-2. User learning curve

Table 5-2. Learning Stages and User's Need

Learning stages	User's needs
Entry-sensorimoter-trust	Clear, concrete, uncluttered information and structure Control of the results of actions Explanations of the results of actions
Beginner-preoperational-self-confidence	Help in creating appropriate, simple schemata Encouragement to ask questions Many simple examples
Intermediate-concrete operational-initiative	Insight into the logic structure of the information Open-ended examples that they may extrapolate from their own environment Safety net as they explore
Power-formal operational-competence	Systemic understanding of the information Opportunities to apply analytical and problem-solving skills Frameworks that launch original thought about, and application of, the information
Outsider	Recognition that not every user wants to traverse the user curve Modular components of information that enable them to come to information over and over Reliable maps that guide them to just the information they want

The first line from the outsider shows the level of performance skill. The second line shows the learning level and the third line shows the respective psychological changes. There are five profiles in this learning curve, and, as in the new entry moment, learning starts from the sensory-motor stage. The users need to believe that they can learn. These five profiles and respective user's need is summarized in Table 5-2 according to Coe (1996).

Flexibility refers to the multiple ways of exchanging information between the user and the system. The system can take the initiative actions, although the user may take the initiative action as well. This point is extremely important for a dialogue system design that addresses who shall take the initiation—the system or the user. A flexible system should allow the users to perform the task in different ways that fit into their own preference and mental model. Multimodal interaction is regarded as one of the best choices to achieve the best flexibility and robustness. Task migratability concerns the transfer of control for the execution of tasks between the system and the user. This means certain control can be carried out by both the system automatically and by the user as well. Substitutivity allows the equivalent values of input and output to be arbitrarily substituted for each other. For example, a time can be said as 1.5 hour, or 1 hour and 30 minutes, or 90 minutes. The system should also be able to be automatically modified based on its knowledge of the user and the environment. The user-initiated modification called adaptability and the system initiated modification called adaptivity (Dix, 2003).

Robustness refers to the possibilities of error detection, error tolerance and error recovery. The system shall provide the necessary support to the users and the system performance shall be transparent so the user can interpret the situation of the performance.

5.3.3   Usability Evaluation

To be able to evaluate the usability of a system, we need to have a useful theory for evaluation. The theory of evaluation should provide a basis for choosing the scenarios on which an interface would be tested, in selecting the personnel who would be used as test persons and in defining the measures that are suitable to use as metrics for evaluation. The following questions would need to be carefully handled when any evaluation is going to be carried out:

a) What are the characteristics of the overall system performance? How should it be measured? Should we measure the human behavior, or system output, or some other appropriate measures? At what level should they be taken?

b) How should we choose the test persons? Should they be naïve people who are then trained extensively, or experts on existing systems who are trained to overcome their background knowledge and habits?
c) How much training is required to arrive at a stable performance where we can be sure that we are evaluating the properties of the interface and not just the learning and adaptive behavior of the operators?
d) How detailed and near fidelity should the scenarios be? What should their properties be? Which properties of the interface should be used? How can context be specified?
e) If general principles are being evaluated, what is the minimal set of applications that are needed in the evaluation?
f) How should we measure team performance? How can cooperation and communication be measured effectively?

When carrying out the usability evaluation, one should understand that users rarely work in isolation interacting solely with an interaction system. People who use the system to support their work may at the same time interact with peripheral devices (printers, card readers, etc.) with help systems and user support, with other users on the network, and with colleagues or customers either face-to-face or via the telephone. The balance and nature of these interactions is task-dependent; it varies from workflow to workflow. In short, evaluation of the performance of a system or product in the hands of its users must reflect the wider context in which the system will be used. It is also essential to evaluate how successfully task goals are achieved, both in terms of quality and quantity, as well as the cost of achieving those goals, which may be measured in terms of time or money (Macleod, 1996).

The evaluation of usability should be carried out through the entire design process, from the planning of the design to the final products produced (as was indicated in Figure 4-8). The usability evaluation includes the following objectives:

a) Discover the potential flaw in design concept by user trial
b) Compare with existing usable products to set up the baseline criteria for new development
c) Determine whether predefined criteria have been met.

Usability testing is an important issue in the UCD process. Usability testing is a process of learning from users about a product's usability by observing their using the product. The primary goal for a usability testing is to improve the usability of a product. The participants to the test should represent the user group since they will do the real tasks that the product is designed for. It normally requires a multiple observation and measurement records with a team of observers and analyzers to interpret the data. The usability testing can be carried out in a special usability lab, in a normal testing room and in the field. Barnum's book *Usability testing and research* provides details about how to carry out the usability testing (Barnum, 2002).

Heuristic evaluation, which was developed by Jakob Nielsen (1994), has been mostly used to evaluate the usability of a system. This heuristic list was derived from a factor analysis of 209 usability problems. Nielsen claims that this list of heuristics provides a broad explanatory coverage of common usability problems and can be employed as general principles for user interface design. Heuristic evaluation is a method for structuring the critique of a system, so it is suitable to be used for evaluating the early stages of design, as well as different stages of the design and the function of the final product. Nielsen's ten heuristics are:

a) *Visibility of system status*: Always keep users informed about what is going on through appropriate feedback within reasonable time.

b) *Match between system and the real world*: The system should speak the user's language, with words, phrases and concepts familiar to the user, rather than system-oriented terms. Follow real-word conventions, making information appear in natural and logical order.

c) *User control and freedom*: Users often choose system functions by mistake and need a clearly marked emergency exit to leave the unwanted state without having to go through an extended dialogue. Supports undo and redo.

d) *Consistency and standards*: User should not have to wonder whether words, situations or actions mean the same thing in different contexts. Follow platform conventions and accepted standards.

e) *Error prevention*: Make it difficult to make errors. Even better than good error messages is a careful design that prevents a problem from occurring in the first place.

f) *Recognition rather than recall*: Make objects, actions and options visible. The user should not have to remember information from one part of the dialogue to another. Instructions for use of the system should be visible or easily retrievable whenever appropriate.

g) *Flexibility and efficiency of use*: Allow users to tailor frequent actions. Accelerators—unseen by the novice user 4 may often speed up the interaction for the expert user to such an extent that the system can cater to both inexperienced and experienced users.

h) *Aesthetic and minimalist design*: Dialogues should not contain information that is irrelevant or rarely needed. Every extra unit of information in a dialogue competes with the relevant units of information and diminishes their relative visibility.

i) *Help users recognize, diagnose and recover from errors*: Error message should be expressed in plain language (no code), precisely indicate the problem and constructively suggest a solution.

j) *Help and documentation*: Few systems can be used with no instructions so it may be necessary to provide help and documentation. Any such information should be easy to search, focused on the user's task, list concrete steps to be carried out and not be too large.

Usability evaluation can mean rather different things to different people. It can focus on narrow or broad issues. They may assess conformance with guidelines, reflect expert opinion or assess performance in the hands of users. Their scope may range from some micro-issues such as, the study of individual interaction with detailed aspects of a user interface to evaluating the use of a system to support work in an organizational setting. A successful interaction with interface objects is essential for effective and satisfaction between an individual user and a computer system or an interactive product. The experience of evaluating usability shows that even quite small changes in presentation and feedback—both in the design of interaction objects and the manner of their incorporation into dialogues which support work tasks— can greatly influence the quality of interaction (Macleod, 1996).

By comparing different categories proposed by Don Norman (Norman, 1988), Ravden and Johnson (1989), and by Dix (2003), as well as the *Heuristic evaluation* from Jakob Nielsen (1994), with the six categories in ISO/IEC 9126-1, we can have one impression that usability evaluation is a complex task. The coupling between different functional requirements and the design parameters that can satisfy these functional requirements are not simple. It is understandable since human cognition is not simple, and usability reflects most of human cognition aspects.

Lo and Helander (2004) proposed a decoupling method by applying axiomatic design theory (Suh, 1990). Axiomatic design theory provides a framework for characterizing component and relational complexities in design. The principle concepts of axiomatic design can be summarized as follows:

a) Design domains are used to group together different types of attributes in design.

b) In a domain, the attributes form a hierarchical structure.

c) Decision-making in design is perceived as a mapping process between attributes that belong to two adjacent domains.

d) Design equations are used to represent this mapping process between domains.

A design equation can be written (Suh, 1990):
$$\{FR_i\} = [A_{ij}]\{DP_j\}$$

Here, $\{FR_i\}$ is a vector of functional requirements, which represent design goals. $\{DP_j\}$ is a vector of design parameters, which represent a design solution, and $[A_{ij}]$ is a design matrix. A design matrix for $\{FR_2\}$ and $\{DP_2\}$ is the following form:

$$[A] = \begin{bmatrix} A_{11} & A_{12} \\ A_{21} & A_{22} \end{bmatrix}$$

If the value of the design matrix is 0, it signifies no mapping or coupling between the respects $\{FR_i\}$ and $\{DP_j\}$. Otherwise, a mapping exists.

If we look back to Nielsens's heuristics, we can break it into the [FR] and [DP] language (Lo, 2004) as shown in Table 5-3.

Table 5-3. Reproducing the Nielsen's heuristics into FRs and DPs languages

Functional requirements (FRs)	Design parameters (DPs)
(FR_1)= Visibility of system status	(DP_1)= Feedback within reasonable time
(FR_2)= Match between system and the real world	(DP_2)= Speak the user's language
(FR_3)= User control and freedom	(DP_3)= Emergency exit
(FR_4)= Consistency and standards	(DP_4)= Follow platform conventions
(FR_6)= Recognition rather than recall	(DP_6)= Make objects, actions, and options visible
(FR_7)= Flexibility and efficiency of use	(DP_7)= Allow users to tailor frequent actions
(FR_8)= Help user recognize, diagnose and recover from errors	(DP_8)= Error message should be expressed in plain language (no code), precisely indicate the problem, and constructively suggest a solution

Heuristic 5 (error prevention) is a function requirement (FR_5), but its design parameter is not made clear by Nielsen, and Heuristic 10 (help and document) is a design parameter (DP_{10}), but its function requirement is not clear.

The design matrix is proposed by Lo (Lo, 2004) as Figure 5-3. The explanation of this equation: to make the system visible (FR_1), we need to have an appropriate system feedback (DP_1) and to make objects, actions, and options visible are also belong to visibility (DP_6). At the same time, the type of language used for feedback (DP_2) should be the same as the platform conventions (DP_4). Error message (DP_8) is one type of feedback closely related to the language expression.

This equation is not empirical testable, as it is not based on any design context. For any specific design, the matrix may be different, as the specific (FRs) and (DPs) can differ in detail. The analysis should keep track of the multiple relations between the design goals and design solutions.

$$
\begin{Bmatrix} FR_1 \\ FR_2 \\ FR_3 \\ FR_4 \\ FR_6 \\ FR_7 \\ FR_8 \end{Bmatrix} = \begin{pmatrix} A_{11} & A_{12} & O_{13} & A_{14} & A_{16} & O_{17} & A_{18} \\ A_{21} & A_{22} & O_{13} & A_{24} & A_{26} & O_{27} & O_{28} \\ O_{31} & O_{32} & A_{33} & O_{34} & A_{36} & O_{37} & O_{38} \\ O_{41} & A_{42} & O_{43} & A_{44} & A_{46} & O_{47} & O_{48} \\ O_{61} & O_{62} & O_{63} & O_{64} & A_{66} & O_{67} & A_{68} \\ O_{71} & O_{72} & A_{73} & O_{74} & O_{76} & A_{77} & O_{78} \\ A_{81} & A_{82} & A_{83} & A_{84} & O_{86} & O_{87} & A_{88} \end{pmatrix} \begin{Bmatrix} DP_1 \\ DP_2 \\ DP_3 \\ DP_4 \\ DP_6 \\ DP_7 \\ DP_8 \end{Bmatrix}
$$

Figure 5-3. **The design equation from Nielsen's Heuristics**

There are more and more studies related to usability design principles and evaluations for different interactive speech systems (Dybkjaer, 2001; 2004; Gaizauskas, 1998; Hone, 2001). We will discuss this issue in respective chapters.

The term "usability testing" can mean different things to different people, depending on their training and experience, and depending on different stages of the design iteration process. As such, it is important to know that usability testing is not a *function testing or reliability testing validation testing*. A functional testing is to verify if the users are able to perform certain tasks. A reliability testing is to verify that the product performs as designed, and validation testing is to verify that the product performs without errors or "bugs." Although these tests must be performed to assure the accuracy and validity of the product, none of these types of tests affirms a match with the users' wants, needs and desires (Barnum, 2002). Unfortunately, many reported usability tests of speech interaction systems are actually the functional, reliability and validation tests.

5.4   HUMAN NEEDS AND SATISFACTION

Satisfaction is one of the important parameters that measure the usability of different products. Satisfaction can be understood on different levels. The first level is the comfortable feeling when using a certain product or working with a certain interface. The second level is the pleasant, enjoyment and happy feeling when using the product.

It is human nature that people are never satisfied with what they have. The dissatisfaction becomes the initial power and motivation for improving the existing design and searching for a new design. Motivation is a need, desire or incentive that energizes behavior and directs it toward a goal. Maslow (1970) described a "hierarchy of human needs" with five nirvana named physiology, safety, belonging and love, esteem and self-fulfillment. Coe (1996) categorized them into three supersets of needs: basic needs, psychological needs and self-actualization needs. This forms a pyramid or hierarchy of needs as shown in Figure 5-4.

On the left side of this pyramid, human needs are divided into three different levels. As soon as people have fulfilled the needs in the lower level of the hierarchy, they will then want to fulfill the needs higher up. Inside the pyramid, there are some suggestions about how to fulfill those needs when designing the interactive system as it was illustrated inside the pyramid.

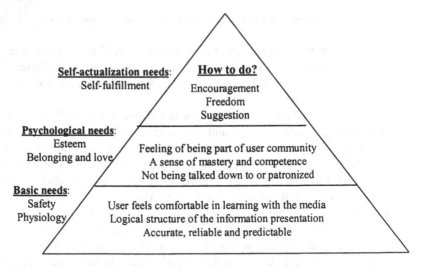

Figure 5-4. The hierarchy of user's needs (I)

The pyramid indicated that when people get used to having something, they then start looking for something more. This can be explained by I Ching philosophy. The world is constantly changing. The changes can be caused by internal factors of human and external factors. For every individual person, the knowledge, the experience and the skill of performance is changing; the self-consciousness based on self-understanding, self-recognition during interaction with other people, the technology and the environment is also changing. Technology is constantly developing, and the environment is also changing all the time. Living in this constantly changing world, it would be very strange if people could be satisfied by what they already have! The motivation of looking for new things to fulfill the new needs is actually the initial power for the development of everything we live on in this world, good or bad.

5.4.1 Trust

Trust is one of the issues related to usability. It is one of the properties of satisfaction. From a social perspective, trust, in general, is a mental state, an attitude toward another agent (usually a social attitude). Trust has internal and external characteristics. The internal characteristics include: ability/competence, self-confidence, disposition as willingness, persistence and engagement. The external characteristics include: opportunities and interferences. If we said the truster (x) has a goal g) that x tries to achieve by using (y) as the trustee, x has four basic beliefs (Falcone, 2002):

a) *Competence belief*: Y has some function that can produce/provide the expected results to reach the goal g.
b) *Disposition belief*: Y will do what x needs.
c) *Dependence belief*: X needs y and relies on y.
d) *Fulfillment belief*: X believes that g will be achieved.

Most of the studies about trust are related to automation or social/organizational issue (Gallivan, 2001; Harrington, 1999; Muir, 1996; Tan, 2002). There is very limited research work related to human-computer interaction. Moray (1999) has summarized some of the main findings on trust between humans and systems:

a) Trust grows as a positive function of the reliability of the system. It is also a function of trust in the recent past.
b) Trust declines if the human disagrees with the strategy adopted by the system and when errors or faults occur.
c) Trust can be partitioned: Operators are often able to assign differential trust to different parts of the interaction according to how the different subsystems behave.

5.4.2   Pleasure

In modern life, the satisfaction in usability concept does not satisfy people anymore. As Jordan (2000) pointed out: People are no longer pleasantly surprised when a product is usable, but are unpleasantly surprised by difficulty in use. Satisfaction in the usability definition (ISO DIS 9241-11) is referred to as "comfort and acceptability." It is concerned with avoiding negative feelings rather than producing overtly positive emotion (1998). This is the basic level of satisfaction. The high level of satisfaction needs self-fulfillment. Self-fulfillment creates the pleasure and happiness. It is possible that usability and pleasure in product use are directly equated. If not, then the usability of user-centered designed products will fall short of offering optimal experiences of the user (Jordan, 1998).

Most of the products were designed for accomplishing certain tasks. If a product does not contain appropriate functionality, it cannot be usable. Appropriate functionality will not ensure usability, and usability will not ensure pleasure. If a product is not usable, then it is most unlikely to be pleasurable. Usability is the basic requirement. The relationship between the functionality, the usability and the pleasure of a product can be shown by another pyramid suggested by Jordan (2001) in Figure 5-5. This pyramid is comparable with the one Coe suggested in Figure 5-4.

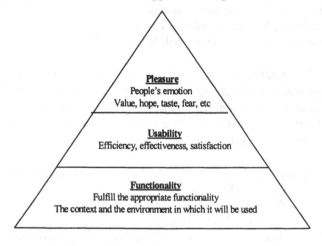

Figure 5-5. The hierarchy of user's needs (II)

What is pleasure? What aspects affect the pleasant feeling? How does one design product to stimulate people's pleasant feelings? Tiger (1992) has developed a framework for addressing the pleasure issue in general:

a) *Physiological pleasure*: This is the pleasure derived from the sensory organs. It has strong connection with touch, taste and smell as well as feelings of sexual and sensual pleasure.
b) *Social pleasure*: This is the enjoyment derived from the company of others, including friends, lovers, colleagues or even like-minded people. It might also include a person's relationship with society as a whole.
c) *Psychological pleasure*: It is gained from accomplishing a task more than from usability approaches. It pertains to people's cognitive and emotional reactions.
d) *Ideological pleasure*: It refers to the pleasures derived from "theoretical" entities such as books, music and art. It pertains to people's values.

The four-pleasure framework is not a theory of pleasure. It does not explain why people experience pleasure. This framework breaks down the issue of pleasure into four different sections. This helps the designer to consider the full spectrum of the sorts of pleasures that a product can bring (Jordan, 2000). This framework may not cover all types of pleasure. There is a certain pleasure that a specific product provides to the user that may not be able to be categorized into these four categories. It is also not a suggestion that all products should provide all four types of pleasure.

The pleasure feeling comes from the interaction between humans and the products or the systems. When it comes to the context of design, an approach should take from both a holistic understanding of the people who the products or the system are designed for and the product benefits' specification according to the four-pleasure framework. Based on these analyses, the specification of product design properties can be identified.

The aspects of analyzing people's characteristics according to the four-pleasure framework are summarized below (Jordan, 2000):

a) *Physio-characteristics*: These characteristics have to do with the body. The analysis can take a different approach, such as special physical advantage and disadvantages; muscular-skeletal characteristics; external body characteristics (such as height, weight, and body shape, etc.); body personalization (such as hairstyle, clothing, tattooing, etc.); and physical environment and physical dependencies, including the reaction to the physical environment.
b) *Socio-characteristics*: These have to do with a person's relationship with others. The analysis aspects can be sociological characteristics, social status, social self-image, social relations, social labels, social personality traits and social lifestyles.

c) *Psycho-characteristics*: These refer to the cognitive emotional characteristics of people. The analysis can take approaches such as special talents and difficulties, psychological arousal, personality traits, social-confidence, learned skills and knowledge.

d) *Ide0-characteristics:* These characteristic have to do with people's values. The analysis approach can be personal ideologies, religious beliefs, social ideology, aesthetic values and aspirations.

In the context of products, physio-pleasure relates to the physical characteristics of the products that people can feel, for example the tactile and olfactory properties. Socio-pleasure can refer to the social interaction a product can facilitate. Psycho-pleasure might include issues relating to the cognitive demands of using the product and the emotional reactions engendered through experiencing the product. Ideo-pleasure refers to the aesthetics of a product and the values that a product embodies (Jordan, 2000).

Jordan also studied different characteristics of the properties of the products that are associated with emotional feeling of pleasure by comparing many different consumer products and similar products with different designs. The results are shown in Table 5-4 (Jordan, 1998).

Table 5-4. Emotion and properties associated with pleasure products

Properties	Emotions
Features/functionality	Security/comfort
Usability	Confidence
Aesthetics	Pride
Effective performance	Excitement
Reliability	Satisfaction
High-status product	Entertainment
Convenience	Freedom
Size	Sentiment/nostalgia
Brand name	

To ensure the pleasure with products, the designers should understand the user's requirement, not only the requirement due to the usability and cognitive issues, but related to the pleasure feeling. They should also be able to link the product properties to pleasure responses. The quantification measurement of pleasure in the products is another challenge for the researchers and designers (Jordan, 2001).

Of course, there are many excuses for modern interactive products that are hard to identify as far as its users, the use environment and defining of functions. The design of such products is only for the purpose of making people feel happy. These designs are called "incomplete design" (Tannfors, 2004), or "design of useless products[4]." Normally, these are multimodal and multimedia interaction products. Speech, sounds, music, pictures, and colors are commonly used in the design. But I will not discuss this type of product in this book.

5.5 USER-CENTERED DESIGN

The user-centered design (UCD) is a design process that can assure the usability of the products. It has its focus on the user. All potential users of the proposed information system have the opportunity of being actively involved, either directly or indirectly, in the whole analysis, design and implementation process. Practically speaking, the UCD can change during the design process from soliciting user opinion to designing to accommodating work activities and the capacity of people who are going to use the system (Noyes, 1999). So, generally speaking, the UCD process places the user in the center of the design and takes the workplace, environmental factors and organization into consideration.

A good example of a UCD approach to speech interface system design is reported by Holzman (2001) on the design of the speech interaction system for the Trauma Care Information Management System (TCIMS). The TCIMSConsortium focused on mobile, field-based components and communication of patient data and medical logistic information among field medics. They followed the user-centered system design processing from the earlier beginning of analysis of all the possible users/environments/tasks to make up the design concepts. Then they discussed the concepts with potential users. They applied the users' feedback to finalize the design, then implemented them as prototype and evaluated them by the user. They made solid consideration of every detail from the hardware, functionality, multimedia input/output, communication and speech-audio user interface for different users. Even the evaluation methods have also been carefully designed, and the preliminary test results showed high usability and acceptance by different users.

5.5.1 What is UCD process?

A UCD process has early focus on users and the tasks that involve understanding the user in the application context. Usability testing is done

---

[4] www.digitlondon.com and www.digitfeed.com

from beginning to the end. Empirical measurement from usability testing can help designers have a better understanding of users and their performance, such as ease to learn, to use of the products, or their experience towards satisfaction and pleasure. A UCD is an iterative design process that fixes the problems found by the user in usability testing. There are a few principles in the UCD process that insure the usability of the products/software. These are (Maguire, 2001):

a) Understanding and specifying the context of the user and the task requirements, allocation of function between user and system according to the human's capacities and limitations and the understanding of the special demands of the task. At this point, the designers should actively involve the user in the design process.
b) Specifying the user and organizational requirements.
c) Iteration of design solution: using a user's feedback from early design solutions. Simple paper mock-ups of screen layout to prototypes are important for the success of the design, and this process can be iterated for many times until the satisfaction criteria is matched.
d) Multi-disciplinary design teams: The development is a collaborative process that benefits from the active involvement of various parties, each of whom has insights and expertise to share.
e) Evaluate designs against requirements.

The Context of Use Analysis (CUA) and the usability evaluation methodologies are at the heart of the UCD process (Maguire, 2001). CUA is a useful method in the early stages of development, such as requirement specifications and interface design work. Usability evaluation tools are useful for the product/prototype evaluations. A fully user-centered design process will demonstrate a number of qualities (Smith, 1997):

a) User engagement in the design process: This means that users can communicate and negotiate with the developer. The user has real decision-making powers.
b) A socio-technical design process: The job satisfaction is at the center of consideration, and it contains iterative and evolutionary approach.
c) Appropriate user-centered tools implemented: Here the priority is given to user interface design.
d) The ability to react to changing needs: This can include all the levels of needs, as we discussed in Section 5.4.

ISO 13407 provides a framework for user-centered development activities that can be adapted to numerous development environments. By this standard, there are five essential processes that should be undertaken in order to incorporate usability requirements. These processes are plan the UCD design process, understand and specify the context of use, specify the user and organizational requirements, produce design and prototypes and carry out user-based assessment. The process is carried out in an iterative fashion until the particular usability objectives have been attained. Usability testing is performed after every stage of the process. The UCD lifecycle process description is shown in Figure 5-6.

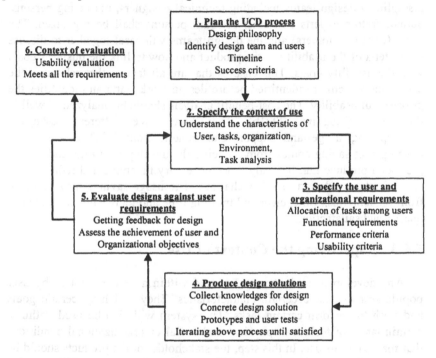

Figure 5-6. The interdependence of user-centered design activities. Inside the window of each process, what people do in this process is described.

There is very rich information available about different designs and evaluation methodologies from different website. The practical guidelines for ISO 13407[5] and Handbook of User-Centered Design[6] were published with the INUSE and RESPECT projects. Maguire (Maguire, 2001a; 2001b) discussed in detail what people should do in each step of the UCD process.

---

[5] http://www.ucc.ie/hfrg/emmus/hftoc.htm
[6] http://www.ejeisa.com/nectar/inuse/6.2/c...

Here I will not repeat the available information in detail; I will just give a short summary of the process and discuss some argumentation that has appeared recently.

5.5.2   Planning the UCD Process

This is a strategic activity. The business goal should be clearly set up. The stakeholders of the product will be identified here. A multiple disciplinary design teams including technical designers, marketing persons, human factors experts and other relevant persons shall be organized. The users (or the customers) and the design teams will meet together to discuss the criteria of the usability of the product and how to link the business goals with the usability goals. This ensures that are all factors that related to the use of the system are identified before design work starts and identifies the priorities of usability. The cost-benefit analysis should be analyzed as well in this stage (Maguire, 2001b). There are three different savings—development, usage and support can benefit from UCD process. The development saving comes from reducing the development time and cost to product a produce that has only those necessary functions and reducing the cost of future redesign or other changes. The usage savings may be made from reducing task time, error and training time and increasing productivity, quality and satisfaction.

5.5.3   Specifying the Context of Use

Any developed product is to be used within a certain context by user populations that have certain characteristics. They will have certain goals and wish to perform certain tasks. The system will also be used within a certain range of technical, physical and social or organizational conditions that may affect its use. In this step, the stakeholder of the products should be identified. The user analysis should be carried out here (we will discuss this in detail in section 5.2 in this chapter). Task analysis (Baber, 2001; Chipman, 2000; Diaper, 1990; Flach, 2000; Hackos, 1998; Kirwan, 1992; Luczak, 1988; Schraagen, 2000; Sheridan, 1997) and context-in-use analysis (CUA) (Maguire, 2001a) should be carried out as well. CUA is a structured method for eliciting detailed information about a system in the context of use. This information is a foundation for later usability activities, particularly user requirements specification and evaluation. CUA is a serial of methods including surveys, observations, interviews, questionnaires, etc. A specific design questionnaire that is used can provide details of the characteristics of the users, their tasks and their operating environment. Stakeholders attend a facilitated meeting, called a Context Meeting, to help complete this detailed questionnaire.

- **Specify the user and organizational requirements**

Requirements elicitation and analysis is crucial in this process. ISO 13407 (ISO, 1999) provides the general guidance on specifying user and organizational requirements and objectives. It states that the following elements should be covered in the specification:

a) Identification of the range of relevant users and other personnel in the design.
b) Provision of a clear statement of design goals.
c) An indication of appropriate priorities for the different requirements.
d) Provision of measurable benchmarks against which the emerging design can be tested.
e) Evidence of acceptance of the requirements by the stakeholders or their representatives.
f) Acknowledgement of any statutory or legislative requirements, for example, health and safety.
g) Clear documentation of the requirements and related information. Also, it is important to manage changing requirements as the system develops.

There are some general methods that can support the requirement specification studies (Maguire, 2001b). These methods are interview users, focus group discussions, and scenarios of use, personas, existing system and competitor analysis, task or function mapping and allocation of function. The usability criteria shall be clearly identified here.

- **Produce design solutions**

All design ideas will go through iterative development as they progress. Paper prototype, mock-ups and simulation of the system are necessary. They allow potential users to interact with, visualize and comment on the future design. Major problems with the design can be identified and changes to the design may then be made rapidly in response to user feedback. This helps to avoid the costly process of correcting design faults in the later stages of the development cycle. There are many methods, including techniques for generating ideas and new designs (brainstorming and parallel design), use of design guidelines and standards (to ensure compliance with legal requirements) and techniques for representing future systems (storyboarding, paper-based prototyping, computer-based prototyping, Wizard-of-Oz and organizational prototyping) (Maguire, 2001b).

- **Evaluate design against user requirement**

Designs should be evaluated throughout development. There are two main reasons for usability evaluation. a) To improve the product as part of the development process by identifying and fixing usability problems; and b) to find out whether people can use the product successfully.

Problems can be identified by many methods as well. User-based methods are more likely to reveal genuine problems, but expert-based methods can highlight shortcomings that may not be revealed by a limited number of users. User-based testing is required to find out whether people can use a product successfully.

User-based testing can take place in a controlled laboratory environment, or at the users' work place. The aim is to gather information about the users' performance with the system, their comments as they operate it, their post-test reactions and the evaluators' observations. Evaluation studies can be carried out in three levels of formality when performing: participative (least formal), assisted (intermediate) and controlled evaluation (most formal). A participatory approach is appropriate to understanding how the user is thinking. Questions may include their impressions of a set of screen designs, their impression of what different elements may do their expectations from the result of what their next action will be and their suggestions about how individual elements could be improved. The objective of an assisted approach is to obtain the maximum feedback from the user while trying to maintain as realistic an operational environment as possible. Thinking aloud protocol is used where the user is requested to perform tasks. The evaluator only intervenes if the user gets stuck. A controlled user test is required to find out how successful users will be with the full working system, as closely as possible replicating the real world in the test environment. The controlled user test can be used to evaluate whether usability requirements (effectiveness, efficiency and satisfaction) have been achieved (Maguire, 2001 b).

- **Context of evaluation**

Before a product is released to the customer, the final usability evaluation needs to be carried out. One of the main duties in the evaluation is to compare if all the usability criteria are met in the product.

A good product shall be delivered with good support. If a product is used in ways that were not envisaged in design, or through inadequate user requirements specification, it can be rendered unusable. The support is related to both its supplier and its customer. The support includes: assisting the installation process and training the user to maintain the system and provide technical support, documentation and online help, initial and continued training and learning for users, capturing feedback on the product in use, providing of health, safety and workplace design activities and considering the environmental factors. Some products are much more likely to be associated with major organizational changes than others. Organizational changes, including the physical workstation and environment issues, may need to be made at the time of product implementation (Maguire, 2001 a).

5.5.4 User Analysis

UCD requires users involved in the design process from the beginning to the end of the project development. To be able to formulate a realistic usability product plan, it is necessary in the earlier stages to define who the target users are and for what kinds of tasks these users will wish to use the product/software.

There may be a wide range of different users. Any employee or customer of the organization who will be directly or indirectly affected by the system should be considered the users (Smith, 1997). So the user group can be categorized as follows:

a) *Manager user* who is responsible for the system operating inside an organization system. He is more interested in the efficiency and the benefit of the entire system. Manager users are responsible for identifying the need for new or revised systems and for their initial specification.
b) *End user* is the one who directly uses the system. The end user is a crucial element in the successful acceptance of information systems within the organization.
c) *Customer users* are those who are immediately affected by the inputs to, and output from, an information system or part of the system.
d) *System user* is the person who either generates or manages the application itself rather than using the software system to support an organizational role.

In recently years, people have used *stakeholders* to describe different type of a user group. Stakeholders are those who influence or are affected by the system but not the actual user. The above four types of users are part of the stakeholders. The others are recipients of information, marketing staff, purchasers and support. If the user population is composed of more than one user type, then an analysis should be completed for each type. Relevant characteristics of users also need to be described. These can include knowledge, skill, experience, education, training, physical attributes and motor and sensory capabilities (Maguire, 2001a). Different user population may have different shapes, sizes, psychological and cognitive capacity with different expectations. The analysis of a user's personal attributes should include the issues of age, gender, physical capabilities and limitations, cognitive capabilities and limitations, attitude and motivation. Users have different skill levels. The analysis of the user's experience, knowledge and skills should include the issues of product or system experience, related experience, task knowledge, organizational knowledge, training, input device skills, qualifications and language skills. A novice user has little application knowledge but considerable system knowledge. A casual user, however, has considerable application knowledge but limited system knowledge, whereas an experienced user has considerable knowledge of both the application area and the system itself.

The interactions between user and system can be identified by their task complexity, the frequency of use and their adaptability (Smith, 1997). What needs to be known about the user before design can be made? The knowledge about the user can be divided into three categories (Hackos, 1998):

a) How they define themselves (jobs, tasks, tools and mental models)
b) How they differ individually (personal, physical, and cultural characteristics, as well as motivation)
c) How they use products over time and the choices they make about the levels of expertise they want or need to achieve (stages of use)

Thomas and Bevan published a handbook of *Usability context analysis* (1996). They provide very handy methods to help specify in a systematic way the characteristics of the user, the tasks they will carry out and the circumstances of use. It focuses on knowing the users, knowing their goals and knowing the circumstances of the system use. At the same time, we shall analyze the user group (the stockholder of users) from the four-pleasure framework's perspective (see section 5.3 in this chapter).

5.5.5   User Partnership

It is impossible to create user-centered information without having an ongoing dynamic relationship with a user. Building the partnership includes the understanding of the user's cognitive processes of sensation and perception, learning and memory, problem solving, accessing information and acting. A designer should have some clear answers to the questions: What is the user's motivation? What is the user's expectation of this system? What kind of problem-solving strategies does the user need to apply, and what kind of obstacles may exist? Does the user need to apply different learning process and styles? What kind of feedback does the user need?

Normally, the technical designer does not have first-hand information of the user. Coe (1996) has recommended several methods for the engineers to build up the user's partnership. These methods include site visits, competitive benchmarking, brainstorming, mind mapping, storyboarding, paperwalks, draft and prototype reviews and usability testing. Site visits give the inside information about the user, physical and psychological environment, organization condition and needs and problems, etc. The competitive benchmarking is to make the comparison of products with the competitor's. Brainstorming and mind mapping are group problem-solving sessions. Storyboarding is visualizing information contents and layout. Paperwalks test a storyboard with the user to validate the design. Draft and prototype reviews provide the user with the information they have gone through in the earlier stages.

5.5.6   Usability Requirements Analysis

As was illustrated in Figure 5-6, before producing the design solution of a system, it is important to specify the requirement of the design. The design requirement study is not only for functional purposes. To satisfy the usability test, the designers should have a clear picture in mind what the critical success factors for the designing system are. To achieve the critical success factor, the system has to meet the aims of buyers and users.

The analysis of the user's aims, user's mental models and what the role of the technology/product is should take place in the beginning of the design process. The requirements for the design can be identified by this analysis. Usability evaluation requests should be identified as well. These requirements are derived from the aims of the particular application and formulated for this particular situation. When the usability requirements are defined, they should be transferred into the design technical languages. All of these processes require multiple disciplinary teams, and normally the potential users can be involved to a certain degree.

The analysis of the design requirement should be based on the application context. Maguire (Maguire, 2001a) recommended a five-step process for the analysis of design requirements in design context.

a) *Step 1: Describe the product or system.* It is important to have a high-level understanding of the product and the reason for its development.
b) *Step 2: Identify users and other stakeholders.* This will help ensure that the needs from all kind of users are taken into account of and, if required, the product is tested by them.

c) *Step 3*: Describe the context of the use. The components of context of use analysis include four different aspects: 1) the description of the system includes its purpose, the main application areas, the major functions and the target market; 2) user analysis includes the experience, domain knowledge and performance skill; 3) a list of tasks; and 4) environmental analysis includes not only the technical environment, but also the physical and organizational environment. This description will help the usability analysis in the design context.

d) *Step 4*: *Identify important usability factors*. The identification activities have to be based on the earlier three steps of the analysis and find out the critical components at will affect the usability of the product.

e) Step 5: *Document potential requirements or test conditions*.

5.5.7   UCD Process in Practices

Philips Company has reported that one of their projects has developed a standardized approach to the evaluation of user interface for Philips' Business Groups (van Vianen, 1996). The product creation process (PCP) is the center part of the approach. PCP contains four main phases as below. The strengths and weaknesses of some testing techniques are listed in Table 5-5.

a) *Know-how phases*: The purpose in this phase is to gather and generate new ideas for concepts. It will systematically study the existing problems, test the concepts to answer specific questions, and apply know-how in several fields. The techniques used in this phrase are 1) user workshops discussing the general issues related to the products and re-design specifications; 2) focus group to have a deep insight discussion of the product; and 3) a usability test of the existing products for their effectiveness, efficiency and satisfaction.

b) *Concept phase*: It will encompass carrying out work for future product generations. Here the ideas and concepts are selected. It will determine the usability problems and acceptance criteria and answer specific questions. Some informal tests and usability tests for the new concepts will be carried out.

c) *After product range start (PRS)*: It will form the specification of the rough product range and test the final concept in detail and includes verification of implementation. Here the usability test will also be carried out.

d) *After commercial release (CR)*: Here the product is released for delivery and distribution on the market. Again, the inventory of problems and remarks, confirmation of usability criteria and detailing of problems will be documented and usability tests will also be carried out.

IBM Consulting has a history of applying human factors knowledge to their product design back to the '50s (Ominsky, 2002). They focus on the concept of UCD and the key pieces that make the success of the UCD approach which are:

a) Early and continuous focus on the user for all the possible aspects that may have impact to the design.
b) Design the total user experience with multidisciplinary team. The total user experience includes all of what users will see, touch and hear.
c) Relentless focus on the competition.
d) Frequent usability evaluation at all stages of design.

Table 5-5. Overview of the strengths and weaknesses of some testing techniques

Technique	Strengths	Weaknesses
User workshop	User involvement Intense and effective initial concept generation Combine evaluation/solution generation	Much effort required and expensive, but provides more in-depth information at an earlier phase
Focus group	Can discover the unexpected Revealing and prioritizing issues that are important to users	Poor basis for decision making Information mainly from opinion leaders Little measurable data
Usability test	Can measure task performance User involvement	More expensive and time consuming, but provides more in-depth information at an earlier phase
Informal test	Quick	Measurements can be unreliable Users often not from target group
Inventory of usage	Objective information Easy to process	Need for special equipment
Inventory of comments/remarks	Input from users based on end product	Self-selective Dependent on motivation

The multidisciplinary team includes the total user experiences leader, marketing specialist, visual designer, HCI designer, user research specialist, technology architect, service and support specialist and user assistance architect. To be able to perform the UCD approach, they developed fifteen work products and educated the other design team members on the advantages of performing UCD activities up front in the design process so that maximum value could be attained and rework due to poor design minimized. The fifteen work products can be summarized into three categories as shown in Table 5-6 (Jordan, 1998).

Table 5-6. **Work products for UCD in IBM**

UI Planning	UI Design	Usability Evaluation
• Usability design and evaluation plan • User profiles • Usability requirements • UCD approach to use case modeling • Use case validation report	• UI conceptual model • UI prototype • UI architecture • UI design guidelines • Visual resources • UI design specification	• Current solution evaluation • Early usability evaluation • Usability test plan • Usability test report

The differences between these two approaches are that Philips' PCP has tried to involve the users physically in all process, while IBM's approach is focusing on understanding the user.

5.6   SOCIAL TECHNICAL ISSUE

Social and organizational factors can make or break the deployment of information and communication technology. Any computer system may interfere with the existing authority and social relationships within an organization. There are many different technologies and analysis methods that one can use and Dix (2003) has a short discussion on some of the methods in Chapter 13 of his book *Human-computer interaction*. The selected discussion includes social-organizational approaches to capturing stakeholder requirements, including social-technical models, soft-system methodology (SSM), participatory design and ethnographic approaches.

Social technical models focus on representing both the human and technical sides of the system in parallel to reach a solution that is compatible with each. Methods vary, but most attempts to capture certain common elements as in (Dix, 2003):

a) Why the technology is proposed and what problem it is intended to solve
b) The stakeholders affected, including primary, secondary, tertiary and facilitating, together with their objectives, goals and tasks
c) The changes or transformations that will be supported
d) The proposed technology and how it will work within the organization
e) External constraints, influences and performance measures

SSM models the organization, of which the user is part, as a system. Participatory design sees the user as active not only in using the technology, but in designing it. Ethnography, on the other hand, views the user in context, attempting to collect an unbiased view of the user's work culture and practice.

5.7   ADAPTIVE AND INTUITIVE USER INTERFACE

Here I would like to discuss two types of interfaces, one the adaptive interface and another, the intuitive interface. These two types of interfaces are highly usable, but the development process may not necessarily follow the UCD process. One thing is in common, however, and that is to have a deep understanding of users, especially the end users.

Adaptive User Interfaces (AUIS) are meant to be adaptive to differences or changes that exist or take place within the user population of a computerized system (Browne, 1993). The main goal of adaptation is to present an interface to the user that is easy, efficient and effective to use. They are a promising attempt to overcome contemporary problems due to the increasing complexity of human-computer interaction (Dieterich, 1993). The AUIS makes the complex systems usable, as it presents what the user wants to see. It fits heterogeneous user groups. It simplifies and speeds up the performance process. It also takes increasing experience into consideration (Dieterich, 1993).

There are a number of fundamental questions that are frequently asked when it deals with the rationale for an adaptive interface behavior (Dieterich, 1993; Kühme, 1993):

a) What might the interface need to be adaptive?
b) Who should adapt and what should their role be in the adaptation process?
c) What levels of the interaction should be adapted (for example presentation or functionality)?
d) What information should be considered when looking for opportunities for adaptation?
e) What goals should be furthered by the chosen level of adaptation when triggered by the information under consideration?
f) When should the changes be made?
g) What are the usability problems that ought to be addressed by building adaptive interfaces?

There are four reasons and benefits that have been identified for adaptive interface design. These four reasons and their respective benefits are listed in Table 5-7 (Kühme, 1993).

Table 5-7. The reasons and benefits for user adaptive interface design

Reasons	Benefits

User with different requirements:
  a) Individual cognitive abilities
  b) Personality
  c) Preferences and desires
  d) Knowledge and interests
  e) Interact with task differences
	a) May allow more users to access the systems' functionality
	b) Higher degree of satisfaction
	c) Less learning efforts

User with changing requirements:
  a) User's development from novice to expert
  b) Changes in perception, understanding and the use of the system
	a) Optimal support during the learning process from novice to expert system user
	b) Continuously used with maximal efficiency

A user works in a changing system environment: Changes in a system environment are routinely caused by almost every interaction with the system.	a) Maintains interface consistency across the continuous modification of the system b) The effort of individualizing a user interface can be saved across version changes, leading to a higher efficiency of system use

A user works in a different system environment:
  a) With different applications
  b) One application in different environments
	a) User interface in a consistent way across system and application boundaries
	b) Lower error rate and higher efficiency

The mechanisms of adaption may be expensive, but they are programmable. There are two different ways to approach the adaptive design (Benyon, 1993):

a) A single design that will suit everyone. The design makes an average user have better understanding and exploits the system. It normally provides a variety of help and assistance for users so that they can obtain advice appropriate to their needs, when and where they need it. The problem with this approach is that the instruction can be very complicated and a user may have difficulties using it, or remembering how to use it.

b) The computer can adapt itself to individuals or groups of users. The system is supplied with a suitable theory of interaction and how interaction can be improved. The computer is in a position to change its functioning, its structure or the representations provided at the interface to better match the preferences, needs and desires of the users.

The information of individual users is the essential element for the adaptive design. People differ from one another. Three important questions that arise from it are (Benyon, 1993) how do people differ? What differences are useful? Which ones are stable and which ones really have an impact on the interaction? The differences between AUIS and intelligent interfaces is defined by Dieterich (1993) as: an intelligent interface is the integration of an AUI, both with an intelligent Help System, making context-sensitive and active help available, and with an Intelligent Tutoring System, supporting the user in learning the use of the system.

An intuitive interface is built on existing general and technology-specific user knowledge. The goal of the design is to minimize the gap between the user's existing capabilities and the system. It requires minimal learning for a given task complexity (Bullinger, 2002). It uses natural human modes of expression such as speech and gesture. It minimizes the gap between the users' intentions and goals and the interaction features and affordances provided by the system. For reducing that gap, it takes more fully into account the user's natural modes of expression and perceptual capabilities. At the same time, those usability requirements, such as efficiency of interaction, transfer of learning to new applications, or standardization of interaction techniques, will still be critical factors, thus making the system more intuitive to use (Bullinger, 2002).

5.8   USAGE-CENTERED DESIGN

There is a debate for involving the user in the design (Eason, 1995): users may not be aware of alternatives to current technology; users may expect the new system to be simply an improved version of the old one; users may not be able to step back from their work practices to see how technology can change the way they work; and/or they might not be familiar with the design methods used or the technology; and they may simply feel over-awed by the design process (which leads to their feeling unqualified to comment). It is hard to have the user contribute to the quality of the design solution. At the same time, it is difficult to collect all the information about the users.

Instead of focusing on the user, Rakers (2001) suggests focusing on the roles, goals and responsibilities people have. Rakers writes "Users have roles, these roles are associated with responsibilities, these responsibilities request actions to be taken, goals to be reached, and boundaries and limits to be set and adhered to." This idea leads to the usage-centered approach of the design.

Compared to the UCD process, usage-centered design does not have practical theories and methodologies in the design process and evaluation. The design goal and the process are different from usage-centered design and UCD. Here the usability is the fundamental, but not the final goal of the design. The final design goal will be to fulfill the higher level of human needs as shown in Figure 5-4, which are the psychological needs and self-actualization needs. How does one approach the final goal? Coe's (1996) suggestion about how to do so is purely from a psychological perspective. How should a designer start his work? How does one measure and evaluate the design process, and how does one know that final goal is reached?

5.8.1   Creative Design

As we indicated earlier, the UCD approach works well for improving the existing products in the usability aspects. It has its limitations toward the innovative design. Innovative design, in turn, requires the coexistence of a set of favorable factors that relate to ability, possibility and feasibility (Bullinger, 2002). As Bullinger indicated, ability subsumes all factors that influence people and their innovation attempts such as knowledge, imagination, motivation, creativity and the absence of psychological barriers. Possibility refers to the supportive nature of organizational and other environmental factors such as corporate structure, leadership, reward schemes and working environment. Feasibility includes technological as well as economical factors.

The creative processes can be segmented into four more-or-less distinguishable phases (Bullinger, 2002). Each of these phases has different requirements:

a) *Preparation*: The definition of the focus for change is made. Cognitive aspects identify situations where an improvement might happen. Tools and methods are needed to structure the issue. Information has to be easily accessible via different channels, and communication has to be facilitated.
b) *Incubation*: This is the approach to the issue via conscious and subconscious processes.
c) *Illumination*: This is the spontaneous generation of possible clues to the issue. There can be the freedom to cross-perceived borders.
d) *Verification*: This is intellectual clarification, or whether the illumination output respects the initially stated requirements and formulation of a

concept. It requires visualization and discussion facilitation, support of cooperative work, feasibility analysis and documentation.

The creative and innovative design is most likely to happen in the earlier stages of the design process when the design idea is forming. It often ends up in at the prototype level. Before it finalizes into the product, the usability concept and the usability design principles and process should be applied.

5.8.2 Design Criteria and Process

What are the general criteria for a good interface? Flach pointed out that "a good representation must present the deep structure in a way that it can be directly perceived and manipulated. The topology of the representation should allow direct perception of the deep structure (meaning, affordances). Direct manipulations of the representation should allow direct interaction with the deep structure of the problem" (Flach, 1998). In other words, the users and the systems are the partner relationship. The users should understand directly how the system is working and the results he or she can expect.

If a new design is being offered, the comprehensive interface design process from the usage-centered approach should involve the following steps:

a) Identify top-level task requirements (e.g. mission analysis, or work analysis).
b) Analyze and model the task (allocation of function, i.e., what the user will do and when; what the system will do and when—but not how).
c) Determine what communication needs to take place between humans and the system.
d) Develop recommendations for interaction requirements (dialogue) based on the type of communication and context.
e) Develop initial recommendations for interface technology, based on type of interaction requirements.
f) Discuss with technical experts to understand the possibilities and constraints of different technologies.
g) Develop initial design specifications for the interface content based on detailed task analysis, human factors knowledge and predictive models of human performance.
h) Produce a rapid prototype of the design.
i) Evaluate the design with user trials to establish how the user performs his allocated functions.
j) Re-iterate as required to achieve the required human performance.

For different interaction systems, the details of the design criteria and process can be somehow different. Different human factor books provided different interface design methods/process. The summary of different methods based mainly on Hackos's book (1998) is shown in Table 5-8. The detail of each method can also be found in many HCI books, so I will not go into detail about how it works.

Table 5-8. Interface design methods/process

Interface design method	Brief definition
Qualitative usability goals and measurable objectives	List of the goals you need to achieve to ensure that your users find the interface usable. Quantitative measures of the goals, as needed
Objects/actions: Nouns/verbs	Lists of nouns and verbs that represent objects you need to create and actions you need to support
Metaphors	Conceptual models that reflect how your users will think about the new interface
Use scenarios	Brief narrative descriptions of the users, tasks and environments as they will look in the new design
Use sequences	Ordered lists of the tasks as they will be performed by the users in the new design
Use flow diagrams	Drawings of the specifics of a task, including objects, actions and decisions, as it will occur in the new interface
Use workflows	Drawings of the major tasks of a larger activity as they will be performed by an individual user or by groups of users
Use hierarchies	Tasks arranged in a hierarchical arrangement to show their interrelationships in the new interface
Storyboards	Scripts and illustrations that illustrate the new design in the context of users, tasks and environments
Rough interface sketches	Preliminary and very rough sketches of screens, menus, dialogue boxes and more in preparation for paper prototyping
Video dramatizations	Video dramatizations of the new work environment with the new product in place

5.8.3 Ecological Approach

A primary goal of ecological interface design (EID) is to make the intentional and causal links explicit in the interface. In other words, the surface structure of the interface should directly reflect the deep structure of the work domain as reflected in relations along the diagonal of the abstraction/decomposition space. Ecological approach to the interface design is regarded as one of the most developed theories on usage-centered approach to the design (Flach, 1998).

EID theory did not address on the questions pertinent to interface design such as context sensitivity, visual momentum and dialogue. So it did not provide any other methods than what we learn from our earlier sections. It has tried to characterize the meaningful distinction within a work domain in two dimensions: *abstraction* and *level of decomposition*. Rasmussen (1986) distinguished five different levels for characterizing complex social-technical systems with respect to the abstraction dimension. The relationship between different levels of decomposition can be represented as a triangle in Figure 5-7. Categories at higher levels within this hierarchy provide the rational context for understanding why lower level constraints are meaningful. Categories at lower levels in the hierarchy provide the physical details of how constraints at higher levels can be implemented (Flach, 1998).

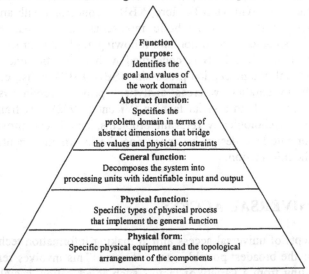

Figure 5-7. Rasmussen's abstraction hierarchy represents a nested hierarchy in which relatively few categories provide a meaningful characterization of the function purpose of a complex system.

This theory has been developed for an interface design framework for complex human-machine systems (Vicente, 1992). Most successful cases, or practically used EID cases, were applied to the industrial control room where complex tasks were performed (Burns, 2000; Christoffersen, 1998; Howie, 1998; Terrier, 1999; Torenvliet, 2000), and some applications to interactive microworld simulation (Reising, 2002).

Any interface design for complex systems started with the first question: What is a psychologically relevant way of describing the complexity of the work domain? This requires a representation formalism for describing the work domain's constraints (Vicente, 1992). It defines the informational content and structure of the interface. The second question is: What is an effective way of communicating the information? This requires a model of the mechanisms that people have for dealing with the complexity. This would provide a basis for determining the form that the information should take, the idea being that information should be presented in a form that is compatible with human cognitive and perceptual properties. These two questions define the core of the interface design problem (Vicente, 1992).

The abstraction hierarchy (AH) developed by Rasmussen (Rasmussen, 1986) is a useful framework for representing a work domain in a way that is relevant to interface design. It distinguishes the perception processing into three levels: skill-based behavior (SBB); rule-based behavior (RBB); and knowledge-based behavior (KBB). Here, KBB is concerned with analytical problem-solving based on a symbolic representation, whereas RBB is concerned with situation recognition and following applicable procedures to solve the problem, while SBB is concerned with automatic, unconscious actions. In general, the perception processing (SBB, RBB) is fast, effortless and proceeds in parallel, whereas analytical problem-solving is slow, laborious and proceeds in a serial fashion (Vicente, 1992). AH framework represents the psychologically relevant problems, and it can provide the foundation for interface design by specifying the information content and structure of the information.

5.9 UNIVERSAL ACCESS

The concepts of universal access aim at making information technology accessible for the broadest possible range of users. This involves removing barriers resulting from a variety of factors such as education, social status, level of income, gender, disabilities or age (Bullinger, 2002).

In the late 1990s, a group at North Carolina State University proposed seven principles of universal design. These principles give us a framework in which to develop universal design. These were intended to cover all possible areas of design and be equally applicable to the design of interactive systems (Story, 1998):

a) *Equitable use*: The design does not disadvantage or stigmatize any group of users. The user's appropriation, security, privacy and safety provision should be available to all.
b) *Flexibility in use*: The design accommodates a wide range of individual preferences and abilities through choice of methods of use and adapt to the user's pace, precision and custom.
c) *Simple and intuitive use*: Use of the design is easy to understand, regardless of the knowledge, experience, language or level of concentration of the user.
d) *Perceptible information*: The design promotes effective communication of information regardless of the environmental conditions or the user's abilities.
e) *Tolerance of error*: The design minimizes the impact and damage caused by mistakes or unintended behavior.
f) *Low physical effort and fatigue*: The design can be used efficiently and comfortably.
g) *Size and space for approach and use*: The placement of the system should be such that it can be reached and used regardless of body size, posture or mobility.

The universal assess interface differs from the classical HCI is the user group. For universal assess interface, the users are no longer homogeneous, with certain skill in knowledge and performance. Here, the users include all potential citizens, including the young and elderly, residential users as well as those with situational or permanent disabilities (physically and probably mentally as well). This brings a new challenge to the interface design for the following aspects (Akoumianakis, 2001):

a) No single design perspective, analog or metaphor, will suffice as a panacea for all potential users.
b) Design will increasingly entail the articulation of diversity for all potential users or computer-mediated human representation to describe the broader range and scope of interactive patterns and phenomena.
c) The developmental theories of human communication and action may provide useful insight toward enriched and more effective methodological strands to interaction design.
d) The designed functionality can be executed in different contexts, which arise due to different machine environments, terminals or user conditions.

e) Universal design lacks the solid methodological ground to enable early assessments of quality attributes pertinent to universal access, comprehensive evaluation techniques or formal methods to specify what universal access implies in certain application areas.

From the traditional approaches in user modeling, cognitive engineering and task analysis, it is hard to match the requirement of design "for all." The requirement of developing a user model of "design for all" should be able to capture (Stary, 2001):

a) Different sources of knowledge
b) Procedure of task accomplishment
c) Individual skill or performance
d) Dynamics of interaction
e) Adaptation in the context of human-computer interaction

Stary (2001) suggested two ways to define user groups: functional and individual perspective. A functional description of a user might deviate from an assumed best practice description and lead to individual sub-processes for task accomplishment, however, resulting in the requested output of work. Coupling the user context with the task context requires the use of particular relationships, "handles". A user handles a certain task in a certain functional role. The level of experience and skill development is captured through personal profile. In this situation, a typical user-centered design process does not fit into the design requirement while usage-centered approach probably provides the better solution.

Inclusive design is another approach, beside "design for all," to universal design. It is based on four premises (Benyon, 2005):

a) As we change physically and intellectually throughout our lives, the varying ability is a common characteristic of a human being.
b) If a design works well for people with disabilities, it works better for everyone.
c) Personal self-esteem, identity and well-being are deeply affected by our ability to function well in the society with a sense of comfort, independence and control.
d) Usability and aesthetics are mutually compatible

A total inclusion is unattainable for many different reasons (technical or financial). It is just a matter of minimizing the inadvertent exclusion. During the user characteristics analysis, it is important to make a clear distinction between fixed characteristics and changing ones. In each of these two categories, the common and rare incidence is distinguished. For each incidence, different alternative solutions will be suggested and balanced during the design. During the design, some existing systems should be tested, and people with special needs should be included. The designer should consider whether new features affect users with special needs both positively and negatively. The special needs users should be included in the usability tests. Normally these special needs users are disabled people (Benyon, 2005). Often, the universal access interfaces are multimodal.

5.10   ETHNOGRAPHY METHOD FOR CONTEXTUAL INTERFACE DESIGN

Most of the speech-related interactive system is contextually dependent. To shorten the gap between the designer and the user and to find the way for designs to be "realistic," designers must find ways of understanding work in practice. Ethnography methodology has been interesting many HCI designers.

Ethnography is regarded as a useful tool to study the social context of HCI. The aim of the design ethnographer is to describe and interpret cultures for the purpose of designing a future tool that will change the culture studies (Mantovani, 1996). It attempts to understand human behavior through different techniques such as observation, interview, material gathering, desk research, conversation analysis, etc. Ethnography is a key empirical data-gathering technique (Macaulay, 2000). It is not a theory but a set of methods (Hammersley, 1995). The study is always carried out in the field setting.

Since the ethnography methods involve collecting qualitative data—typically the verbal description of some individual's behavior—to use this data for the design is a tricky issue. The only way to solve the problem requires an ongoing dialogue between designers and ethnographers, so the ethnographer himself becomes the link between the field setting and the designer (Macaulay, 2000).

Different social life theories are needed for the ethnography study. Theory is used as the tool to provide the opportunities from intuition to insight (Macaulay, 2000). Ball and Ormerod (2000) have suggested that there are ten features that prototypical ethnography would be characterized by:

a) *Situatedness*: Data are collected by a participant observer who is located within the everyday context of interest (e.g. a community of practitioners).

b) *Richness*: The observer studies behavior in all manifestations, such that data are gathered from a wide range of sources including interviews, team discussions, incidental conversations and documents as well as non-verbal interactions.

c) *Participant autonomy*: The observees are not required to comply in any rigid, pre-determined study arrangements.

d) *Openness*: The observer remains open to the discovery of novel or unexpected issues that may come to light as a study progresses.

e) *Personalization*: The observer makes a note of his own feelings in relation to situations encountered during data collection and analysis.

f) *Reflexivity*: The observer takes a reflective and empathetic stance in striving toward an understanding of the observee's point of view, the observer takes account of, rather than strives to eliminate, his own influences to the behavior of the observees.

g) *Self-reflection*: The observer must acknowledge that any interpretative act is influenced by the tradition to which he belongs.

h) *Intensity*: Observations must be intensive and long-term, such that the observer should become immersed in the ongoing culture of the observee's environment.

i) *Independence*: The observer must not be constrained by pre-determined goal-set, mind-set or theory.

j) *Historicism*: The observer aims to connect observations to a backdrop of historical and cultural contingencies.

Ethnography method is a time-consuming work. It normally needs months or even years to observe and understand the work. At the same time, the people who perform the ethnography method should have special training the skill of observation and how to interpret what he or she saw toward interface design. The results from the ethnography method cannot be the only guide to the interface design.

5.11   CONCLUSION

In this chapter, we discussed the concept of usability and different definitions of usability from standards. The analysis of human needs indicated that human beings would never be satisfied with what they have, and they will always look for something new. Nowadays, usability of a product is the essential requirement for the success of its market. To approach the usability requirements, a user-centered-design (UCD) approach has been recommended and broadly used. What should be done in each step of the iterative life cycle of UCD process has been externally discussed.

Applying a UCD process probably cannot satisfy all possible design requirements, especially for innovative, intuitive and universal design. Different methods that can be used for usability tests and evaluations, included in the design process have been discussed. What we did not discussed here is the application of different design theories to the speech interaction system design. This part of the discussion will be appeared in later chapters.

Chapter 6

HUMAN FACTORS IN SPEECH INTERFACE DESIGN

6.1   INTRODUCTION

A speech interaction system is a computer interaction system that allows people to exchange information deliberately or to communicate with a computer system and at least part of their tasks performance through some form of speech input and output. Such systems can be unimodal or multimodal. The unimodal system uses only speech for input/output of the information while multimodal uses additional modalities such as a graphical speaking face or keyboard, etc., for information exchange. A speech interaction system will at least consists of recognizing speaker utterances (automatic speech recognition system, ASR), interpreting them (language understanding) with respect to the application, deriving a meaning or a command (speech generation) and providing consequent feedback to the user by a speech prompt (synthetic speech) or other modality of output or a system action.

Human speech production is a complex process. It starts with an intention to communicate after various levels of mental process that translate an idea into sequences of motor neuron firings and, in turn, active muscles that generate and control acoustic signals. The outcome of this process may be perturbed by so many factors that it is probably rare for two utterances to be so similar that no difference could be detected. Much (perhaps all) of the variability of speech carries information about the state of the speaker. Speech is normally regarded as the most natural way to exchange information. Children learn to speak first before learning how to read and write. The speech utterance contains very rich redundant information that written text cannot compare with. This redundant information is generally useful in social interactions between humans but is difficult to be "understood" by machines.

Among the speech technology, the speech recognizer and synthetic speech are the two components of many speech interaction systems that are directly interacting with human users. These two technologies stand in the front line of human computer interface. Behind them, there are different integration technologies such as spoken language dialogue systems, language understanding, language generation and translation.

There are a very limited number of human factors and usability studies related to the application of speech interaction system design. Most of the studies on human factors on the ASR system were reported in then '80s or early '90s. The results from these studies have strongly impacted to the human-friendly design of the ASR system. Currently, most speech systems are designed by computational linguists and computer scientists who are mainly concerned with the properties of the language models used in the systems. These designers pay much more attention to one or more of the following: the accuracy, coverage, and speed of the system, and usually believe that an improvement in any of these factors can be a clear win for the interface design (James, 2000). From an investigation of large amounts of publications, one can easily find that there are a lot of papers with usability as one of the keywords. The "usability" in these papers is not the same concept as we discussed in Chapter 5. Naturally, if a system has improved speech recognition accuracy, coverage and speed of the system, it has the potential impact to improve its usability since the functionality of a system is the basic requirement to usability. Still, these factors belong to the reliability of the system, rather than the usability of the system.

In the work of design of speech interfaces, there are a few major problems that cause the low usability of the system. People who are currently working on the design of speech interfaces are, in most cases, not interface designers and, therefore, do not have enough experience in human cognition and usability issues. Sometimes, they are not experts in speech technology either, and they cannot discover all the possibilities of speech technology. Speech, as an interface modality, has vastly different properties other than modalities, and, therefore, requires different usability measures (James, 2000), and there are not yet universal usability measurement methods for speech interaction systems. The naturalness and intuitive interaction of speech requires new design theories and design process guides. Due to these problems, the low usability of speech interaction systems is the major obstacle in applications.

Speech interaction system design is a multidisciplinary science. We will approach this subject from a broad scope.

6.2 THE UNIQUE CHARACTERISTICS OF HUMAN SPEECH

In Chapter 2, we discussed the cognitive psychology of human speech and language understanding. The characteristics of human speech and comprehension have a strong impact on interface design. It differs considerably from the traditional modalities of keyboard and mouse, media of dual tone multiple frequencies, computer-console and other visual-mechanical interfaces. The characteristics of human speech and comprehension can be summarized as follows:

The language process is complicated. In speech, the fundamental sounds-phonemes combined to form the smallest meaningful unit of words-morphemes. A string of words that follows certain grammar makes up the sentence and has certain meaning. Spoken language differs from written language in that the vocal intonation (prosody) can change the literal meaning of words and sentences. Spoken language contains much redundant information, including the speaker's feeling and intention, rather than the words itself. The efficacy of human-human speech interaction is wrapped tightly with prosody. The human voice has evolved remarkably well to support human-to-human interaction. We admire and are inspired by passionate speeches. We are moved by grief-choked eulogies and touched by a child's call as we leave for work (Shneiderman, 2000).

There are three main levels of analysis in the comprehension of sentences: syntactical (grammatical) structure, literal meaning and interpretation of intended meaning. Comprehension would be impossible without access to stored knowledge. There are three types of systems in the brain that process language. One is to receive non-language representations of sensory input and provide a way of organizing events and relationships. Another one represents phonemes, phoneme combinations and syntactic rules for combining words. The third neural system interacts with the category system and speech system to form the concept (Kolb, 2000). Language is a relatively automatic process because it comes from repetition of well-learned responses or the mimicking of input. At the same time, language representation changes with experience.

Speech is acoustic and sequential in nature. Physiologically, speech perception follows the basic mechanism of auditory perception. The three fundamental physical qualities of sound—frequency, amplitude and complexity—are translated into pitch, loudness and timbre as the qualities of sound perception. Speech contains numerous hints, such as pitch, intonation, stress and timing, to sentence structure and meaning. Spoken sentences can be an ambiguous and unclear signal, depending on the speaker's sound quality, language capacity and accents, and it is strongly affected by the environmental noise. Spoken word is spread out in time. The listener must hear and comprehend all information one element at a time. Speech information is transient a word is heard and then it ends. Similarly, spoken input must be supplied one word, one command, one number or one datum at a time. This requires the user to remember and order the information in a way that suitable for the machine to understand. As this, speech interaction is time consuming and has high memory demand. It is not like the visual display of information that can be continuously available. A visual display allows the user to search back to information whenever necessary. Spoken information is temporal in nature.

There is a cognitive resources limitation for problem-solving when speech input/output consumes short-term and working memory. The part of the human brain that transiently holds chunks of information and solves problems also supports speaking and listening. Speaking consumes the same cognitive resources as problem-solving, and it is difficult to solve complicated problems and speaking at the same time. Therefore, working on tough problems is best done in quiet environments—without speaking or listening to someone (Shneiderman, 2000). Hand-eye coordination is accomplished in different brain structures, so typing or mouse movement can be performed in parallel with problem solving. Proficient keyboard users can have higher levels of parallelism in problem solving while performing data entry. Product evaluators of an IBM dictation software package also noticed this phenomenon (Danis, 1994). Danis's report indicated that keyboard users could continue to hone their words while their fingers output an earlier version. In dictation, users may experience more interference between outputting their initial thought and elaborating on it.

The human head and brain are uniquely evolved to produce speech (Massaro, 1997). When humans hear speech or find themselves speaking, they immediately change their processing strategy from "Is this a person?" to "What should I know about this person?" to "How should I respond?" (Nass, 2000) and "How much does this person understand me?"

The human brain will act the same when confronted by speech technologies. Because human-directed responses are automatic and unconscious (Reeves, 1996) and require continuous reminders to be extinguished, humans do not have the wherewithal to overcome their fundamental instincts concerning speech. Thus, individuals behave toward and make attributions about voice systems using the same rules and heuristics they would normally apply to other humans (Nass, 2000).

Speaking interaction has social aspects. In most situations of human-to-human communication, listeners use more than just acoustic information to interpret a spoken message. The history behind the conversation, the context during the conversation and the familiarity with the speaker and the subject are all aids to the perception. When available, visual cues also assist such as lip reading. Lip reading provides the main cues for deaf people and it also provides important cues for normal conversation to solve the confusions in the acoustic information. The research from social-psychological approach showed that people treat media the same way as they treat other people (Reeves, 1996). Speech interaction systems can provide the user a strong feeling; when the system answers questions or reacts to the spoken sentences, it becomes a social actor. The user is building a relationship with the system. The relationship will affect the user's perception of the usability of the system. The user would automatically expect that the system would understand him or her, just as another person should understand him or her. The personal feelings will be involved in the interaction between humans and the speech system.

These five characteristics of human speech will strongly affect the speech interaction system design. We will discuss them in more detail in the following sections.

6.3   HUMAN SPEECH RECOGNITION SYSTEM

A brief desciption of speech recognition systems in statistical modeling paradigm is shown in Figure 6-1. The speech signals are received in a microphone. They are converted into digital format by use of an analog-digital converter. The feature of the speech signal will be analyzed, and the results of the feature analysis will be classified into patterns. There is a sequence of stored patterns that have previously been learned. These stored patterns may be made up of units as words, or shorter units as phonemes (the smallest contrastive sounds of a language). The comparison of the sequences of features with the stored patterns is to find the exact matches. After the pattern classification, the language processing will carry out its job and turn the speech signals into the text of the sentences. The fundamental of speech recognition mechanism implies the difficulty of speech recognition study: the speech signal is highly variable due to different speakers, different speaking rates, different contexts and different acoustic conditions.

Modern linguistics has developed over many years, but the most successful computational linguistic applications use very little modern linguistic knowledge (Huckvale, 1997). For example, the linguistic view of phonetic segments is that they have realized identifiable and distinguishable properties or *features* by which they may be recognized. The engineering view is that segments are associated with regions in a multi-dimensional space formed from a fixed set of continuously valued observational parameters. Huckvale (1997) indicated that this shift from specific features to generic parameters has been necessary owing to two facts: (1) the enormous variability in the realizations of segments according to producer, context, environment and repetition, and (2) the imperfections of pattern recognition systems for feature detection. Together, these mean that the presence or absence of a detected feature cannot reliably indicate the presence or absence of an underlying segment. As each observation is made up from a large number of features, pattern recognition is hard in higher dimensional spaces.

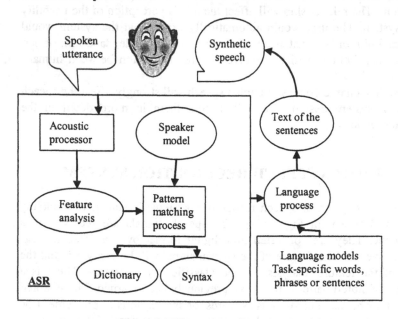

Figure 6-1. The process of speech recognition

The reasons to make such gaps are because of the different goals between speech engineers and the linguistician (Huckvale, 1997). The goal of the speech engineer is to reach the high performance of the speech system. The performance is measured by the degree of success at communication and how readily and accurately speech can be communicated. The underlying structures of the speech are only interesting to the extent that they make communication more accurate and more reliable. So if the regularities and constraints in language can be captured and expressed in a mathematical formalism that can exploited to make man-machine communication practical, then it doesn't matter to the engineer that such a formalism allows all kinds of unknown phenomena to occur, or is beyond human cognition.

6.3.1   Automatic Speech Recognition System

Automatic speech recognition (ASR) technology makes human speech signals useful in carrying out different activities. Upon detection and recognition of a sound or string of sounds, the recognizer can be programmed to execute a predetermined action. Consequently, many groups of individuals have benefited from using ASR in human-machine interaction, in human-to-human communications, and as a tool to control their immediate environment.

Automatic speech recognition is the capability of a machine to convert spoken language to recognized words. These words are processed by an understanding stage and results into one or more actions. Here are some of the dimensions in which ASR differs:

a) *Degree of speaker independence*: the user, who speaks to the recognizer, trains a speaker dependent recognizer. Usually speaker-dependent recognition systems can have high recognition accuracy. But the training process can be difficult and time consuming. Speaker-independent recognizers attempt to maintain recognition accuracy independent from the user. The user does not need to train the system, but the accuracy was usually less than a speaker-dependent system. Another one in between is the speaker adaptive system. The factory trained it before it was delivered to the user, and the system can automatically adapt to the speaker's characteristics and reduce the training requirements during use.

b) *Vocabulary size and complexity*: Normally, ASR has been classified as "small vocabulary" and "large vocabulary" systems. The small vocabulary ASR is defined as all words in the vocabulary that must be trained at least once. Large vocabulary ASR system recognizes sounds rather than the whole words. It has a dictionary that contains the pronunciation of all words in the vocabulary. When other factors are equal, a larger vocabulary is likely to contain more confusing words or phrases that can lead to recognition errors.

c) *Speaking rate, co-articulation*: Speech sounds can be strongly affected by surrounding sounds in rapid speech. Words in fluent speech usually have worse recognition rate than isolated words. In isolated speech, the speaker must leave distinct pauses between each word. It is very unnatural for the speaker but can be useful for command and control or under high noise conditions.

The most popular ASR systems are statistical modeling paradigms in nature. In the past ten years, the ASR technology has made tremendous progress, and most of the ASR systems in the market can recognize continuous speaking of sentences. The accuracy in the testing laboratory environment can reach 99 to 100% accuracy. Many ASR systems can have high tolerance on the variances of human speech, with different accents and with high noise.

Speech recognition has also made dramatic improvements, though these improvements are harder to quantify. Church (2003) describes a transition from small vocabulary isolated word template matching in 1962 to large vocabulary continuous speech recognition and interactive dialogue systems in 2002. In the past fifteen years, the automatic speech recognition technology, even for large vocabulary systems, has made consistent error-rate reduction. The great advances in large vocabulary speech recognition systems over the past two decades were largely due to the power and flexibility of the HMM framework. HMMs could simultaneously exploit the continuous increase in available training data and computing power. Large-vocabulary speaker-independent continuous speech recognition (LVCSR) is now available in reasonable price in many computer stores. Large vocabulary continuous speech recognition (LVCSR) in an ideal condition, such as read, clean, monotonous and native speech, has realized its performance of approximately 98%, and research targets have been shifted to real-world speech recognition. Recognition of spontaneous, noisy, expressive and/or non-native speech is considered challenging task, and a lot of effort is being made to improve the performance. Most applications of LVCSR systems are used for handicapped users and some professional users. These include lawyers and doctors who dictate their documents, since these people normally have slow typing skills and are not good at spelling the words.

All successful ASR applications have a number of features in common: First, they have tasks that demand small vocabularies; Second, they have dedicated users; Third, in terms of safety, the application is non-critical (Noyes, 2000). Small vocabulary voice command-and-control is becoming a familiar feature for users of telephone-based interactive voice response (IVR) systems.

However ASR advances have slowed down quite a bit in recent years. This statistical modeling paradigm of ASR and is also called knowledge-ignorant models because it has nothing to do with linguistic knowledge and has more or less reached its top level. These systems are often overly restrictive, requiring that their users follow a very strict set of protocols to effectively utilize spoken language applications. Furthermore, the ASR system accuracy often declines dramatically in adverse conditions to an extent that an ASR system becomes unusable (Lee, 2004). Even a state-of-the-art ASR system usually gives much large error rates, even for rather simple tasks operating in clean environments. If it is in a noisy environment, the recognition accuracy will still decrease dramatically.

It seems the past successes of the prevailing knowledge-ignorant modeling approach can still be further extended if knowledge sources available in the large body of literature in speech and language science can be objectively evaluated and properly integrated into modeling. The future development will be knowledge-supplemental modeling of the ASR, which means to integrate multiple knowledge sources and to overcome the acoustic-mismatching problems and "ill-formed" speech utterances contain out-of-task, out-of-grammar and out-of-vocabulary speech segment. The new approach will give a better understanding of human speeches, than just acoustic signals. It provides an instructive collection of diagnostic information and potential benefits for improving our understanding of speech, as well as enhancing speech recognition accuracy.

6.3.2　Speech Feature Analysis and Pattern Matching

The first step of speech recognition is feature analysis (see Figure 6-1). Feature analysis distills the information necessary for speech recognition from the raw speech signal. Feature analysis is one of the most important parts of speech recognition. Through feature analysis, the speech signal can be coded as a computer signal. In lifetime application of the ASR system, the feature analysis should discard irrelevant information such as background noise, channel distortion, speaker characteristics and manner of speaking.

The second step of the speech recognition process is pattern matching (Figure 6-1). Pattern matchers take into account both the variability of speaking rates and the constraints of correct language. The way of pattern classification has been regarded as the philosophy of the speech recognition system (Roe, 1993). Some knowledge about pattern classification is also important for studying the physical and psychological aspects of speech recognition. There have been four classes of pattern matchers for speech recognition: template matchers, rule-based systems, Artificial Neural Networks (ANN), and Hidden Markov Model (HMM) systems.

The principle of template matching is to store examples of the speech patterns called templates that consist of sequences of feature vectors for each speech pattern. Unknown speech features are compared to each of the templates to find the closest match.

Rule-based systems set up a series of criteria in a decision tree to determine which of the units of language is presented in the speech signal.

For large complex speech recognition tasks, it is difficult to create sets of rules that generalize well across the many variations in the speech signal.

HMM systems are very successful speech recognition algorithms and are used in virtually all applications except where low cost is the overriding concern. HMMs can be defined as pairs of discrete-time stochastic processes. HMMs use dynamic programming techniques reminiscent of template matching to achieve time normalization. The principle advantage of HMM systems over template-based approaches is that HMMs retain more statistical information about the speech patterns than templates. HMMs retain information about the complete distribution of features present in the training data. This translates to greater discrimination power (Roe, 1993).

The basic idea underlying Artificial Neural Network (ANN) research is to take limited inspiration from neurobiology data about the human cortex in order to develop automatic models that feature some "intelligent" behavior. The fact that the basic problem of ASR is that of pattern recognition suggests the use of ANNs in this area. The idea that the brain cortex is able to recognize speech and that ANNs are crude models of the cortex is not in itself a valid justification since ANNs and the cortex are known to operate differently. There are several good reasons in favor of using ANNs[7]:

a) ANNs can learn from examples, a feature particularly crucial for ASR
b) ANNs, unlike HMMs, do not require strong assumptions about the statistical properties of the input data
c) The ANN formalism is in general intuitively accessible;
d) ANNs can produce highly non-linear functions of the inputs
e) ANNs have highly parallel and regular structures making them especially suitable for high performance hardware implementation

In this approach, pattern classification is done with a multilayer network of perceptions with each link between perceptions assigned a weight determined during a training process. Each perception finds the sum of its inputs partially clips it with a sigmoid function and passes the result through the links to the next layer of the network. It has proven difficult for neural networks to achieve the level of time alignment of the speech signal that HMMs have attained (Roe, 1993).

The dictionary (Figure 6-1) contains the vocabulary of all words that ASR has been designed to recognize. The words are stored as sequences of

---

[7] Course material of international summer school neural nets "Speech processing, recognition, and artificial neural network" given by Professor Jean-Paul Haton. 1998, Oct 5—14 in Vietri, Italy.

phonemes, and a phoneme database is used that contains statistical descriptions of all phonemes, including variation in the speaking characteristics among different speakers.

The syntax (Figure 6-1) contains a description of valid combination of words to sentences. Thus, a complete description of all valid sentences is available for the pattern-matching process.

The speaker model (Figure 6-1) describes how to interpret words as phoneme sequences and how to consider variations in the speech characteristics. Speaker models are typically based on HMM or ANN.

A parallel computational approach to increase the accuracy of an ASR system is to combine speech recognition and language processing. In the language process, the system possesses a knowledge base with domain concepts, hierarchical goal trees and a user model. Based on the actual context and the dialogue history, it can use its higher-level knowledge to generate a set of predictions about probable sentences.

6.3.3 Speech Synthesis

The technology generating the speech output is called "speech synthesis". There are two very different categories for speech synthesis: the first one is called "canned speech" because the output speech is generated on the basis of pre-stored messages. With this type of synthesis, very high quality speech can be obtained, especially for quick response applications. This technique, however, requires the use of large memory and is not flexible. The second one is "text-to-speech synthesis." It allows the generation of any message from text. This generally involves a first stage of linguistic processing, in which the text input string is converted into an internal representation of phoneme strings together with prosodic markers, and the second stage of sound generation on the basis of this internal representation. The sound generation can be made either by rule, typically using complex models of the speech production mechanism (formant synthesis), or by concatenating short pre-stored units (concatenative synthesis). The speech quality obtained with concatenative synthesis is generally considerable higher; however, the memory requirements are larger than with formant synthesis. The current technology in text-to-speech synthesis already yields quality that is close to human speech when considering the word intelligible, but it is still inferior to human speech in terms of naturalness and expression of emotion.

6.3.4   National Language Processing

Language process is the computer technology that understands the meaning of a sequence of words. Language is so fundamental to humans, and so ubiquitous, that fluent use of it is often considered almost synonymous with intelligence. Many people seem to think it should be easy for computers to deal with human language, just because they themselves do so easily. Actually, for computers, it is one of the most difficult tasks. Natural language (NL) understanding by computers is a discipline closely related to linguistics. It has evolved to incorporate aspects of many other disciplines, such as artificial intelligence, computer science and lexicography.

One way to illustrate the problems of NL processing is to look at the difference between the fundamental goals of an ASR system and an NL system (Bates, 1994). In an ASR system, there is well-defined input and output. The input is a speech signal, and the output is a word string. It is fairly easy to evaluate the quality of different SR systems. In an NL process, it is extremely difficult to have precisely the characteristics of the input/output. They are varied, hard to specify, difficult to get common agreement on and resistant to the development of easily applied evaluation metrics (Bates, 1994).

The input language might be a sentence at a time or multiple sentences all at once. It might not be sentences at all in the sense of complete grammatical units but could be fragments of language or a mixture of sentences and fragments. The input might be grammatical, nearly grammatical or highly ungrammatical. It might contain useful cues like capitalization and punctuation or, particularly if the input comes from a speech processor, all punctuation and even the sentence boundaries might be missing. The ultimate output from a NL system might be an answer from a database, a command to change some data in a database, a spoken response or some other action on the part of the system. But these are the output of the system as a whole, not the output of the NL processing component of the system. This inability to define output for an NL process system will cause problems in a variety of ways. One of the reasons that NL is challenging to computational linguists is its variety (Bates, 1994). Not only are new words frequently introduced into any natural language, but also old words are constantly reused with new meanings (not always accompanied by new morphology).

A spoken-language-understanding system is the integration of speech recognition with natural-language-understanding systems. If spoken language is the input to the system, then an appropriate response will depend on the meaning of the input. In this way, spoken-language-understanding systems could be created by the simple serial connection of a speech recognizer and a natural-language-understanding system. There are certain reasons that make up the differences between spoken language understanding system and general NL system (Moore, 1994). Spontaneously spoken language differs in a number of ways from standard written language. If the natural language system were not adapted to the characteristics of spoken language, even a speech recognizer could not deliver a perfect transcription to a natural-language-understanding system and performance could still be wrong. Spoken language contains information that is not necessarily represented in written language, such as the distinctions between words that are pronounced differently but spelled the same, or syntactic and semantic information that is encoded with different prosodic characteristics in speech. It is high interest of to extract this information to solve certain understanding problems more easily with spoken input.

6.3.5 Language Modeling

Study of language theory includes acoustic, phonetic, phonological and semantic structures of speech as well as the mathematics of formal languages. There are two fundamental tasks in applying language modeling to speech recognition (Roe, 1993): a) The mathematical description of the structure of a language, its valid sequences of words or phonemes; and b) Given such a mathematical description of the language, the compute shall efficiently calculate the optimum sequence of symbols from the acoustic pattern classifier that meets that mathematical description.

Of the three stages of speech recognition—feature analysis, pattern classification and language modeling—language modeling is the weak link, and recent research in computational linguistic and statistical language models have made very good progress as a scientific base from which to attack these formidable problems.

6.4 HUMAN FACTORS IN SPEECH TECHNOLOGY

Regarding the speech interaction system design, the human factor or HCI designers have different focuses compared with speech technology engineers. Human factors or HCI designers are concerned with such issues as information grouping and differentiation, consistency with the way users perform their tasks and clear specification of the purpose of each interface element. The influences theories are from cognitive psychology and social sciences, while speech technology engineers are more focused on corpus analysis, recognition accuracy, speech auditory signal analysis, etc. The main theories and research methodologies are linked with theories of language, which are an attempt to explain (among other phenomena) what makes a discourse (or dialogue) coherent, socially acceptable and appropriate for the reader/listener at hand, and how various communicative functions are realized in language at the various levels of the linguistic systems (e.g., pragmatic, semantic, lexical and syntactic). There is very limited collaboration, cross-references or even mutual knowledge between the HCI and speech technology.

Human factor specialists can contribute to the development of speech interaction system design application in several ways: a) to explore the possibilities of what the technology can do to the people in the using context. This may include integrating the speech technology into existing work practices and identifying the tasks that might be more suitable for speech interaction. This means considering not only mobility requirements and physical constraints on the use of hand and eyes, but also the psychological demands imposed by other task components. b) To design the speech user interfaces, with the aim of improving the usability. Specific topics under this heading include the selection of task vocabularies, the design of system feedback and specification of dialogues for error correction. c) Operational use of speech recognition equipment in order to gain a better understanding of processes that affect the performance of these devices. The ability to avoid or to remedy the causes of poor performance is critical. Even the best-designed application will fail if the user is unable to achieve acceptable levels of recognition accuracy. d) Integration with other technologies for accomplishing the work. In this situation, a speech interaction system is part of the large interaction systems. The human factor specialists can help the designers to collect the job and the task requirements and the design criteria for the entire system, as well as for the speech interaction system. e) Development of the usability design guidance. In this aspect, the combination of cognitive psychology, social-psychology and the knowledge about users in their application environment is important for the designers to select and to adapt the design guidelines into their specific design context. f) Performance evaluation of the system. The evaluation of the system performance toward the user satisfaction can be a very difficult issue, and there is not a practical theory for it right now.

6.5   SPEECH INPUT

Earlier this chapter, we discussed the differences between human speech and written text. Many studies also indicated that as soon as people start to communicate with a system via speech, they would automatically treat the interaction media as a social actor. Still, there are many differences between human-to-human communication and human-machine communication. Compared with human to human communication, the assessment to human-computer spoken language has revealed that spontaneous speech has been more discontinuance and that self-repairs are substantially lower when they speak to a system, rather than to another person (Oviatt, 1993). The degree of discontinuance is related to an utterance's length. When an utterance is spontaneously spoken, it may involve false starts, hesitations, filled pauses, repairs, fragments and other types of technically "ungrammatical" utterances. Speech recognizers and natural language processors have difficulties accepting these phenomena. These have to be detected and corrected.

In general, a speech recognition system involves the system that understands a natural language, but it normally does not cover the entire language. It is a trade-off problem here during the application. Normally, a system that can handle a large vocabulary has less recognition accuracy and is sensitive to the speech variance and environmental noise. A speech recognition system that handles small vocabulary can have better recognition accuracy and more robustness. Users will employ utterance outside the system's coverage. It is difficult to give sufficient data on which to base the development of grammar and templates that can cover all the possible utterance that a cooperative user may generate.

The determination of the system's language coverage is influenced by many factors in the situation. This question is context-dependent and designers should investigate it through the following aspects (Oviatt, 1993)

a) Clearly define, or select a relatively "closed" domain, whose vocabulary and linguistic constructs can be acquired through iterative training and testing on a large corpus of user input.
b) The system should easily let the users discern the system's communicative capabilities.
c) The system should guide the users to stay within the bounds of those capabilities.
d) The system should indicate clearly what level of task performance users could attain.
e) The designer should have a clear understanding of what level of misinterpretation users will tolerate, and what level is needed for them to solve problems effectively, and how much training is acceptable.

6.5.1   Human Factors and NLP

Ozkan and Paris (2002) have pointed out that the NLP disciplines do not collaborate with human-computer interaction (HCI) studies. They discussed several application areas and methodologies issues. The inclusion of HCI perspectives and methods increases the overall cognitive and contextual appropriateness of NLP systems. For example, the task analysis, contextual inquiries and other traditional HCI techniques, which investigate the goal, relevance, groupings and structure of information queries, complement the NLP technologies of generating information in a correct and relevant manner. The communication between human to human and human to machine is different. The interaction between human to human and human to machine will also be very different. Human behavior towards a computer and human expectations from a computer are also different. So the text in the corpus should not be the only objects of study for the design of the NLG system.

The evaluation methodology in NLP has been traditionally to assess the readability and fluency of automatically generated tests. It is not enough for the usability consideration. It would be useful to follow the principles of user-centered design process and perform task-based evaluations, thus evaluating the generated texts in the context of a task. This would allow researchers to measure the actual impact of the texts on their intended users.

6.5.2   Flexibility of Vocabulary

Letting the user adapt to the speech system is not a good design of the system. There are big individual differences of how quickly the user can learn and adapt to the system language. It would affect the user's perception of the usability of the system. Nowadays, most applicable speech interaction systems have limited vocabulary size such as a couple of hundred words combined into different sentences. The vocabulary selection and the syntax design can be crucial for the success of such a system.

The vocabulary should be selected in such a way to avoid possible misrecognitions of ASR but allow the certain robustness level of the system and to demand the minimal amount of training of the user to remember the vocabularies. Users have problems recalling correctly all vocabulary items (Taylor, 1986). This aspect may be independent of environmental stressors and attributable solely to the (inadequate) strategies individuals employ to recall words from their short-term memory (Levelt, 1993).

Jones et al., (1992) reported their visit to thirty worksites where ASR was applied. ASR was being used either to replace an existing input medium, or to run in parallel with that medium. In these sets, selection of vocabulary items was based on established forms, which generally meant those used with manual input devices (either dedicated keys, key combinations or fully transcribed commands). They found that compatibility between voice and manual input systems is clearly desirable, in so far as it can be achieved, not least because of problems that could be encountered if it is necessary to switch between the two. The reason was that the resulting vocabularies were not designed for maximum discriminability by the ASR system. Trials frequently indicated a need for vocabulary modification, typically in one of two circumstances: a) where there was a high incidence of confusion errors (e.g., between "*next* and *six*"), a new word would be substituted for one of the conflicting items; and b) where users found an alternative word more "natural" for a particular function. Many experienced users had become skilled at substituting alternatives for vocabulary items that were poorly recognized.

The variability of word selection by humans is a fundamental fact of human behavior (Furnas, 1987) as far as the language regulation is permitted. A speech interface must be designed with this variability in mind to improve user performance and increase user acceptance. This is not to say that a speech interface should include thousands of commands. It is not technically or economically feasible to build an interface capable of dealing with all words and their variations.

6.5.3 Vocabulary Design

In order to narrow the size of a speech-interface vocabulary, the application and the domain must be considered in the design (Sheniderman, 1992). Even the earliest speech interface studies suggest that the spoken commands must be designed using words from previously spoken interactions. In addition to domain specificity, the vocabulary of the speech interface should reflect the user's expectation of an appropriated model of the dialogue (Baber, 1989). Transcription conventions have been developed to capture a domain and task-specific vocabulary (Rundicky, 1990; 1989). Wizard of Oz (WOZ) processes were usually used to capture the vocabulary of people performing a specific task.

Dillon (1997) studied the user performance and acceptance of a speech-input interface in a health assessment task. This study examined their performance and acceptance of the interfaces. They selected novice and expert nurses to perform a hands-busy and eyes-busy task using a continuous speech-input interface. Three factors were found to improve the user's performance: expertise in the domain, experiences with the interface and the use of a small vocabulary. Experience with the interface corresponded with a higher degree of acceptance. Available vocabulary size and level of expertise did not affect acceptance.

6.5.4   Accent

Accent is a challenging problem in speech recognition. It is one of the most important factors that create undesirable variability in speaker independent speech recognition systems. Speaker accent is an important issue in the formulation of a robust speaker independent recognition system. Second-language learning requires a modification in the patterns of intonation, lexical stress, rhythm and grammar, as well as the use of additional distinctive phonemes, which are perhaps unfamiliar to the speaker.

The person with a foreign accent can be aided in processing deviations in articulation, rhythm, voice and symbolization. Hansen and Arslan (1995) conducted a series of experiments whose results motivate the formulation of an algorithm for foreign accent classification. A useful application for such an algorithm would be for selection of alternate pronunciation models in a multiple entry dictionary for a speech understanding system. An accent sensitive database is established using speakers of American English with foreign accents with an accuracy of 81% in the case of unknown text, and 88.9% assuming known text. Finally, it is shown that as accent sensitive word count increases, the ability to correctly classify accent also increases, achieving an overall classification rate of 92% among four accent classes.

The causes of accent to recognition problems are complicated. Accent induced by foreign language has a lot to do with the person's language knowledge, the degree of influences from other language, the speech task and the environmental settings (Chen, 2003). The recognition errors due to the accentual pronunciation may be reduced by introducing additional algorithms or databases (Hansen, 1995). The other types of recognition errors due to the wrong pronunciation or wrong express in the sentence can only be solved by the system design. In this aspect, the human factor specialists can help with the vocabulary selection, the syntax design and even introduce the user-mental models, domain task knowledge and system modeling to overcome the problems.

6.5.5   Emotion

The efficacy of human-human speech interaction is wrapped tightly with prosody. People learn about each other through continuing relationships and attach meaning to deviations from past experiences. Friendship and trust are built by repeated experiences of shared emotional states, empathic responses and appropriate assistance. "Emotion in speech" is a topic that has received much attention during the last few years. The purpose of these studies is to dictate the hidden information in human speech and to increase the recognition accuracy.

Human emotions normally include love, sadness, fear, anger and joy/happiness as basic ones, and some people add hate, surprise and disgust, and distinguish "hot" and "cold" anger. Quite some research effort is now being put into a field that is called "affective computing" (Picard, 1997). The goal in affective computing is to design ASR and TTS related algorithms (e.g., agents) that understand and respond to human emotions.

Although the emotion may manifest itself on the semantic level, the emotion content is to an important extent carried by prosodic features. A large number of studies investigate the relation between acoustic features of utterances and the emotion. It appears that a number of "basic" emotions such as anger, sadness and happiness can quite well be described in terms of changes in prosodic factors: pitch, duration and energy (Cowie, 2003; Scherer, 1981). But automatic detection of the emotion state from the speech signal is not straightforward. For example, recent studies show that the triplet happiness, sadness/neutral and anger can be distinguished only with an accuracy of 60–80% (Li, 1998; Whiteside, 1998). When more emotion tags are to be recognized (some studies distinguish eight or more different emotions), the detection results decrease substantially; depending on the task, the performances range from 25% to about 50%. These results are largely dependent on the size of the acoustic datasets and on the way the emotion classifier is trained.

A number of studies narrow down "emotion" to "stressed" in the sense of stressful (Zhou, 1998). In that case, the task is not to recognize the emotion itself, but to binary classify utterances as stressed or "not stressed." With an optimal choice of the features used for classification, such stress detection may improve recognition score. In another example, Huber, et al., (Huber, 2000) simplifies the classification of emotion to the binary question whether one particular emotion (anger) is present or absent. They report the highest classification (anger versus neutral) result of 86% for acted speech, but a more realistic test using non-acted speech (in a Wizard of Oz setting) yielded a classification rate of 66% correct, which shows the sensitivity of performance results on the paradigm of the test.

A very pragmatic issue is the usefulness of emotions to ease the human–machine communication. For example, in dialogue systems it may be very fruitful to be able to detect "moods" such as frustration, irritation or impatience, such that more appropriate dialogue handling can be chosen or the call be redirected to a human attendant. The attempts made so far to integrate prosodic information and speech recognition have been unsuccessful for improving recognition accuracy. Prosodic information has been incorporated into a variety of ASR-related tasks such as identifying speech acts, locating focus, improving rejection accuracy (confidence modeling), locating disfluencies, identifying user corrections in dialogue, topic segmentation and labeling emotions (Bosch, 2003).

6.6   ASSESSMENT AND EVALUATION

There are some standard methods for assessing various components involved in language engineering. The standard methodology includes how to get enough speaker samples from the representative population, how to compare different ASR systems and how to design the experiments and statistically analysis the data (Gibbon, 1997). There are also the standard methodologies developed to assess different recognition systems, speaker verification systems, synthesis systems and the interactive systems (Gibbon, 1997). The assessment methodologies addressed mainly technical requirements. For instance, the assessment of interactive systems are mainly for the three categories of users: First is for the engineers to compare the existing interactive dialogue systems in order to select the best one for their particular purpose; second is for the system engineers to gauge the performance of existing systems for diagnostic purposes, in order to improve their performance; and third is for the designers to learn how to go about designing an interactive dialogue system from scratch (Gibbon, 1997).

Here the assessment and evaluation methods we try to address are from the end user's point of view, taking the human factors and human cognition and human behavior, the effects from the social and organizational issue and the environment into consideration. The complexity of the human-computer interface, the subtle role of speech and language processing, the complicated users' variability and application environment and the task specifications have been and continue to be the difficulties for the application. The difficulties may come from the following aspects:

a) The field conditions are very different from laboratory conditions. The results from a simulated environment in the laboratory can hardly be compared directly to the field application.
b) Lack of agreed-protocols for specifying the different field systems
c) Lack of the "standard" to select appropriate "off-the-shelf" HCI components according to the specific purposes.
d) Lack of a clear understanding of the implications of the performance of a system component on overall system effectiveness.

e) Lack of the agreed-upon evaluation methods to measure the overall effectiveness.

The evaluation of speech interface is not only the quality of the speech recognition and verbal communication itself; it also evaluates how well the task is performed and the costs to the user in terms of effort, motivation and affect.

According to the UCD process, the assessment of the speech interaction system should be carried out in the earliest stages of the design process as possible. The assessment should also be carried out before the design is finished. Then the simulation for the assessment should also be modified according to the possible changes introduced by different factors.

It is necessary to identify the critical features for the device and the tasks in the simulation. Identify the potential user group, and the specification of the users related to the performance of the speech interaction system will be understood as early as possible. The assessment system should also provide the evaluation criteria.

Life and Long (1994) presented a speech interface assessment method (SIAM). The method is intended to support evaluation of speech-based computers. Normally, people start the design by simulating the target system where the speech interaction system is going to apply. The simulated dialogue is then based on an analysis of a current version of the task and on judgment of how this would likely be changed by the introduction of the speech interface. These judgments are difficult, and the longer the lead-time before the system enters service, the greater the danger of mismatch between the simulated dialogue and the final implementation.

SIAM is a procedure to perform the usability evaluation of the interaction system. First is to make the preliminary problem specification that contains the statement of technical issues to be addressed in the assessment. Second, a preliminary system specification is generated by a target system analysis and applying different task analysis methods. Here the task and the user are identified. The purpose is to set up the assessment scope. Third is specifying the model of device-user interaction. Here the configuration of the diagnostic table for the assessment is built up. Fourth is specifying the empirical evaluation. Here it requires designing an experimental process and all the details for carrying out the experiment. Fifth is gathering the interaction data, which is measured during the experiment. Normally the data includes objective data, such as time to finish the task and the error rate, and subjective comments include how much effort the subjects have applied to the interaction system and their satisfaction level, and even compare the alternative systems that the subjects had experienced before. Life studies on a few assessments concluded that the structured evaluation methods provided could improve the quality of speech interface evaluations by leading the user systematically through the process of evaluation and the help to ensure that important steps are appropriately covered. This procedure potentially offers a mechanism by which relevant knowledge may be assessed by non-specialists. But the assessment quality still depends considerably on the ability of the assessor to interpret procedures and sources of discipline knowledge appropriately.

The novel aspect about SIAM is that a) the structured method can make the evaluation knowledge explicit easily; b) non-specialists can use it. The evaluation may be conducted before the system is specified; the evaluation will give the estimation of task quality and costs to users and, it can give the diagnoses of failures to meet the performance desired of the target system.

Heuristic evaluation is a popular discount usability method that is commonly used for different user interface evaluation. In a heuristic evaluation, an interface design is critiqued based on a checklist of design guidelines or heuristics. Nielsen's ten usability heuristics (Nielsen, 1994) are too general to be used for speech interface evaluation. In section 5.3.3, Chapter 5, we made some discussion on Nielsen's ten usability heuristics. This heuristics evaluation needs to be further developed based on different application specification. There are some of such heuristics and design guidelines for spoken dialogue systems (Dybkjaer, 2001; 2004; Suhm, 2003). The detail of usability evaluation for different speech interaction systems will be discussed in the latter chapters.

6.7   FEEDBACK DESIGN

In human-human communication, listeners are not passive; they provide a number of feedback cues to speakers, which help to determine whether the message is being received and understood. These cues can be verbal or nonverbal vocal cues (utterances such as "Uhuh," "Mmm," "Yes," "Pardon," etc.) and another range of non-vocal cues (eye contact, facial expressions, nods of the head, and postures). Although these cues are important features for good communication, none are essential since verbal communication is still possible when some cues are eliminated as on the telephone, or where communication is restricted to typed messages (Chapanis, 1975). In a similar way, it has been suggested that "redundant" feedback could be provided in speech systems to facilitate communication between man and machine (Rodman, 1984).

In general, for any human computer interaction system, the feedback design is one of the important parts of the entire system, and it will directly affect the user's satisfaction of the system. The feedback design should fulfill the following requirements:

a) Let the user notice that the system understands his requirement and the required action is now carried out by the system
b) Indicate where the user is now in his navigation process
c) Diagnose the errors user made and provide the suggestions for correction

The feedback design can make the system transparent for the user and let the user feel that he is in control of the system, so to increase the user's confidence on his performance. For speech interaction systems, as ASR systems are far from perfect, one extra important function for the feedback design is to let the user determine whether an utterance has been recognized correctly, and the required action will be carried out by the system (Jones, 1992).

6.7.1 Classification of Feedback

Some form of feedback is generally desirable to verify voice input. Theoretically, feedback should provide adequate error-detection facilities without overloading the user with information and without interfering with speech data entry. In general, ASR feedback can be classified in three ways:

a) *Modality of presentation*: visual and auditory are the most common alternatives.
b) *Timing in relation to each utterance*: concurrent and terminal are the most common. Concurrent feedback (or primary feedback) occurs when a system responds immediately to a recognized command. In terminal feedback (or secondary feedback), the system repeats commands back to the user for verification before any action is taken.
c) *The amount of specificity provided*: may range from the simple symbolic nonverbal indicator (a tone, for example) to verbal feedback (such as the word recognized by the ASR device).

Secondary feedback is the feedback after correcting the error, and it is generally preferred when accuracy is particularly important, either because actions are safety-critical, or recovery from an error is difficult (Jones, 1992). Secondary feedback on ASR system can be either visual or auditory modalities, or can involve both modalities simultaneously.

The selection of what type of feedback is often determined by the particular ASR applications. Which form it takes depends on what is essentially a multi-component task. Even in a simple data-entry task, the user performs both the primary task of entering speech data, which has to be coupled with the secondary task of monitoring feedback. Feedback may interrupt working memory.

6.7.2   Modality of Feedback

A major ergonomic problem when designing automatic speech input systems is in determining the most effective sensory mode for feedback.

The basis of a "Stimulus-Central Processor-Response (S-C-R) compatibility model of dual task performance (Wickens, 1982) suggests that visual feedback should follow manual data entry and auditory feedback should follow speech data entry. Speech feedback should be preferred for speech input, and it has the added advantage of allowing the operator to concentrate on the data entry process; he or she does not need to monitor a visual display terminal while entering information.

Unfortunately, the argumentation goes into different directions, depending on the application circumstances and what aspects researchers are emphasized. In a report of avionics applications, Welch (1977) indicated that the transitory nature of auditory feedback makes it too difficult to use in practical situations. He therefore used auditory feedback only to augment visual feedback. Also, in a study reported by Martin and Welch, (Martin, 1980) users had difficulty with voice feedback, a result which contradicts the prediction made by Wickens and Vidulich (1982). Martin and Welch (1980) found that auditory feedback was too slow to be of value in speech data entry tasks and that it actually increased error rates, not only during speech data entry, but also during manual data entry. The reason for increasing errors by auditory feedback is possibly that the auditory output interfered with the user's attempt to formulate the next utterance properly (Hapeshi, 1989). It seems that voice feedback may present serious difficulties for the users.

According to McCauley (1984), visually presented feedback is the best because the information was constantly available, but did not interfere with the task; the user simply looked at the screen when information was wanted. Auditory tones may be sufficient in applications that have a rigid syntax or a small vocabulary (Lynch, 1984). For more complex vocabularies and syntax, tones would be too vague to be useful. In these cases more detailed feedback is necessary, which could be either visual or spoken or a combination of tones and visual feedback. The study from Gould, et al., (1983) simulated an accurate, large vocabulary word-processing system in which the relatively few errors that would occur could be corrected later using the syntactic and semantic context available in natural languages. With most data entry applications, errors would need to be corrected immediately; therefore, subjects would have to attend to the feedback. In this case, visual feedback has an advantage over voice in that it can remain displayed to be scanned by users when necessary, while voice feedback is transient and hence places heavy demands on a short-term memory.

Jones and Hapeshi (1989) studied different types of feedback combined with different primary tasks of data entry. This study provides clear evidence that when speech data entry from short-term memory is required, verbal feedback presents a significant burden to processing capacity. Concurrent verbal feedback, whether visual or auditory, will result in difficulties for human short-term memory in addition to slowing data throughput. Terminal visual and auditory feedback can impair speech recognition performance, particularly during isolated speech recognition. This suggests that where speech data entry from short-term memory is required, nonspecific concurrent feedback should be used. The present results also show that for feedback in any ASR application, there will be an inevitable speech-accuracy trade-off (Jones, 1989). Where error detection and correction is essential, and hence specific information must be provided to serve as a basis for action, the constraints demonstrated in this study will still apply.

Above studies were carried out fifteen to twenty years ago, the results are still valid for today's application. Voice feedback is not always good as it may overload the working memory and interfere with concurrent tasks. Voice feedback is slow and transient as well. If it is available, visual feedback is preferred. Even though speech feedback may interfere with the presentation of other communications or with the next spoken input, the feedback mode selection will probably be determined by the particular circumstances of the application. For example, in the aviation or vehicle applications, visual feedback may not be proper because the pilot's eye can be busy with other tasks the vibration and acceleration of the airplane can cause difficulties for the pilot to read the information on the screen in time. Visual feedback in such environment and task specifications may lose its advantages. In some other applications, especially for speech dialogue system via telephone, visual feedback is not available.

6.7.3 Textual Feedback or Symbolic Feedback

It is well-established cognitive knowledge that decisions for negative information take longer time to interpret than those for positive information (Ratcliff, 1985). Clark and Clark (1977) suggest that decisions for negative information involve the construction of the positive instance followed by its negation. The data support this theory. The use of textual feedback resulted in faster reactions to both correct and incorrect feedback. In ASR use, incorrect feedback corresponds to device misrecognition. Users must recognize and correct such misrecognitions in order to maintain system performance. A laboratory study (Baber, 1992) indicated that textual feedback has a consistently faster reaction time than symbolic feedback. There was no learning effect on textual feedback performance. This was constant across all item types and all trials. However, there was a definite learning effect for symbolic feedback. Therefore, textual feedback is better suited to ASR feedback than symbolic feedback.

Baber, et al., (1992) reported that verbal decisions, such as error handling, are best supported by textual feedback. This means that device errors would be most effectively dealt by the operator if feedback is textual. If it is available, textual feedback use in conjunction with symbolic feedback resulted in the best levels of performance. The use of symbols is argued to provide system feedback related to the process display (and presumably in line with subjects developing "mental models" of the process). In this way, feedback is integrated into the task. Text provides additional validation of commands and scope for error handling.

6.8   SYNTHESIZED SPEECH OUTPUT

The technology generating the speech output is called "speech synthesis." There are two very different categories for speech synthesis: the first one is called "canned speech," because the output speech is generated on the basis of pre-stored messages. With this type of synthesis, very high quality speech can be obtained, especially for quick response applications. This technique, however, requires the use of large memory and is not flexible. The second one is "text-to-speech synthesis." It allows the generation of any message from text. This generally involves the first stage of linguistic processing, in which the text input string is converted into an internal representation of phoneme strings together with prosodic markers and the second stage of sound generation on the basis of this internal representation. The sound generation can be made either by rule, typically using complex models of the speech production mechanism (formant synthesis), or by concatenating short pre-stored units (concatenative synthesis).   The speech quality obtained with concatenative synthesis is generally considerably higher; however, the memory requirements are larger than with formant synthesis.

Synthetic speech is becoming a critical component in different military and industrial warning systems, feedback devices in aerospace vehicles, education and training modules, and aids for the handicapped consumer products and the products that concern functional independence of older adults. It is also an important part of the speech dialogue systems.

6.8.1   Cognitive Factors

Research on natural language comprehension demonstrates that the first step of speech perception is word recognition and lexical access. Further steps are sentence comprehension (with its syntactic and semantic aspects) and text comprehension. At this point, the listener's characteristics are taken into account, for example, the linguistic ability involved in segmenting and analyzing speech into appropriate morphemic and syntactic units, content-related knowledge and the capacity to deploy appropriate comprehension strategies, motivational and attention processes. At the same time, the external factors such as test properties (e.g., length, complexity, content) and acoustic properties (e.g., speed, pitch, etc.) (Samuels, 1987).

There are a few cognitive factors that would affect speech listening. We will focus on three factors: perception, memory and attention (Delogu, 1998).

Perception: Speech perception requires an analysis and decoding of the relevant information available in the sound pattern. The perceptual process includes: a) the determination of the presence or absence of acoustic and phonetic features, b) their extraction from the acoustic wave, and c) their combination into recognized phonemes or syllables.

Speech prosody has a potential contribution to perceptual segmentation. Some studies show that if sequences of words or syllables lack prosodic continuity, recognition of the message may be hampered. Studies have also shown that local speech rate has an important effect on the identification of vowels (Nooteboom, 1977).

Memory: Here the most important factor is the limitation of working memory (WM). WM allows the temporary storage of information essential to the current task. WM has limited resources allocated to various tasks and sub-tasks. The elements that are stored (and that can be rehearsed) are not always the same as those that have been actually heard. This process is called re-coding and allows WM to store more information than it could if the stored information were to be kept in memory untransformed. The initial words of a sentence must be remembered in some form until the final words are uttered in order to understand the entire sentence.

Attention: Speech perception requires selective attention. Although people are not consciously aware of the process by which they perceive speech, and they are not aware of making any effort in this process, there are limits to their ability to recognize spoken language. In order to recognize words or phonemes, they must listen selectively to a single channel of speech.

6.8.2   Intelligibility

The pragmatic value of synthetic speech, or text-to-speech (TTS) systems, is mainly determined by two factors: the intelligibility of the speech that is produced and the naturalness of the speech output (Bosch, 2003). The comparisons between natural and synthetic speech are basically focused on two parts: single-word intelligibility and word or sentence comprehension. A typical procedure (modified Rhyme test) is to ask listeners to identify the word they heard from among a set of words differing by only one phonetic feature (House, 1965). Logan, et al., (1989) results showed that the synthetic word only makes about a 2% error rate higher than natural speech. Even with equal single-word intelligibility, people take longer to comprehend synthetic speech.

In an experimental study, Simpson (1975) was able to show that as long as synthetic speech messages were presented as sentences (and contained linguistic cues), intelligibility was higher than a terse two- or three-word format. It was found that airline pilots rated the intelligibility of phoneme-based synthesized speech warnings as equal to or greater than normal Air Traffic Control radio communications (Simpson, 1980). Using an experimental synthesized voice callout system in an airline cockpit, Simpson (1980) found that pilot performance improved during difficult segments compared with their performance during human pilot callout. Therefore, it appears that the crude and artificial sound of synthetic speech may actually improve performance in applications with competing voices when there are sufficient linguistic and contextual cues to aid intelligibility.

6.8.3   Comprehension

In the last decades, there is evidently improvement of segment intelligibility and the smooth concatenation of the synthetic units in the design of TTS systems; as a result, the TTS word intelligibility has substantially improved during recent years. However, less success has been achieved in making the synthetic speech more natural. Even in modern TTS systems, there is quite a long way to go in order to improve the prosody and naturalness of the synthetic signal.

In a series of experiments Luce, et al., (1983) found that free recall of spoken word lists was consistently poorer for synthetic speech compared to more natural speech, and in ordered recall this difference was substantially larger for the early items in a memorized list than for later items. Luce, et al., also found that when digits were presented visually for retention prior to the spoken word lists, fewer subjects were able to recall the digits when the word lists that followed were synthetic. According to the authors, these results indicated that difficulties observed in the perception and comprehension of synthetic speech is partly due to increased processing demands in short-term memory.

The process of speech perception includes the detection of acoustic–phonetic cues to form/activate words, a grammatical analysis to form well-formed sentences, semantic determination and disambiguation and, on top of this, the pragmatic use of prosodic cues (Bosch, 2003). Compared with human speech, the synthetic speech still lacks the prosodic information. We are highly sensitive to the variations and intonations in speech, and, therefore, we are less tolerant to the imperfections in synthetic speech. We are so used to hearing natural speech that we find it difficult to adjust to the monotonic non-prosodic tones that synthesized speech can produce. In fact, most speech synthesizers can deliver a certain degree of prosody, but to decide what intonation to give to the sentence as the feedback information, the system must have an understanding of the domain tasks.

Research has demonstrated that compared with a natural voice, speech produced by TTS synthesizer places an increased burden on perceptual and cognitive resources during the comprehension process. Experimental studies showed that if the subjects were asked to repeat the synthetically produced messages from TTS, they were less accurate and take a longer time to understand the message. The results of the comprehension tests by Delogu, et al., (1998) suggested greater workload demands associated with synthetic speech, and also confirmed that on occasion contextual and linguistic bias may affect the results. Subjects listening to synthetic passages are required to pay more attention than those listening to natural passages (Delogu, 1998).

A number of factors could account for the processing speed differences between synthetic and natural speech (Paris, 2000): a) At the phonemic level, synthetically generated phonemes are "impoverished" relative to natural speech because many acoustic cues are either poorly represented or not represented at all (Pisoni, 1981). In other words, there are differences in the amount of information conveyed by natural and synthetic speech at the phonemic level. Normally natural language cues, such as contextual cues, become more important when bottom-up cues are less robust during language comprehension (see Chapter 2). b) At the prosodic level, such as intonation, voicing, stress, durational patterns, rhythm, etc., prosodic cues provide perceptual segmentation and redundancy, speeding the real-time processing of continuous speech. Studies show that prosodic cues guide expectancies, cause search processes to end when contact is made between an acoustic representation and a cognitive representation and influence the actual allocation of processing capacity in terms of power, temporal location and duration.

Prosody in TTS systems is generally limited to the addition of pitch contours to phrase units marked by punctuation. Paris's et al., study (2000) also indicated that prosodic cues provide a powerful guide for the parsing of speech and provide helpful redundancy. When these cues are not appropriately modeled, the impoverished nature of synthetic speech places an additional burden on working memory that can exceed its capacity to compensate under real-time demanding circumstances.

6.8.4   Emotion

People's emotional states contribute strongly to the prosody of human speech and provide different semantic meaning to the utterance. Pitch is the most relevant acoustic parameter for the detection of emotion, followed by energy, duration and speaking rate (Picard, 1997). It was found that in a number of cases, speaking rate, segment duration and accuracy of articulation are useful parameters to determine the emotion state of the speaker. To simulate emotion states in synthesis, one usually modifies pitch, segment duration and phrasing parameters to create the desired emotive effect.

6.8.5   Social Aspects

Reeves and Nass (1996) study the social aspects of humans toward media, and their study showed that people interact with media in the same way that they interact with other people, and this media become social actor. Speech recognition systems are one such media with which people interact adhering to the rules of socio-psychological interaction.

Kotelly (2003) in his book *The art and business of speech recognition* has described some factors that are related to the social aspects of a speech recognition system, mainly toward the dialogue system design. These social aspects includes a) the flattery comments, for example, the computer gives a comments like "good job" to the user, which would affect the user's task performance. The question is how to design such flattery comments and how often they should appear; b) designing the dialogue in such a way to encourage the user to try harder; c) should we anthropomorphize our system, and how far we can go? In here, instead of using "the system will read . . ." such machine sentence, using "I will read . . . ." This may mislead the user that the system has feelings. d) Should speech-recognition systems be humorous? Human is fine, but it should be in the right context and done in the right way to the right person with the right timing. These can make the design more difficult. Furthermore, people sometimes see humor in a commercial setting as an indication that the company doesn't take its business or its customers' needs seriously. One more social factor that has received some extra attention is whether we use a male or a female voice for a system. We would like to discuss this factor in more detail.

The attributions that listeners make to male and female computer voices may affect how the listener processes the intended message produced by the TTS system. In particular, if the message is one that is intended to influence the listener, not just inform the listener, then these attributions may have a major impact (Mullennix, 2003). There is substantial evidence indicating that social influence is affected by interactions between the gender of the listener and the gender of the speaker. This factor is found in listening to human speech. Some studies indicate that people apply similar gender stereotypes to male and female computer voices (Mullennix, 2003; Nass, 1997).

Before we discuss this aspect, we can give a brief overview of what human listeners attribute toward human speech and synthetic speech. Some studies suggest that there may be some cross-gender differences in preferences for particular types of voices, and perhaps some differences in how men and women perceive human and synthetic voices. Mirenda, et al., (Mirenda, 1989) found that male and female listeners preferred human male and female voices over synthetic (DECtalk DTC01) male and female voices. Carli, et al., (1995) found that likeableness was a more important determinant of influence for female speakers than male speakers, while Carli (Carli, 1990) showed that female speakers who spoke more tentatively were more influential with male listeners and less influential with female listeners than those who spoke assertively. These findings indicate that gender is an important factor. It was also found that there is a tendency for females to be more easily persuaded than males, especially when the speaker is a female (Mirenda, 1989). Some view this gender difference as a tendency for some males not to be persuaded regardless of the situation (Burgoon, 1998).

A Mullennix, et al., (2003) study showed that listeners view human and synthetic speech quite differently. They found that listeners focused on differences in the quality of speech and found the female synthetic voice less pleasing, while female human speakers were more truthful and more involved than the synthetic speaker and that the message uttered by the human speaker was more convincing, more positive, etc. The study also found that, as indexed by attitude change on the comprehensive exam argument, that a female human voice was more persuasive than a female synthetic voice.

The Mullennix, et al., (2003) study suggests that there some differences and some similarities exist between male and female human speech and male and female synthetic TTS speech in terms of social perception and social influence. They confirmed that human and synthetic speech is gender stereotyped in a similar fashion. In their study, it appears that when listening to a persuasive appeal, female human speech is preferable to female synthetic speech and male synthetic speech is preferable to female synthetic speech.

The Mullennix, et al., (2003) study also showed that a male synthetic speaker was regarded as more powerful, softer, less squeaky and as speaking more slowly than the female synthetic speaker. These particular differences are perceptual in nature and are most likely due to differences in synthesis quality between male and female voice. The message produced by the male synthetic voice was rated as more favorable (e.g., good and more positive) and was more persuasive, in terms of the persuasive appeal, than the female synthetic voice. Thus, the differences between the voices are not completely due to perceptual factors, as higher-level perceptions of the message and persuasiveness are also affected.

Reeves and Nass (1996) report that when listeners are exposed to male and female computerized voices, evaluations from male-voiced computers are viewed as "friendlier" and are taken more seriously than evaluations from female-voiced computers. Also, they showed that both male and female listeners rated a female-voiced computer as more knowledgeable about love and relationships, while a male-voiced computer was rated as more knowledgeable about technical subjects (Reeves & Nass, 1996, p. 164).

Although the technology behind TTS is advancing rapidly, with better and more natural-sounding computerized speech soon available, the issue of whether we gender-stereotype speech emanating from a computer is still a viable and important issue.

6.8.6   Evaluation

The evaluation of TTS should include both intelligibility and naturalness measurements. There are a few commonly used methods for measuring the intelligibility of the synthetic speech. The modified rhyme test (MRT) (House, 1965) and the diagnostic rhyme test (DRT) (Voiers, 1983) are still the most frequently used methods for assessing the intelligibility of TTS systems. However, several objections have been made to MRT and DRT such as their not being representative of running speech and not appropriate for diagnostic purposes, and causing response biases and perceptual confusions (Carlson, 1989). As a consequence, new tests have been developed such as: the Standard Segmental Test designed within the Output Group of the Esprit Project SAM (van Bezooijen, 1990), the Cluster-Identification-Test (CLID) developed at Ruhr-University in Bochum (Jekosch, 1992), the Bellcore test corpus developed at Bellcore (Spiegel, 1990; Van Santen, 1993) and the Minimal Pairs Intelligibility Test developed at AT&T Bell Laboratories (van Santen, 1993). Nusbaum developed an interesting methodology for measuring the naturalness of particular aspects of synthesized speech, independent of its intelligibility, making it possible to separately assess the individual contributions of prosodic, segmental and source (Nusbaum, 1995).

The evaluation of synthetic speech that is only concerned with particular aspects of synthetic speech examined under special conditions is not enough. Subjects should be asked to evaluate the cognitive aspects as well, such as memorize, comprehend or process the content of the message. The consideration should take place on the concurrent cognitive demands of the environment as well as the properties of the spoken text itself and listener dimensions (age, hearing acuity, exposure to synthetic speech and mother tongue). The critical user, tasks and environmental characteristics (i.e., the context of use) have to be identified and described. The experimental tasks and the stimuli should be designed to be as close as possible to the actual applicative situations where users are likely to carry on complex tasks. The tasks would require them to memorize, process and understand longer and more difficult material (Delogu, 1998). In such a perspective, the evaluation of the quality of the synthetic speech can only be determined by taking the whole communication situation into account (Jekosch, 1994).

As we discussed in Chapter 5, there are common evaluation problems that are hard to solve. The assessment of the functional quality aspects of speech also face many practical questions (Möller, 2001):

a) What are the quality features and which quality features are of primary interest?
b) Which type of assessment (laboratory tests, user surveys, field tests) is the most appropriate for the assessment task?
c) Is the test setting representative of the present and future application? These include the test material, the test subjects and the assessment tasks.
d) What should be measured and how should it be measured? What is the internal and external validity of the study?

6.9   ERROR CORRECTION

Whenever ASR is used in operational systems, some recognition failures and user errors will inevitably occur. Effective provision for error detection and correction is therefore an essential feature of good system design. For ASR to be usable there must be a quick and easy way to correct recognition errors. The error corrections for dictating documents, for giving comments or using in the dialogue system are very different. User acceptance of speech technology is influenced strongly by the speech recognition accuracy, the ease of error resolution, the cost of error and their relation to the user's ability to complete a task (Kamm, 1994).

The error happens when there is a mismatch between speech recognition algorithms and human models of miscommunication, which results in the user being unable to predict when or why a system recognition failure will occur. One of the functions to design the system feedback is intrinsically linked to error correction. The feedback design can help the user detect the error during speech inputting. Error detection does not guarantee error correction, given the potential problems surrounding feedback monitoring. Without feedback, people have no guidance in organizing their repeat input to resolve the error reliably.

6.9.1   Speech Recognition Error

There are three types of failure of speech recognition between human-to-human communication (Ringle, 1982):

a) *Perceptual failure*: in which words are not clearly perceived, are misperceived are misinterpreted.
b) *Lexical failure*: in which the listener perceives a word correctly but fails to interpret it correctly.
c) *Syntactic failure*: in which all words are correctly perceived and interpreted but the intended meaning of the utterance is misconstrued.

There are many differences between human-to-human communication and humans speaking to the computer. A human being is used to speaking to another human being but not talking to a machine. It normally takes some time for the user to adapt to talking to the machine. Human beings may turn to speaking un-naturally, especially when an error occurs. Human speech can be varied due to tiredness, illness, emotional reasons, etc.

There are three major causes of recognition errors: acoustic confusions between vocabulary items; the discrepancy between the training and applications context; and the individual characteristics of the user (Hapeshi, 1988). For those systems that are using limited vocabulary with a trained user, there are a couple of ways to limit the recognition error. For example, one could minimize acoustic confusions by the careful selection of a vocabulary containing acoustically distinct items (as in the ICAO alphabet) and by introducing a rigid syntax where errors are most likely to occur or where they may have serious consequences. One can also train the template in a similar environmental condition as it is intended to be used; both physical factors (background noise, vibration, etc.) and psychological factors (task, stress, workload, etc.) should be taken into consideration. In recent years, there has been a quick technical development regarding how to handle the recognition errors due to the environmental factors and stress. I discussed these other chapters, so I will not repeat here.

The types of user characteristics that are most likely to affect recognition accuracy are knowledge and experience of the ASR system, user attitudes, and user motivation. All of these can be improved with user training and practice. Other factors that will determine the acoustic characteristics of the user's voice, such as sex, age, the existence of speech impediments, etc., could be accounted for with appropriate hardware and software adjustments (gain controls, feature detecting algorithms, etc.).

There are four possible types of recognition errors that often happen (Baber, 1993; Jones, 1992; Karat, 1999):

a) *Substitution*: The ASR device wrongly identifies a word that is apparently spoken correctly, or the input utterance is matched against the wrong template.
b) *Insertion*: Some extraneous sound, such as a cough, is matched to an active template and produces a response.
c) *Rejection*: The ASR fails to respond to the user's utterance, or the input utterance is not matched against any stored template, and therefore ignored.
d) *Operator errors*: The user's input or response is inappropriate to the task being performed.

Nowadays, manufacturers claim similar figures for recognition accuracy. It seems that commercially available ASR systems have probably reached an asymptote in terms of their performance capacities. This implies that the current technology has reached a point at which only minor adjustments can be made to improve it.

Errors are, by definition, unpredictable and, as such, can never be wholly obliterated, although their incidence of occurrence can be reduced. There are two distinct approaches to cope with errors: one is on the assumption that errors can be beaten into submission with more powerful technology; the other assumes that errors will stubbornly remain a part of ASR, perhaps migrating from one form to another but remaining nevertheless. The second approach leads to the proposal that we should be endeavoring to produce error-tolerant systems. In other ways, the requirement for error correction does not necessarily imply that speech technology is inferior to other input devices. After all, most forms of computer input devices have the potential to produce incorrect input, either through user error or technical malfunction.

It is just a matter of which type of input produces fewer errors than the other type under the circumstances and the user's attitude to accept such errors.

6.9.2   User Errors

Not all the misrecognitions of ASR are errors from the imperfection of an ASR system it may due to the fact that, the user does not speak the utterance correctly. User error is noticed by many studies. Bernsen, et al., (Bernsen, 1998) regarded this user error as "by-products" of the dialogue system. They classified some series of interaction problems as user error rather than dialogue interaction design error. In their study, they realized that user error has received comparatively little treatment in the literature on interactive speech systems. They have argued that it needs to have a balanced viewpoint between "the system must be able to deliver what the user wants no matter how the user behaves" and "users just have to get used to the system no matter how stupidly it behaves." Users make errors during spoken interaction and some interaction problems are the compound effects of interaction design errors and user errors. This makes the evaluation of the dialogue system difficult to separate errors made solely by users from compound errors and those from pure errors of dialogue interaction design.

Normally we can distinguish two different types of human errors. One is call "slips" and another "mistake." A slips error occurs when people have the right intention, but failed to do it right. It normally is caused by poor physical skill or inattention. Mistakes happen when people have wrong intentions. They are caused by incorrect understanding of the information, situation, etc., "slips" errors often happen in the speech of the user. Such error could result from a number of factors when people speak to computers:

a) The restrictions imposed by the speech recognition devices to the user's speech styles.
b) The user applies illegal words to issue an utterance.
c) The user's response might be adversely affected by the feedback provided by the device.

Many studies have indicated that training has very limited effects to improve the user's speech habits. Human speech and language adaptation occurs only within natural limits. Many of the features of human speech production are not under full conscious control, such as disfluencies, prosody and timing (Oviatt, 1998). There are studies indicating that the knowledge of the cognitive factors that drive disfluencies makes it possible to design corresponding interface techniques that minimize the occurrence of the hyperarticulates. Hyperarticulate speech refers to a stylized and clarified form of pronunciation, which has been observed informally in connection with the use of interactive speech systems. As hyperarticulated speech departs from the original training data upon which a recognizer was developed, it degrades the recognition rate.

Oviatt's et al., study (1998) indicated that the user's profile of adapted speech to a computer differs from that during interpersonal hyperarticulation. For example, the pattern of hyperarticulation to a computer is somewhat unique: users did not alter their amplitude when resolving error with the computer and change in fundamental frequency was minimal. Speakers can adjust their signal characteristics to accommodate what they perceive to be specific obstacles in their listener's ability to extract lexical meaning. The cause of the adaptation is the speaker's expectation, believing, experience and understanding of how the system works. Oviatt developed a CHAM model (computer-elicited hyperarticulate adaptation model). According to this model, she proposed several possible ways to improve the performance of current spoken language system on hyperarticulate speech:

a) Training recognizers on more natural samples of user's interactive speech of systems, including error resolution with the type and base-rate of errors expected in the target system
b) Adjustment of durational thresholds or models of phones since duration adaptation is the primary change that occurs during moderate hyperarticulation
c) Design a recognizer specialized for error handling, which could function as part of a coordinated suite of multiple recognizers that are swapped in and out at appropriate points during system interaction
d) Use of a form-based interface with content-specific input slots
e) Develop adaptive systems designed to accommodate differences in a system's base-rate of errors and individual differences in a user's hyperarticulation profile

Under high workload condition, different types of user errors may occur (Chen, 1999; 2003; 2000) such as insertion, submission, or mis-pronunciation of words within a sentence. Besides these, speakers typically increase both amplitude and variability in amplitude while simultaneously speaking at a faster rate and with decreased pitch range. In stress situations, noteworthy adaptations the speaker makes include an increase in fundamental frequency and change in pitch variability. The relation between the speech signal and intended phonemes is a highly variable one, which is not entirely captured by positing a constant mapping between phonemes and physical acoustic or phonetic characterizations, nor by factoring in local coarticulation effects (for review see [Oviatt, 1998]).

6.9.3   User Error Correction

For speech interaction systems, not only high recognition accuracy, but also adequate error correction is crucial to realize productivity gains. Regardless of the intelligence of the error handling software in an ASR system, facilities for feedback and user-error correction must be provided. Some studies on dictation programs found that users spend 66% of their time on correction activities and only 33% on dictation and that one third of the time used was spent on simply navigating from one location to another. This means that performing a standard dictation task is much slower than using mouse and keyboard (Karat, 1999). One study on using voice commands in navigation showed about 17% of all navigation commands fail (Sears, 2003).

When we discuss the error correction, we have to discuss it according to different systems. The error correction for a spoken dialogue system is very different that for a dictating system. For a speech command-control system, since each command issue will normally trigger a respective function, the error correction design is based on the specification of the system performance. Normally, before the command is executed, a proper feedback, either auditory or visually, will be available for the user to confirm that the command is correctly issued and understood by the system. The error correction for speech dialogue system design is crucial. We will discuss this aspect in Chapter 7.

Many detailed analyses of users' error correction patterns revealed that the potential productivity gain of using speech dictation is lost during error correction. Baber and Hone (1993) discussed the problems of error correction in speech recognition applications in general terms. They pointed out that interactive error correction consists of two phases: first an error must be detected; then it can be corrected. Jones (1992) indicated that most error-correction procedures were based on one or more of the following strategies:

a) *Immediate re-entry*: An incorrectly identified word is repeated where feedback is provided by a visual display, and the previous response is overwritten.
b) *Backtrack and delete*: A sequence of words is entered as a continuous string, and a "scratch-pad" feedback display may be edited by means of a backtracking command, which deletes the terminal item each time it is invoked. This arrangement has the disadvantage of deleting correct responses that occur after an error.

c) *Backtrack with selective editing*: In this case, editing commands allow the user to position a cursor over an incorrect entry. This prevents the drawback identified in the previous method, but requires a relatively complex editing dialogue since provision must be made for insertion, deletion or reentry of individual items.

d) *Cancellation and re-entry*: A single command is used to delete one or more responses, which are then reentered from the beginning.

A number of strategies can be employed for the confirmation process when using speech as data input. The simplest is one in which the number is repeated back to the user by synthetic speech and the user says "no," "wrong" or "correction" and repeats the number back if it is not right. Ainsworth (1988) studied this kind of strategy and showed that in order to minimize the time taken for recognition, confirmation and correction if necessary, the optimal number of digits spoken before feedback is given depends on the recognition rate and on the time taken for recognizing and synthesizing utterances. The above strategy can be improved by moving each order from the active vocabulary as soon as it is known to be an incorrect response (repetition-with-eliminative strategy). This allows the correct digit to be determined more rapidly (Ainsworth, 1992). This strategy may be useful in the cockpit or other applications when speech input is a short command.

Another system was also tested in Ainsworth and Pratt's (1992) study. In this system, it can be removed when it is necessary for the user to repeat the digit after an incorrect recognition attempt. The system can simply suggest the next most likely digit, and so on, as determined by the recognition algorithm (elimination-without-repetition strategy). This saves the time taken by the user in repeating the digit, but the effects of odd noise remain, and it removes the possibility for the user to produce a more typical pronunciation on his second or subsequent recognition attempt. In addition, a second interactive correction method could be chosen from a list of alternative words.

6.10   SYNTAX

Syntax was likely to be utilized in most of the applications. Words were grouped into major functional subdivisions. Many of the system dialogues encountered during the survey were organized in terms of a tree-structured syntax. This is a convenient and well-established method of improving recognizer efficiency. Potentially confusable pairs of items can also be used within the same vocabulary if the items concerned are located in different branches of the syntax tree. The main advantages of the tree-structured syntax are an overall improvement in recognition accuracy, as well as the freedom to use potentially confusable vocabulary items in different sections of the dialogue. The main disadvantage is that misrecognition of key words can leave users "lost" (Jones, 1992). Users could lose track of their current position within the syntax structure; command words were used at times when the corresponding templates were not active, making correct recognition impossible.

Restrictions on word order were sometimes difficult to remember, especially under conditions of high workload or stress (Jones, 1992). For example, a command like "set up radio channel" could also be delivered as "radio channel set up." If system designers specify a syntax with strictly-defined rules for word order, without establishing which order is most natural for the user population, this kind of problem is likely to occur. More generally, users expressed a preference for greater flexibility in the dialogue structure, feeling that it would allow quicker and more efficient interactions with the system.

For small vocabulary systems, the vocabulary selection and syntax design directly affects the system performance (Thambiratnam, 2000). There are different ways to improve the performance of the voice inputting system by the design of the syntax. It depends on where the system is used who is using it. Many studies on speech interaction system performance did not give a clear description about how the vocabulary was selected and how the syntax was designed. The results of bad system performance can be questioned on different levels: whether it is an ASR system problem, or the selection of the vocabulary, or the syntax design, or the integration with the other parts of the system. A careful vocabulary selection and syntax design may change the system performance and user's acceptance dramatically.

6.11   BACK-UP AND REVERSION

The term "back-up" in ASR systems refers to any input devices provided as an alternative to the speech input. It is a kind of multimodal interaction system when the users can have another modality for the data or comment input. The difference between the back-up design and multimodal interface system is that the alternative input device is designed for using when the speech input does not function. If the speech input device is working, a user is less likely to use the manual input system.

Currently, there is no speech recognition system that can reliably achieve 100% accuracy, even with a modest vocabulary. It sometimes may require an alternative input device parallel to the speech input. Manual input also has a certain error rate in both human and machine performance. It requires a balance between manual input error, speech recognition error and how easy it is to make the error correction.

Hapeshi and Jones (1988) have pointed out several reasons for the need of alternative methods of input:

a) *Recognition and user errors*: Even if ASR devices were very reliable, there would still be a need to allow for occasional recognition errors. If no back-up system is provided, the user may need to continue simply repeating the utterance until a correct match is made.

b) *Safety*: For some applications, such as in aviation and other transportation systems, both speed and reliability are essential. Back-up systems are always provided even when the systems are very reliable in case of breakdown or damage. If an error-prone system, such as ASR, is included, then back-up is needed as a safety measure.

c) *Initial user training period*: During initial contact with the speech-input device, users are not able to achieve high accuracy and productive work. A back-up system may be essential during a user's early experience of the system when error rates are likely to be high and confidence is most vulnerable.

d) *Individual preferences*: Some users may prefer to use voice input all the time, others are less likely, especially if they find the system prone to recognition errors. Also, there may be occasions when even the highly motivated user may suffer from vocal fatigue. Therefore, providing a back-up input mode would allow for individual preferences or idiosyncrasies in performance, and give users the option of using voice input or keyboard input.

Generally, the major reason for providing back-up is to allow for error correction in more effective ways. In most of the cases, the alternative input is the manual model. Under certain conditions the user will be forced to switch to the alternative input mode, at least temporarily. Hapeshi and Jones (Hapeshi, 1988) discussed the problem caused by switching from voice input to manual input:

a) *Detracts from the advantages of voice*:

The advantages of using voice input when the eyes and hands are busy on a concurrent task are lost when there is a need to switch to manual input. If visual feedback is provided, the operator has to use the hands for manual input and eyes for keyboard and visual display. When using speech input systems as remote control, then a portable manual back-up system has to be provided also. This is expensive and impractical. In some applications, voice input was safer than manual input systems (Coler, 1984). If the user has to switch from voice input to the manual input, may by encountered many other problems.

b) *Extra user training required*:

Voice is a natural mode of communication, so in general, training is relatively easy. If a back-up facility is necessary, additional training must be provided. In ASR systems, users can achieve a reasonable level of proficiency in fifteen minutes to a few hours (Lind, 1986). Even if users were proficient in using the back-up facility, there may be some loss of skill if back-up is used only occasionally. Therefore, the back-up mode is likely to be much slower and prone to errors.

c) Coding system different:

In ASR systems, a single utterance can be used to replace a sequence of keystrokes, but it can cause problems if there is a need to use a back-up facility. In general, if users are allowed to adopt names with which they are familiar, then they may forget the code sequences necessary for keyboard entry. In avionics, for example, pilots may be encouraged to enter waypoints using a name rather than geographical points (Taylor, 1986a; 1986b). However, if voice input fails, then geographical points need to be recalled and entered on a keypad; if pilots have come to rely on the convenience of names, they may find difficulty in reverting to the older method.

Most of the studies about using "back-up" systems are from the '80s. As far as the speech recognition error is exists, an alternative way of input would always be considered in the design to make sure the system is working. Nowadays, people are more interested in multimodal input rather than the "back-up" system design. But the knowledge gained from either "back-up" system design can still be interesting.

6.12   HUMAN VERBAL BEHAVIOR IN SPEECH INPUT SYSTEMS

The ideal situation for a speech interaction system is that the system would be able to recognize any spoken utterance in any context, independent of speaker and environmental conditions, with excellent accuracy, reliability and speed. For a variety of reasons, today's automatic speech recognition systems are still far from reaching the performance of their human counterparts. Humans possess a much more sophisticated sensory system for the interpretation of acoustic signals. Humans use additional sources of information; humans have access to much broader and sophisticated sources of knowledge about the actual context, including common sense and;

humans generally keep track of the actual context as well as of the history of the spoken dialogue (Schaefer, 2001).

User's attitudes toward the speech interaction system can change, and this can affect the system performance. Nelson (1986) reported that in the initial stages of using a speech-input system in an inspection environment, errors were due to giggles, coughs and general self-consciousness on the part of the users. There was noticeable improvement in performance at the end of a week, after which voice input was as fast as manual input. After six weeks, the voice input was on average 30% faster than manual input.

Computers can be viewed as a special type of interlocutor. Humans may have a model of what they believe a computer is able to process. This model can affect their verbal output. The feedback the user perceived from the computer speech interaction system may also affect the user's verbal behavior, and hence affect the performance of the system. A few studies tried to address this subject. Most of these studies have employed variants of what is termed the "Wizard of Oz" paradigm.

An overview of the results of Chin (1984), Richards and Underwood (Richards, 1984a; 1984b), and Amalberti, et al., (Amalberti, 1993) is as follows. In the human-computer condition, as compared to the human-to-human condition, the subjects' verbal exchange is characterized by:

a) *Fewer dialogue control acts*: fewer read backs, fewer acknowledgements and fewer speech acts designed to sustain or terminate exchanges.
b) *Less dialogue structure*: less thematic cohesion, fewer and less varied request markers, but an increase in explicit communication planning.
c) *A trend towards use of "standard" forms* (i.e. utterances adhering more closely to written language): fewer indirect requisitions, fewer incomplete sentences and more well-formed requests.
d) A trend towards simplification of language: fewer referring expressions (anaphora, ellipsis, pronouns), repetitive use of the same syntactic constructions, shorter verbal complements and smaller vocabulary (higher type-token ratio);
e) *A trend toward simplification of the exchange*: shorter exchanges, slower pace, fewer words per utterance, increase in the number of utterances, less utterance chaining (longer pauses, fewer interruptions, fewer overlaps), fewer digressions and comments and fewer polite phrases. The modifications reported tend to simplify the verbal output.

One explanation for the modifications reported above is that they result from the subjects' a priori model of computer competence. Adaptive behavior of this type is, however, not restricted to human-computer dialogue. Human to human verbal exchanges differ as a function of type of recipient. A human's model about what a computer can do may affect their verbal output (Amalberti, 1993).

Amalberti, et al., (1993) conducted an experiment to test whether models derived from the analysis of human-human task-oriented dialogues adequately describe human-computer spoken exchange; whether the design of human-computer interfaces using continuous speech should be based on these models or on different ones; and whether users behave in the same way when they expect (and have mental representations of) computer interaction as compared to exceptions and representations of a human interlocutor. The results from this study indicated that tolerance to voice degradation was higher when the interlocutor was thought to be a computer.

Weegels (2000) studied the human speech behavior towards the computer. His study found that the user's expectations and misconceptions of a voice-operated service might influence their interactions with the system. Users generally have limited understanding of the spoken dialogue systems and the technology involved. This means in designing a system, one cannot count on the user's knowledge of how a system works. People also bring along their communication habits from the experiences with human operators. Weegels (2000) found that subjects alternately draw from human to human language and computer language, and that they readily change their behavior according to the problems encountered at a given moment. These findings implied that developers might adopt at least two strategies to cope with user's habits and expectations: they can either teach users how to handle the system (by giving instruction) or they can teach the system how to handle users (i.e., to provide users with feed-forward information the instant that a problem arises).

6.12.1   Expertise and Experience of the User

The skill of a speech-interface user varies with experience and practice (Leggett, 1984). The needs and abilities of an inexperienced user are quite different from those of an experienced one (Norcio, 1989). For example, an inexperienced user needs a request or prompt that provides a model for the user's spoken utterance (Zoltan-Ford, 1991) and a system response or feedback method that informs the user if the utterance is correct or contains errors. The inexperienced user also needs a rigid interaction style that provides error-free data entry and guides the user through the step-by-step interactive process (Schurick, 1985). This rigid interaction style, though less natural, increases the comfort and usability of the speech interface for an inexperienced user. This aspect still holds the same in different degree depending on how the system is designed for a human's natural interaction. Many spoken dialogue systems still provide certain guidance for the user to use the precise utterance that the system understands. An experienced user does not need this structure and rigidness. The experienced user may find the rigid interface long, boring, poorly focused, ineffective and sometimes misleading (Brajnik, 1990).

A system designer should always consider the different requirements from experienced and inexperienced users. Modern interaction systems normally claim natural interactivity between the user and the system. The natural interactivity implies that the user needs as little time as possible to learn about the system. Due to the limitation of the technology and limited capacity of the system, to avoid the errors caused by the variance of users' performance, the designers should pay more attention to the inexperienced user and ignore that this inexperienced user can soon become an experienced user and would like to interact with the system at another level.

6.12.2 The Evolutionary Aspects of Human Speech

The human head and brain are uniquely evolved to produce speech (Massaro, 1997), and human speech comes from a socio-evolutionary orientation (Nass, 2000). When a human is confronted by speech technologies, the human-directed responses are automatic and unconscious (Reeves, 1996). Nass writes, "Humans do not have the wherewithal to overcome their fundamental instincts concerning speech. Thus, individuals behave toward and make attributions about voice systems using the same rules and heuristics they would normally apply to other humans (Nass, 2000)." This situation would be very true when it comes to the normal people communicating with a computer dialogue system. It may not be the same when people use speech comment to control certain behavior of the system, as the system may not use speech feedback, and the users are normally experts on the domain of the application and are trained to use the system in certain ways.

Nass and Gong (2000) proposed four evolutionary principles related to human speech. First, a primary goal of speaking is to be understood. So when people find themselves encountering comprehension difficulties, humans use "hyperarticulate" speech (such as increased pauses and elongated words). Oviatt, et al., (1998) demonstrated that computers with poor recognition elicit the same adaptive behaviors. Consistent with an evolutionary perspective, participants exhibited the same adaptations humans use with non-comprehending human listeners.

Second, humans use speech only when their interactant is physically close. A human talks only when the person is physically proximate. Physical proximity is important because when another person, especially a stranger, is close, he or she presents the greatest potential for harm (and opportunity);

hence, one is cautious in one's interactions. Consistent with the idea that speech suggests social presence, participants provided significantly more socially appropriate and cautious responses to the computer when the input modality was voice as compared to mouse, keyboard or handwriting. During design, social errors by the computer, regardless of the mode of output, are much more consequential for speech input compared to other forms of input. Negative comments and behaviors are processed more deeply and lead to greater arousal than positive comments (Reeves, 1996). When combined with presence, these negative behaviors seem yet more threatening and consequential.

Third, humans distinguish voices according to gender. Gender is so important that instead of being encoded in terms of the canonical representation of voices, the distinction of male and female in speech is accomplished via unusually detailed and complex auditory psycho-physical processes involving fundamental frequency, formant frequencies, breathiness and other features (Mullennix, 2003).

Fourth, humans have a very broad definition of "speech"— extending to computer-synthesized speech. Human languages seem remarkably diverse. Many languages have sounds that native speakers from other languages cannot distinguish or produce.

6.13   MULTIMODAL INTERACTION SYSTEM

In the error correction and back-up design sections, we discussed the multimodality design for error correction and data inputting. Multimodal interaction system design is a large and very active research area. Due to the limitation of the space of this book, we will just give a brief introduction of multimodal interaction system design and possible problems.

Human communication is inherently multimodal in nature. People naturally communicate by means of spoken language in combination with gestures, mimics and nonlinguistic sounds (laughs, coughs, sniffs, etc.). An important aim of multimodal interaction is to maximize human information uptake. Effective visual and auditory information presentation means that information is prepared using the capacities of the human's sensory and mental capabilities such that humans can easily process the information.

The purpose of multimodal interface design is for adaptive, cooperative and flexible interaction among people. New interface technologies are being developed that seek to optimize the distribution of information over different modalities (Essens, 2001). In this section, we will have some brief discussion about multimodal interaction system interface design. We shall start with definitions for multimedia and multimodal systems. One of the biggest confusions in this area is due to the definition of modality and medium. Different people give different definitions mainly based on the convenience of the specific applications. Some defined it by a set of object instances, some by giving examples and some are between the paraphrases of a term in use in a given context. Here I would like to give some examples of different definitions by different authors in the publications.

6.13.1 Definitions

There are different definitions for medium and for modality. There is no overall agreement in the field on terminology. Therefore, researchers are forced to work around this confusion. Some authors explicitly provide their own definitions, while others provide examples from which the reader can infer implicit definitions. Here we list some examples of the definitions.

Alty (2002) gave the definition of a medium are a *mechanism for communicating information* between a human being and a computer. In physics the medium transfers the message but is unchanged. It is a *carrier*. A medium can support many different ways of communicating, but it does impose some restrictions on what can and cannot be communicated. The mechanism of communication media is defined in terms of *Basic Lexicon*— the collection of different possible building blocks used in communication; *Syntax*—how these blocks can be validly put together to make acceptable messages; *Pragmatics*—commonly accepted conventions; and *Channel*— human sense used for communication.

In Bernsen's definition (2002): a *"medium"* is the physical realization of some presentation of information at the interface between human and system. In this definition, media are closely related to the classical psychological notion of the human "sensory modalities." Psychologists use the term "modality" explicitly in the context of sensory modalities such as sight, hearing, touch, smell, taste and balance. In Bernsen's definition system, the graphical medium is what humans or systems see (i.e., light), the acoustic medium is what humans or systems hear (i.e., sound), and the haptic medium is what humans or systems touch. Media only provide a very coarse-grained way of distinguishing between the many importantly different physically realized kinds of information that can be exchanged between humans and machines.

The term *"modality"* (or *"representational modality"* as distinct from the sensory modalities of psychology), in Bernsen's definition, is "mode or way of exchanging information between humans or between humans and machines in some medium" (Bernsen, 2002). By his definition, the basic properties of modality have two different dimensions:

- *Dimension 1*
a) Linguistic (/non-linguistic)
b) Analog/non-analog
c) Arbitrary/non-arbitrary
d) Static/dynamic
e) Graphics/light, acoustics and haptic (determine scope, i.e., no gesturing robots or smelling computers)

- *Dimension 2*
a) Focus
b) Abstraction
c) Interpretational scope
d) Specificity
e) Already existing system of meaning
f) Freedom of perceptual inspection
g) Limited expressiveness

The combination of the two dimensions of the modality properties comes up with about forty eight unimodals. After functional pruning and pragmatic fusions, it finally comes up with twenty generic unimodal levels for output modality and fifteen unimodal levels for input modality.

Coutaz (1992) used the term media for "physical devices" and the term modality for "a way to use a media." For example, with the pen (input) medium, you can use several modalities such as drawing, writing, and gestures to provide input to a computer system, and with the "screen", the computer can use several modalities such as text, graphics, pictures, and video to present output to the user. This definition is close to Bernsen's definition of modality.

Some researchers consider that media relate to machines while modalities relate to human beings. Maybury defines (1997) the media as both the physical entity and the way to use it, and modalities refer to the human senses, whereas Bordegoni defines, a medium is a type of information and representation format (e.g., pixmap graphics or video frames), and a modality a way of encoding the data to present it to the user (e.g., graphics, textual language, spoken language and video) (Bordegoni, 1997). "Natural language'" has been counted among the media (Arens, 1993) a mode (André, 1993), and a modality (Burger, 1993). Martin, et al., (1998) took another approach to the subject by giving the following definition of modality: A modality is a process receiving and producing chunks of information.

The confusion between modality and medium will also bring confusion between multimedia and multimodal, of course. The difference between multimedia and multimodal is the use of semantic representations and understanding processes and use the term channel interchangeably with modality. Definitions of multimodality can be very general, stemming from theoretical models of human information exchange. Other definitions can be based on a particular application framework. According to Leger's definition, a *multimedia system* should be defined as a system whose architecture allows managing several input (and output) media (Leger, 1998), while according to Oviatt and VanGent's definition (2002), a *multimodal system* processes two or more combined user input modes—such as speech, pen, touch, manual gestures, gaze and head and body movements—in a coordinated manner with multimedia system output. In essence, most people still agree that multimodality is the use of two or more of the six senses for the exchange of information (Granström, 2002).

Gibbon (2000) introduces the following definitions: *multimodal systems* is a system which represents and manipulates information from different human communication channels at multiple levels of abstraction. Multimodal systems can automatically extract meaning from raw multimodal input data, and conversely they produce perceivable multimodal information from symbolic abstract representations. A multimodal system can be either a multimodal interface that has nothing to do with speech, or a multimodal speech system. Obviously, there are forms of multimodality where speech does not play a role at all, for example, conventional keyboard and mouse input in most current desktop applications, pen and keyboard input in pen-based computers such as PDAs (Personal Digital Assistants) and camera and keyboard in some advanced security systems.

A *multimedia system* is a system that offers more than one device for user input to the system and for system feedback to the user. Such devices include microphone, speaker, keyboard, mouse, touch screen and camera. In contrast to multimodal systems, multimedia systems do not generate abstract concepts automatically (which are typically encoded manually as meta-information instead), and they do not transform the information. A *multimodal speech interface* is an interface that combines speech input or output with other input and output modalities. The overall goal is to facilitate human-computer interaction. In part, that can be achieved by using the same communication channels that people naturally employ when they communicate with another human, but trade-offs are necessary to make such interaction feasible with current automatic recognition technology.

A *multimodal speech system* considers that people accompany speech naturally with non-verbal cues, including facial expression, eye/gaze and lip movements. All cues interpreted together ensure fluent human to human communication. Here, we agree with Gibbon's (2000) definition that *multimodal speech systems* (or audio-visual speech systems) are systems "which attempt to utilize the same multiple channels as human communication by integrating automatic speech recognition with other non-verbal cues, and by integrating non-verbal cues with speech synthesis to improve the output side of a multimodal application (e.g., in talking heads)."

6.13.2   Advantages of Multimodal Interface

Human face-to-face communication normally employs several modalities in order to produce and perceive information. There are two primary modes of production: speech and body gestures (Allwood, 2001). The body gestures normally are facial expressions, head movements and manual gestures. There are two primary modes for perception: hearing and vision. In the multimodal interface studies, people also apply a rough category system. Besides speech as one of the interaction modalities, other types of modalities are called gestures (Paouteau, 2001). Combining speech with other modalities is a strategy that is used when developing multimodal systems to offset the weakness of one modality by using the strength of the other (Cohen, 1992).

Multimodal interfaces are expected to be easier to learn and use. They are supporting more transparent, flexible, efficient and powerfully expressive means of human-computer interaction, so users for many applications prefer them. They have the potential to expand computing to more challenging applications, to be used by a broader spectrum of everyday people and to accommodate more adverse usage conditions than in the past (Oviatt, 2002).

An ideal multimodal interface will be able to recognize human speech, gaze, gesture and other natural behavior that represent human emotional states. Such an interface eventually will interpret continuous input from a large number of different visual, auditory and tactile input modes, which will be recognized as engaging in everyday activities. The same system will track and incorporate information from multiple sensors on the user's interface and surrounding physical environment in order to support intelligent adaptation to the user, task and usage environment. This makes up a new interface design called "multimodal-multisensor interface." This interface could easily adapt to the specific environment and user's characteristics. Such an interaction system will be strongly interesting to military applications, as well as different parts of homeland security systems. Future adaptive multimodal-multisensor interfaces have the potential to support new functionality to achieve unparalleled robustness and to perform flexibly as a multifunctional and personalized mobile system (Oviatt, 2002).

There are many advantages to using multimodal interface for human-system interface design (Benoit, 2000; Bernsen, 2001; Oviatt, 2002):

a) *Modality synergy*: Several modalities may cooperate to transfer the information in a more robust manner. It permits flexible use of input modes.

b) *Multimodal interface*: Satisfies high levels of user preference and increases work efficiency and effectiveness in different levels. It has increased expressiveness

c) *New applications*

d) *Freedom of choice*

e) *Naturalness, little or no learning overhead*: It has superior error handling, both in terms of error avoidance and graceful recovery from errors.

f) *Adaptability*: It provides the adaptability that is needed to accommodate the continuously changing conditions of mobile use. It can adapt to different environments.

g) *Accommodation*: It has the potential to accommodate a broader range of user than traditional interface. It can accommodate users with different ages, skill levels, native language status, cognitive styles, sensory impairments and other temporary illnesses or permanent handicaps.

All of these advantages are also the design goals for any multimodal interaction systems. At the same time, if it is not a well-designed multimodal system, then its positive issues may turn into their negative effects. So Oviatt pointed out that "the design of new multimodal systems has been inspired and organized largely by two things: First, the cognitive science theories on intersensory perception and intermodal coordination during production can provide a foundation of information for user modeling, as well as information on what systems must recognize and how multimodal architectures should be organized. Given the complex nature of a user's multimodal interaction, cognitive science has and will continue to play an essential role in guiding the design of robust multimodal system. The second is the high-fidelity automatic simulations that play a critical role in prototyping new types of multimodal systems (Oviatt, 2002).

6.13.3 Design Questions

Human-computer interfaces face several requirements, such as the need to be fast and robust in regards to recognition errors, unexpected events and mistakes made by the user. It was one of the expectations that multimodal interface design can solve the problems. It believed that a user's preference for multimodal interaction would be congruent with performance advantages over unimodal interaction—including faster task completion time and fewer errors. From a system-design point of view, different modes of presentation can complement each other and provide different types of information. They can also reinforce each other and provide more complete information for the user during a difficult task (Granström, 2002). Multimodal constructions also were predicted to be briefer, less complex and less diffluent than unimodal (Oviatt, 1997). The goal of designing multimodal rather than unimodal systems typically includes enhanced ease of use, transparency, flexibility and efficiency, as well as usability for more challenging applications, under more adverse conditions, and broader spectrum of the population (Martin, 1998; Oviatt, 1997). Some essential questions for the design of the multimodal interface are:

a) What is the proper information the operator needs for his communication and situation awareness? This is the question of how to select the contents of the information?

b) How does one select the best modality for a given task in its application context? This is the question of modality allocation.

c) How does one present it in an intuitive way, so it does not have to be deciphered? This is a question of modality realization.

d) How do different modalities interact with each other, and how do the different modalities reinforce or complement one another? This is a question of modality combination.

e) In what situations or contexts is it a detriment to use one or the other modality?

f) How does one design the system so it performs a good match in terms of functionality, usability, naturalness, efficiency, etc. between the communication task and the available input/output modalities?

g) How to evaluate the performance of multimodal interaction system after it is implemented together with other systems?

6.13.4   Selection and Combination of Modalities

Along with the above questions, one of the general questions that a multimodal interface developer would very much like to know is how to combine modalities and why this combination may improve the interaction. What is an appropriate use of multimodality? There is no proper theory to guide the designer in his design. In most of the design cases, people use their intuitive feelings to select the modality and use trial-and-error methods to learn how to combine the modality in the design. As people provide different definitions for the modality, it makes up the biggest obstacle for the researchers to develop any useful cognitive and design theories. The only reason for the designers to design a multimodal interface is because human beings' communication is multimodal in nature. So the guided theory for modality selection and modality combination has to come from human cognitive knowledge. To understand human cognitive behavior toward different modalities and the interaction of different modalities corresponding to the input and output information in the application context probably is the key issue for the development of the multimodal interface design theory.

Martin, et al., (1998) defined the modality as a process receiving and producing chunks of information. He has identified five types of cooperation between modalities:

a) Type 1: *Transfer*: Two modalities M1 and M2 cooperate by transfer when a chunk of information produced by M1 can be used by M2 after translation by a transfer operator. In this case, the user may express a request in one modality and get relevant information in another modality. It may be used for other purposes such as improving recognition and enabling faster interaction. Transfer may thus intervene for different reasons either between two input modalities, between two output modalities or between an input modality and an output modality.

b) Type 2: *Equivalence*: Two input modalities M1 and M2 cooperate by equivalence for the production of a set I of chunk of information when each element of I can be produced either by M1 or by M2. It enables users to select a command with different modalities, it also enables adaptation to the user by customization and it enables faster interaction.

c) Type 3: *Specialization*: An input modality M cooperates by specialization with a set I of chunks of information if M produces I (and only I) and no modality in M_i produces I. One should distinguish *data-relative* specialization and *modality-relative* specialization and *absolute* specialization, which have one-to-one relations. Specialization may help the user to interpret the events produced by the computer. It may also improve recognition (for tourist information, you may need only give the name of the place). It enables an easier processing in other modalities. It improves the accuracy of the speech recognizer since the search space is smaller. It may also enable faster interaction.

d) Type 4: *Consistency*: The same information is processed by the same modalities. This may help the user minimize the confusion and enhance the recognition and reaction.

e) Type 5: *Complementarily*: Different chunks of information are processed by each modality, and they have to be merged. These are systems enabling the "put that there" command for the manipulation of graphical objects. Complementarily may enable faster interaction since the two modalities can be used simultaneously and with shorter messages, which are moreover better recognized than long messages. Complementarily may also improve interpretation.

6.13.5   Modality Interaction

Directly or indirectly, the benefits of splitting information delivery and data command/entry across visual and auditory modalities are often justified in terms of independent information processing. Chris Wichen's multi-resource theory (1983; 2000) is often cited as the cognitive theory for modality selection. We made brief introduction to this theory in Chapter 3. This multi-resource theory is too abstract to apply to the design in different contexts. Besides, it did not apply so well when focused attention on the work is required.

There are not only positive aspects with multimodal interaction; it does have a negative side as well. There are the costs for the shifting between the modalities. Spence, et al., (2001) have demonstrated that reaction time (RTs) for targets in an unexpected modality were slower than when that modality was expected or when no expectancy applied. RT associated with shifting attention from the tactile modality was greater than those for shifts from either audition or vision. The independent nature of the processing in turn assumes there will be no interference between tasks or degradation in performance when information is presented in different modalities, according to Wickens multi-resources theory (2000). Speech-based systems could be introduced with little cognitive cost because the input and output were related via an internal verbal code. There should be little task-interference when other tasks are delivered with visual presentation, encoded in spatial terms and require a manual response.

Actually, perceptual judgments can be affected by expectancies regarding the likely target modality. In other words, people can selectively direct their covert attention to just one modality, and as a result process events more efficiently in that modality than in situations where attention must be simultaneously divided between several sensory modalities or where attention is directed to another modality. Spence and Driver (1997) have given external review of the studies in this topic. Their study also showed that responses were always more rapid and accurate for targets presented in the expected versus unexpected modality. When subjects were cued to both the probable modality of a target and its likely spatial location, separable modality-cuing and spatial-cuing effects were observed.

Some studies (Cook, 1997; Finan, 1996) indicate that there are problems related to memory and workload as well when it comes to the multimodal interaction:

a) Speech presentation may impose a greater memory burden as information presentation is transient.
b) Switching attention between modalities may be slow (Wickens, 2000).

c) Auditory presentation may pre-empt and disrupt visual presentation because attention is initially directed by sound, or differences in the length of the anatomical pathways in audition and vision provide a faster route for auditory processing (Chapter 3).

d) Speech-related information may have a suffix effect. The suffix effect in psychological literature refers to the loss of the most recently presented information from acoustic short-term memory (Baddeley, 1994) when a delay occurs between the final item presented and recall.

e) Restricted vocabulary used is unnatural.

f) When recognizers fail, it adds extra stress to the user.

g) The segregation of the information processing may be advantageous for processing of specific items of information, but it could interfere with the integration of information across modalities.

These problems are a big challenge for interface design. They will remain even if speech recognition accuracy is increased because they are the limits of the human operator to manage multi-modal environment.

6.13.6 Modality for Error Correction

The mode in which error-correction commands should be conveyed to the computer is important. Multimodal strategies have been used to develop feedback strategies used when the speech recognizer errors. Two major ways for error resolution have been identified: repetition of input by the user and selection of an alternative among possible interpretations by the user. These strategies have been further developed in studies and investigations.

Rhyne and Wolf (1993) were one of the first researchers to discuss potential benefits of multiple modalities for error correction and switching to a different modality to avoid repeated errors. For example, the systems require users to make corrections by voice when these are to be done immediately, but when delayed correction is carried out, manual/visual editors are used when available. If there is a high recognition error rate in a system, then error correction by voice may cause as many problems as it solves, and hence some form of back-up facility should be made available where possible.

Oviatt and van Gent (1996) investigated multimodal error correction in a Wizard-of-Oz simulation study. Results demonstrated that a user usually first attempts to correct an error with the same modality (e.g., speech) that caused the error, but will not continue with the same modality indefinitely. Larson and Mowatt (2003) compared four different error correction methods: a) *Voice Commanding*: The subjects selected the misrecognized word and re-dictated the word to correct it; b) *Re-dictation*: The subjects selected the misrecognized word by mouse and re-dictated the word to correct it; c) *Alternates list*: The subjects selected the misrecognized word by mouse and opened the alternates list dialogue, then selected the correct word; d) *Soft-keyboard*: The subjects selected the misrecognized word by mouse and used the keyboard to correct the word.

The results showed that users prefer the alternate list and the preference increased from 8% to 47% by different design settings for different apparatus. Larson and Mowatt (2003) pointed out that "the combination of strong modes, push-to-talk and the "correct <word>" command made the alternates list a light-weight mechanism that users were more willing to use to attempt corrections. It also facilitated mode switching, which is known to be a useful correction strategy. Mode switching was facilitated by allowing participants to re-dictate immediately while the alternate list was still open.

Suhm (2001) studied the multimodal error correction for speech user interface. The user evaluation showed that multimodal correction is faster than unimodal correction by re-speaking. Among multimodal correction methods, conventional multimodal correction by keyboard and mouse input, for skilled typists, is still faster than novel multimodal correction by speech and pen input. With a graphic user interface, multimodal error correction effectively solves the repair problem for speech recognition applications. Multimodal correction by speech and pen input could outperform correction by keyboard and mouse input for all users with modest improvements in recognition accuracy.

6.13.7 Evaluation

Evaluation is a key issue in multimodal interface research, and it is a complicated question. There are many different aspects in the evaluation work that depends on what the purpose of the evaluation is, whether all the design functions work as they should or the usability in its application context. From the functional perspective, the evaluation normally has two parts, one is the evaluation of individual components, and the other is the overall performance of the integrated system with sophisticated perceptive capabilities (Bauchage, 2002).

The evaluation of individual components is carried out normally in the earlier stages of the design when it comes to the selection of the modality for different task performance. There are some common kinds of errors (Paouteau, 2001) for different modalities: speech recognition errors, gesture processing errors, interpretation ambiguities and users' mistakes.

Normally the performance (accuracy, reaction time, etc.) of the individual modules in the system contributes to the total performance of the system but not as simple as just a summary of them. The main difficulty when it comes to the assessment of the overall system performance is that the system has no cues that presumably indicate whether a given input corresponds to one of those cases, or whether the result actually fits in with the user's intention (Paouteau, 2001).

It was pointed out that (Bauchage, 2002) assessing the performance of an intelligent interface must not be restricted to the evaluation of individual modules at certain points in time. Rather, rating a sophisticated system has to consider how it performs throughout an entire interactive session. Beyond evaluating the accuracy of individual modules, the examination of the integrated systems regarding its usability can be important. There is no off-the-shelf technology available to assess—especially the overall performance of the system. There are some usability testing criteria (Oviatt, 1997; Preece, 2002) that proposed in different literatures, such as:

a) The speed of task execution, or the task completion time.
b) The functionality of the system (i.e., how many different tasks can be performed).
c) The quality of the results (i.e., how good the performance in different tasks is).
d) Task-critical performance errors and self-reported and observed preferences.

Modern evaluation of complex human-machine interfaces and intelligent systems has many years history involving a variety of methods (Lindegaard, 1994). Some of the common problems are still not solved, for example, as the performance is normally measured by different user activities, what will happen if different measurements do not agree with each other? Or what if they do not point to the same direction? A lot of measurements are very context-orientated, so how does one compare the studies with each other?

Sometimes, people introduce the weighted factors to give some scores on different performances to make a common evaluation of the performances. The main difficulties and defects of using the weighted sum methods are as follows:

a) It is difficult to select weighted factors that properly represent the measurement purpose and guarantee the independency among factors.
b) Determining weights has the possibility of controversy, depending on viewpoints.
c) The relation between the index and its factors is actually nonlinear.

6.14   CONCLUSION

This is the most in-depth chapter in this book. It tries to cover all the possible cognitive and human factors aspects that relate to the design of speech interaction systems. The speech interface is divided into several components such as voice input and output, feedback design and error correction, vocabulary selection and speech variables.

Multimodal interface is very pertinent to speech-related interaction system, as it can make the interaction more natural and intuitive. It is a huge research topic, with the limitations of this book, we can only touch the surface of this topic.

Chapter 7

THE USABILITY OF SPOKEN DIALOGUE SYSTEM DESIGN

7.1 INTRODUCTION

Highlighting the recent progress of speech technology on the communication systems, information services via telephone is becoming one of the most important application areas of speech technology and has been rapidly developing. Applications via telecommunication and various telephone voice services such as finance, call center and supply chain, tourist information access, ticket booking for airlines, buses, and railways, order-entry, school or medical information access, etc., are more or less working in the real context, or under the investigation. The concept of the every citizen interface in USA rests on the realization that a large proportion of the population lacks the computer literacy, or even the linguistic capability to use conventional, text-based input and output. Speech offers the user different input and output channels and different interaction methods in a complex system that users are normally not used to in the performance of such systems.

All of those automatic information accesses services apply ASR and TTS technologies, combining into dialogues systems, and have some common characteristics. The system can (or should) accept spontaneous speech, spoken by a wide range of users and suitable for a wide range of environments. It is important and essential that everybody should be able to use it without prior instruction and training. Therefore, the user should be able to talk to the system in a natural way, that is, the system must provide a comfortable environment very much like the way he or she would talk to a human operator (Kellner, 1997). However, the human issues related to the use of both speech input and output have not been seriously considered yet, since this area has no long history.

The objective of this chapter deals with speech technology as part of public communication systems and the interaction between the user and the systems. The public communication systems discussed here mainly serve to provide information for public use. There are many books and journal papers discussing how to design the dialogue system (Balentine, 2001; Cohen, 2004) and different respective technologies. Here, we will focus on human factors and the ergonomic issues, especially the usability of the design, regarding interactive dialogue and man-system design on speech applications through communication systems.

7.2   THE ATTRACTIVE BUSINESS

Nowadays the development of various information technologies has changed human life tremendously. Businesses are also facing the overwhelming problem of the massive volume of interactions with customers. To address this issue, companies have turned to automation. The touch-tone interactive voice service systems were invented. Such systems are easy to set up but with very limited flexibility and capacity. The process alienated many customers due to tedious and frustrating push-button interfaces. Web self-service was also invented in hopes that consumers would use the Internet and not the phone, only to be disappointed by the slowness with which the new channel has been embraced by its customer population.

Compared with the above two services, automatic information access via speech provides the most effective and enterprising telecommunications service. Speech enables companies to dramatically save costs. A typical automated speech call center application can cut the cost per call from approximately fifteen to twenty cents. Only speech can deliver the required customer satisfaction[8]. Only speech allows companies to brand and build new voice-driven services that differentiate—through voice interface and design—and deliver potential new revenue sources. The speech information service can be on twenty-four hours every day without any delay of answering the questions. It can be more flexible and adapt to user requirements.

---

[8] http://www.intel.com/business/bss/solutions/blueprints/pdf/emerging_avp.pdf

A market study was carried out by Billi and Lamel (1997) to estimate the usage of operator-based telephone information services, as well as the expected demand and potential for automated services. His data comes from a study of 130 information offices in six countries, where over 100 million calls are handled per year, and there are at least an additional 10 million potential calls that go unanswered. Fifty eight percent of callers wait less than thirty seconds and twelve percent wait more than one minutes. Ninety one percent of the callers ask only for information (ninety seven million calls), with nine percent also making a reservation. It is estimated that over 90% of the information calls could be served by an automatic system that could recognize 400 city names (and over ninety five percent with 500 names). Automatic services could thus provide an economic solution to reducing the waiting time and extending the hours of service availability, particularly in light of the fact that in most countries it is the information provider, not the client, who pays for the call. It was found that there is quite a large variability in the services offered in the different countries.

In the later '90s, many automatic information access systems via telephone were already in the service or in the improvement stages, such as the Jupiter system that provides meteorological information (Zue, 1997); the RaiTel system that provides information on train schedules, fares, reductions of rate and the other services (Lamel, 1997); and the STACC service, which allows students to consult marks (Rubio, 1997). From 1995, more and more companies have developed some of the ASR products (Kitai, 1997). The application area is increasing all around the world, especially in the industrially developed countries.

Japan's first Voice Portal Service began by Japan Telecom Co., Ltd in July 2001 for providing the information and various services through via mobile telephone, PHS, and telephones; it was soon followed by NTT Corporation who started the commercial level services through the Voice Portal in 2002. The NTT system is provides news and weather reports, stock price information, and information for gourmets and so forth.

From the user's perspective, the service must be easy to use even for naïve users, and any normal speaker should be able to interact with the service. Billi made the survey of the usability of the information system by asking twenty different questions among a total of hundred subjects (Billi, 1997). The highest subjective scores were for ease of use, information provided and friendliness; the lowest were for speed, reliability, preference for human operator and need for improvement.

There are some marketing studies[9] showing that using ASR and TTS technology for automatic information service can have multiple benefits for the companies. The benefits include: a) improving the efficiency by handling more interactions without adding additional personnel; b) reducing cost, saving time and increasing the return on investment; c) increasing the flexibility and user's satisfaction.

7.3   ERGONOMIC AND SOCIO-TECHNICAL ISSUES

Although current market trends show that incorporation of ASR technology into existing or new wireless services as a means of replacement for touch-tone input is a natural progression in the user interface (Chang, 2000), it has induced some other aspects of the system that the designer cannot avoid facing. These aspects can be addressed psychologically, ergonomically and socio-technically. System designers should pay more attention on human factors and ergonomic issues other than the hardware interfaces.

Environmental factors are an important consideration when judging the desirability of speech-based interaction. A number of researchers have speculated that the use of speech-based interfaces in essentially public environments (such as those found in terminal rooms, or in the operation of automatic teller machines) would involve a loss of privacy that users would find objectionable. To date, this has mostly been a matter of conjecture. It is certainly conceivable that some initial reticence with respect to the use of speech-based user interfaces in public environments might be present until experience and convention begin to modify user expectations. Many of these questions will of course be answered as speech-based products begin to enter the marketplace.

Although most of the dialogue systems design for speech-based automatic information access, systems are still task orientated, compared with traditional dialogues carried out using button-press and keyboard inputs, speech dialogues may allow the users more flexibility and robustness. It can provide completely different interactive methods closer to human nature; at least this is what we expect from speech interaction systems.

Cameron (2000) analysed the success and failure of a large number of commercial speech systems developed in the USA over the last decade and concluded that people will use speech when:

---

[9] http://www.intel.com/business/bss/solutions/blueprints/pdf/emerging_avp.pdf

a) They are offered no choice.
b) It corresponds to the privacy of their surroundings.
c) Their hands or eyes are busy on another task.
d) It's quicker than any alternative.

The first three reasons relate in varying degree to external constraints on the user. The last one is the speech service designer's expectation, but it has rarely been used.

Hone and his colleague conducted a survey to assess user attitudes toward the use of voice to conduct interaction with bank machines (Hone, 1998) via large-scale survey and a series of focus groups. The study pointed out that privacy (the concern over one's personal financial details being overheard) and security (the fear of potential attackers hearing the user withdraw cash) were the major reasons for the success of the implementation of the application using speech for ATM transactions.

Technology does not exist in a vacuum. It is used within a specific context and is influenced by many factors within the context. Socio-organizational factors can make or break the deployment of the speech technology in communication systems. But there are very few research papers one can find that addressed this issue for the spoken dialogue system application.

7.3.1   The User Analysis

One special characteristic of these applications is that its user population is very large; almost every civilian in the society could be counted as a potential user. For such an automatic system to be accepted by the market, it is crucial that everybody can use it without prior instructions. Thus, the user should be able to talk to the system in a natural way, very much like he or she would talk to a human operator. The big population of users brings complexity and diversity to the user demand, skills or experience, preference etc. The system must accept untrained, unskilled utterances that implicitly contain unexpected forms of speech (e.g., vagueness, breath, too loud or too small, distortion, and some other noises). Therefore, much attention must be paid to dialogue system design that liberates the end users from mental stress together with trustable performance, to satisfy the usability requirement. Many different design questions have to be considered for the designers such as how does one design a system that can satisfy both novice users and expert users of the system? How does one design a system that is suitable for both a native language speaker and a foreigner? How does one design a system that meets the requirements of both young users and elderly users?

Many dialogue system designers have put much effort in the design to fit the novice users but ignore the skilled users. When humans talk to each other, they soon learn about each other's characteristics during the conversation. It is the same for human beings talking to the spoken dialogue system. After the users try a couple of sentences of conversation with the system, they know how the system is working, and the user soon becomes a "skilled" user. When a user communicates with the system by voice, often expects the counter partner to be like a human, even though he or she knows it is a machine on the other side of the telephone line (Reeves, 1996). The user still expects that the system can "learn" something about him or her and not treat him or her as novice user any longer. How does one design the system so the skilled user can feel satisfaction when talking to the spoken dialogue information system? Most of all, the users' knowledge and learning experiences of the dialogue system can be at very different levels. This has set up a new challenge for the designer.

7.3.2   The Variance of Human Speech

With the spreading of speech technologies used in the speech interface in a public information system, there are some technological issues raised. For example, If a dialogue system could recognize spontaneous speech, this means that the system should be able to recognize dialect and intonation, various way of utterances, vaguer speeches, strong breath, some nonsense syllables, irregular spacing between words and so forth, which generalized across in population. The system should be able to handle the various signal strength that influenced from environmental noise and that noises that generated at input treatment of apparatus or caused by vibration. Present speech technologies are not robustness enough to handle all these problems. Besides, there are also features that are deeply related to the end users' attitude to the speech input that may somewhat relate to his personal attributes, for example, users' behavior in terms of the input words that they generate, users' age and education background, users' attitudes against the misrecognition or misunderstanding of the system responses and so forth.

These are the most typical features of the public information system: a very large vocabulary of words from widely different domains; syntactic-semantic parsing of the complex, prosodic, non-fully-sentential grammar of spoken language, including characteristics of spontaneous speech input such as hesitations ("ah", "ehm"), repetitions ("could could I..."), false starts ("on Saturday, no, Sunday") and non-words (coughs, the sound of keystrokes); resolution of discourse phenomena such as anaphora and ellipsis, and tracking of discourse structure including discourse focus and discourse history; inferential capabilities ranging over knowledge of the domain, the world, social life, the shared situation and the participants themselves; planning and execution of domain tasks and meta-communication tasks; dialogue turn-taking according to clues, semantics, plans, etc.; the interlocutor reacting in real time while the speaker still speaks; a generation of language characterized by complex semantic expressiveness and style adapted to situation, message and dialogue interlocutor(s); and speech generation including phenomena such as stress and intonation. The automatic speech recognition engineering has set up a lot of effort to solve these effects from environmental and user various verbal behaviors. Thus the human issues related to the use of both speech input and output have been seriously considered by the researchers and research engineers.

7.3.3   Dialogue Strategy

The human factor issue of the interface design should be as important as the dialogue strategy, but if the strategy is poor, the system will not work sufficiently. Let's examine an example on a website. Suppose on a railway ticket information service, there are three routes to get to the destination Q from original station A and, furthermore, there are two different railway companies J and K who are setting different fares to get there. There may be some selections such as fare minimums, time minimum and exchanging complex minimum that deeply depend on the railway operational time (e.g., morning and evening rush hours or slack day time), if the system cannot give whole combinational information, the end user may not be satisfied. However, if it performs perfectly, the end user may be lost in a maze since the information that serially is given requires too much effort to convert into parallel information in the memory. Visual interaction will not have such problems since the information is always there on the screen. On the contrary, the auditory complex falls under such a circumstance. The significance of dialogue strategy cannot be disregarded from such an ergonomic view. The appropriate strategy may provide the users with better directions in and user's preference.

If one wants to evaluate and judge the usability to introduce an automatic inquiry system, it is important to look at the aspects of the user group. Novice users like the system to have short introductions and simple explanations of its features that they really believe and expect favorable effects from, while real users who are skilled in using the speech system are very careful to decide and mostly want short dialogues and thus demand very short system prompts without lengthy explanations (Kellner, 1997). This is one of the aspects of the dialogue strategy.

7.4   SPEECH RECOGNITION ERROR

Normally the ASR used for automatic information access via speech are required to be speaker-independent and capable of handling large amount of vocabulary, with enough robustness to recognize different possible accents among the user population. It should be adequately and neatly designed together with dialogue system. Some had hypothesized that the success or failure of any automatic information access systems would hinge almost entirely on the ability to achieve adequate speech recognition accuracy. Sander, et al., (2002) discusses the effects of the accuracy of ASR by measuring of Word Error Rate (WER) for some existing spoken dialogue systems. There are two principal efficiency metrics for dialogue system performance: Time On Task (TOT) and User Words On Task (UWOT). Actually, these were less important compared with Task Completion or User Satisfaction, but they illuminate the effects of WER on the conversations between systems and users. The values of these two metrics are highly correlated, but TOT seems to be a more important determinant of user satisfaction since TOT is one of the direct metrics on the users' workload. They focused on the following three questions. a) What is the nature of the relationships between WER and those two crucial metrics? For example, are the relationships linear, and is a linear regression thus appropriate? b) What level of WER is compatible with successful task performance? Does task success "go off a cliff" beyond some level of WER? c) What level of WER is compatible with high user satisfaction? These are important questions because many obvious applications for spoken dialogue systems involve noisy environments and distant microphones, which are factors that substantially degrade speech recognition accuracy. There are two crucial metrics that indicate a success of such system: the ability of the user to successfully perform the user's intended task and user satisfaction.

Some similar findings have been reported by Ebukuro (2000) in his investigation of comparing the false and practical successful applications of ASR systems. Sales amounts of inexpensive ASR products for PC input applications (consumer applications) were losing, together with their reputations, mainly because of their poor performance. Even though some of the ASR systems could recognize very large sizes of vocabulary (around 150,000 words) and furthermore some of the sales talks were emphasizing no preliminary speaker trainings, with the majority of the end users suffering from insufficient performances that had over 10% of WER and the products still requiring enrollments for better performances. However, if some users have strong intentions on using it, one might find a way to discover the improvements, such as selecting and limiting the words for input and never using such a large size vocabulary. If the words are familiar to them, the performance may be somewhat improved. If careful word selection is available for an application, the error rate may be reduced to a few percent.

7.4.1   Error Correction for Dialogue Systems

Given the state of the art of current speech technology, spoken dialogue systems are prone to error, largely because of user utterance that is misrecognized; sometimes that can be because the user did not use the designed utterance. The error correction in dialogue systems is very different compared with speech comments control, data inputting and dictating systems. The dialogue system under consideration employs both explicit and implicit verification questions to find out whether it has understood the user correctly (Krahmer, 2001). An implicit verification question serves two purposes: It attempts to verify whether the preceding user utterance was correctly understood, and it processes the conversation by immediately asking a follow-up question. One example of implicit verification question is "On what day do you want to travel from A to B." In this example, the time and the locations are in question. Explicit verification is solely aimed at verifying that the system's current assumptions are correct. One example of such verification is "So you want to travel from A to B today at 20.00." The advantages and disadvantages of both verifications are showing in Table 7-1.

According to the design principle, one performance for one function, the implicit verification violated this ergonomics design principle. The test results also showed that when the dialogue is on the right track, it is better to use the implicit verification because it is efficient, while when an error is detected, one should switch to explicit verification (Krahmer, 2001).

Table 7-1. Advantages and disadvantages of the two verifications

	Implicit	Explicit
Advantages	Its combination of purposes is efficient. It is possible to reduce the number of questions.	It is easier for the system to deduce whether the verified information is indeed correct. It is normally a form of "yes/no" question.
Disadvantages	When the system makes an error, users become rather confused.	But users do not always answer with a simple "yes" or "no" to confirm or disconfirm the system's assumptions.

7.5   COGNITIVE AND EMOTIONAL ISSUE

7.5.1   Short-Term Memory

Most people may have heard about the famous Miller short term-memory rule (1956) that we can only remember 7 ± 2 chunks of information. For a visually based information interface, this does not cause a problem for users, because information that needs to be remembered can be re-acquired from the screen itself. But this option is not available for user interactions that are purely speech-based. The effect that human short-term memory has on the design of speech-based interaction may bring new research challenges for the researchers. Schmandt (1994) studied these aspects on voice interactions with computers. He found that from traditional menu-based route to "user friendliness" needs to be reassessed because of the transmittal nature of sound; the speech equivalent of traditional menu systems usually involves a sequential presentation of prompts. For prompt sequences of any length, it would be virtually impossible for most users to remember the desired command until the end of the sequence. One innovation that has been tried with moderate success is to make the prompt sequence interruptible. However, even interruptible sequential prompts leave much to be desired. If the user knows what he or she wants to do but does not know the exact syntactic form of the desired command, then the prompt sequence must be endured to its end. If the user is still discovering the functional capacities of the underlying software, then a verbal prompt sequence becomes very unsatisfactory. Such a presentation technique leaves no room for review, comparison and thought. Exploration becomes impractical.

Thus the human factors issue that needs to be addressed for a speech-based public information system is to devise an approach that allows dialogue with users that has the structural advantages of menu interaction and the directness of free-form spoken natural language (Bradford, 1995).

7.5.2   Verbal/Spatial Cognition

User interfaces have traditionally focused on a visual/spatial approach to problem solving with communication systems. Users developed good experience with the wide use of these computer systems. They consequently became visual/spatial thinkers.

In the automatic information access system via telecommunication, all the information is presented verbally. The phenomenon of verbal-based cognition as contrasted with visual-based cognition needs to be examined from a human factors perspective. We need to have a deeper understanding of how people can convert the visual/spatial thinking into verbal/acoustic thinking. How can we identify those users who would function more effectively with a predominately spatial-based interface? Are there problem domains where verbal/acoustic thinkers might perform better than visual/spatial thinkers? What is the population distribution of these cognitive styles, how common each type is and how many users exhibit a mix of preferred styles? Are there aspects of speech-based user interfaces that could present serious problems to users who are predominately visual/spatial thinker (Bradford, 1995)? These, and many other issues, will need to be subjected to theoretical consideration and empirical research.

The results of research work on the verbal/spatial thinking can be very useful for the dialogue system designer. It will directly relate to how to present the information to the users by speech. How does one formulate the sentences so the listeners can easily configure this verbal message into spatial information? For those dialogue systems for which a multimodal interface can be available, and the information can be presented visually on display and people can make gesture input, the study of verbal/spatial thinking may guide to the selection of right modality for right information access.

Visual feedback plays an important role in human-to-human communication. The inherent lack of visual feedback in a speech-only interface can lead users to feel less in control. In a graphical interface, a new user can explore the interface at leisure, taking time to think, ponder and explore. With a speech interface, the user must either answer questions, initiate a dialogue, or be faced with silence. Long pauses in conversations are often perceived as embarrassing or uncomfortable, so users feel a need to respond quickly. This lack of think time, coupled with nothing to look at, can cause users to add false starts or "ums" and "ahs" to the beginning of their sentences, increasing the likelihood of recognition errors. Lack of visuals also means much less information can be transmitted to the user at one time, due to the human short-term memory capacity.

7.5.3   Speed and Persistence

People can absorb written information more quickly than verbal information. Compared with written tests, the speech output is a slow process and has lack of persistence (Schmandt, 1994). These two factors make speech both easy to miss and easy to forget. To compensate for these various problems, Grice presented cooperative principles of conversation (Grice, 1975) that contributions should be informative but no more so than is required. They should also be relevant, brief and orderly.

If the dialogue flow takes a long time to reach the necessary information, the users may get impatient, so the "economical" design of the dialogue system should be important. There are different approaches one can take, for example, one can eliminate entire prompts whenever possible and interleaved feedback with the next conversational move so as not to waste time; shorten output considerably in cases when it is possible; and use conversational conventions to give users an idea of what to say next.

7.5.4   Emotion, Prosody and Register

Human speech often exhibits complex patterns of amplitude and frequency variation. In addition, the duration of phonemes changes, and imbedded silences may or may not be present. These patterns define the "prosody" of language. An important role of prosody is to communicate the structure of language. This includes phrase boundaries and sentence types such as the rising tone that signifies a yes/no question. Prosodic cue is an important cue for language comprehension in human-to-human communication; therefore, it is important for the speech-based communication system to give the system's output a natural cadence.

Prosodic issues lead naturally to the topic of sentence register, which makes it an important aspect for the dialogue structure design for the public communication system. By varying a number of acoustic factors (primarily vowel duration and word intensity), specific words can be emphasis. This is often used to convey semantic significance. The use of register not only helps speakers understand one another, it also serves as a valuable source of feedback that indicates whether understanding has in fact occurred. To date the author knows of no system that takes advantage of register, and much research remains to be done concerning the rules that govern the proper use and understanding of register.

It is desirable that prosodic cues should also be used by the input subsystem, both to improve recognition rates (this requires a capacity to compensate for prosodic modification of incoming speech), as a source of useful parsing information (Okawa, 1993; Price, 1991). In addition, the extent to which prosodic cues are used to control dialogue is not known (this includes such things as turn taking, topic introduction and dialog repair). This is an area where further investigation could lead to much more natural human/machine communication. It is a matter of everyday experience that speech changes under the influence of strong emotion.

Emotion, prosody and register are that three factors make speech language different from written language. Human speech contains much redundant information with very complex variations. The regulations of these changes are not easy to find; therefore, it is not easy for the computer system to modify human speech and generate real human-like speech can have the prosodic information based on the detail context. For speech-based human-computer interfaces, the most important emotion that influences speech is apt to be frustration for the speech recognition. It is now a hot research topic to map the effect of this emotion on synthetic speech generated by the communication system. People are trying to develop the algorithms that will normalize speech modified by emotion. This is essential if recognition rates are to remain high when the user is under stress.

7.6 AFFECTIVE COMMUNICATION

There are enormous differences between a human speaking to another human being and a human talking to a machine. Humans possess a much more sophisticated sensory system for the interpretation of acoustic signals. Humans use additional sources of information, (e.g., visual information such as gestures and speaker physiognomy). Humans generally keep track of the actual context as well as of the history of the spoken dialogue. Humans have access to much broader and sophisticated sources of knowledge about the actual context, including common sense. Humana can accurately recognize isolated digit sequences and spoken letters, short segments extracted from spontaneous conversation and words in nonsense sentences that provide little contextual information. Human can easily recognize speech with normally occurring degradations.

We all know that natural human-to-human communication is an affective behavior or process and it is evolutionary. Affective communication is communicating with someone (or something) either with or about affect. We communicate through affective channels naturally every day. Indeed, most of us are experts in expressing, recognizing and dealing with emotions in these traditional communications. Humans are experts at interpreting facial expressions and tones of voice and making accurate inferences about others' internal states from these clues.

The dialogue system design for human users to communicate with computers via a telecommunication system does not really consider the affective communication nature of human speech. This is a largely untapped research area for the development of speech technology. What roles can affective communication play between human-computer interactions? What kind of new human factors issues were induced by it? What method, if any, can the designers of speech-based communication system use? From anthropomorphism's perspective, should we leverage this expertise in the service of speech-based interface design, since attributing human characteristics to machines often means setting unrealistic and unfulfilled expectations about the machine's capabilities? It is very natural to believe that the communication between the human and the computer should be as similar as human to human. This has been regarded as a natural interaction with computers. The prosody in human speech is so complicated that it is hard for the synthetic voice to modify it, at least in present state of technology level, the question is: shall we make the synthetic voice distinguish from a human voice so it can be easily recognized as machine-talk, or shall we expect that people may get used to this kind of machine-voice and learn how to comprehend the synthetic voice better?

A speech-based public communication system may have to respond to users experiencing strong, negative emotional reactions. One example would be a user, trying unsuccessfully to find information from a help system under deadline, becomes frustrated and upset. The computer should be able to work with the user and help address the user's frustration, if not also the eliciting problem as well. Such a system might need to communicate sensitive information to a volatile user in ways that don't exacerbate a user's already excited state. Affective communication may involve giving computers the ability to recognize emotional expression as a step toward interpreting what the user might be feeling. Researchers are beginning to investigate several key aspects of affective communication as it relates to computers. The speech-based communication system is flexible, and can be adapted to contextual needs

7.7   LIMITATIONS OF SUI

Designing speech user interface (SUI) is substantially different from designing graphical applications, and its design is difficult in general. Designers new to the process often find themselves at a loss because there are no concrete visuals to sketch. One reason is the complexity of speech interface design, aggravated by the subtle and not-so-subtle limitations of using speech as an interface modality, including recognition errors, slowness of spoken interaction and ambiguity of spoken language.

Very few experts in speech dialogue system design understand human factors well, and human factors experts normally have very limited knowledge on how to design a spoken dialogue system. To summarize what we discussed earlier, we can identify six broad categories of speech interface limitations: speech recognition, spoken language, environment, human cognition, user and hardware (Novick, 1999; Shneiderman, ; Suhm, 2003). Table 7-2 shows its limitation categories, their definitions and specific examples for each limitation. While the speech recognition community should be very familiar with the limitations of speech recognizers and the hardware used to deploy speech applications, this taxonomy makes it obvious that there is much more to speech user interface design than the speech recognizer.

Table 7-2. The categories of speech interface limitations

Limitation category	Definition	Specific limitations
User	Differences and preferences between users	Task knowledge, expert/novice differences, speech competence, level of motivation, Modality preferences
Human cognition	Properties of the human cognitive system	(see Table 7-3)
Environment	Influences from the user's environment	Noise, multiple voices, not all addressed by system, interruptions
Automatic Speech Recognizers	Recognition accuracy to variance of human speech	Errors, finite vocabulary, language model, acoustic model, quality of input signal
Spoken language	Inherent in speech and spoken language	(See Table 7-4)
Hardware	Properties of the hardware	Channel bandwidth and quality, microphones, platform

Table 7-3. Limitations of speech interaction arising from human cognition

Cognitive Limitations	Design Problems	Example
Sequential nature of speech	Speech is slow	Ineffectiveness of delivering news or e-mails over the phone
Working memory	Limited "short-term memory" (STM) capacity of 7±2 chunks	Confusion arising from menus with many options
Low persistence	As spoken output is quickly forgotten, users have trouble remembering instructions	Infeasibility of delivering complex instructions
Speech competes with (other) verbal processing	Speaking competes with other verbal processing—if they occur simultaneously	Difficulty of composing text and dictating it (vs. typing)

Table 7-4. Spoken Language Limitations

Spoken language limitation	Specific design problem
"Spontaneous" character	Chatty behavior leads to inefficient communication, not suitable for command and control, data entry, dictation. Some users are surprised by open-ended prompts and confused about how to respond.
Public	Others can hear speech: no privacy and possible disturbance.
Turn-taking	Users abide by turn-taking protocol of human conversation.
Ambiguity	Need to resolve ambiguities to determine meaning, Difficult to describe complex concepts or abstractions, Designers need to be aware that prompts or spoken output can lead to user confusion and misunderstanding.
Anthropomorphism	Users may assign human qualities to speech systems, including intelligence and trustworthiness, Degree of interface personification ("persona"), Complex and multi-leveled nuances of human conversation are not conducive to machines, Raised user expectations lead to backlash when disappointed.

7.7.1 Speech Synthesis

We discussed the issue on synthetic speech externally in Chapter 6. To summarize the discussion, we can say that the problem with speech synthesis is that humans use many different acoustic cues for speech comprehension, while present synthetic speech lacks the necessary prosodic cues, and, therefore, humans are intolerant of imperfections in synthesized speech.

Thus, compared to human speech, synthetic speech is generally not as well-liked and takes longer for people to become accustomed to (Francis, 1999). It is found that the comprehension of messages in synthetic speech was significantly lower than for recorded human speech for all message types (Lai, 2001).

It is known that the prosodic cues provide a powerful guide for the parsing of speech and provide helpful redundancy. When these cues are not appropriately modeled, the impoverished nature of synthetic speech places an additional burden on working memory that can exceed its capacity to compensate under real-time, demanding circumstances. Thus it revealed the importance of higher linguistic cues (prosodic, syntax, semantics) for synthetic speech processing and further implicates the prosodic modeling of TTS systems as the source of a major performance differential between natural speech and synthetic speech processing (Paris, 2000). Therefore, the designers should strive to represent prosodic cues more accurately in order to avoid high cognitive processing costs associated with even the highest quality TTS systems.

Also existing as a transient phenomenon, spoken output cannot be reviewed or browsed easily (e.g., in a graphical user interface, the interface can pop up the dialogue stack to the previous level by simply presenting a new menu), while in conversation, there needs to be something comparable, such as "What now?"

7.7.2 Interface Design

Research has been done to present a method for identifying design guidelines for broad classes of speech interfaces in two steps (Suhm, 2003). First, the solution database of the framework of SUI design is queried to obtain lists of "solutions" to specific design problems of a broad class of SUIs. Second, expert knowledge is employed to reduce this list to a short list of best practices. This method was employed to generate a checklist of ten guidelines for speech dialogue applications for telephone applications as shown in Table 7-5.

7.8 USABILITY EVALUATION

To permit natural communication with humans, the system must be able to recognize any spoken utterance in any context, independent of speaker and environmental conditions, with excellent accuracy, reliability and speed. This is the ideal requirement for the speech system performances. However, for a variety of reasons, today's automatic speech recognition systems are far from reaching the performance of their human counterparts. These differences induced different challenges for the success of the automatic information access system.

There are very few reports one can find in the literature about usability evaluation and users' satisfaction of these automatic voice service systems. Different improvements have been made since they appeared on the market. The efforts are mainly on the increased accuracy of ASR performance and the intelligibility of synthetic voices, as well as adjusting the dialogue strategies due to the better understanding of human verbal behavior. There is no systematic human factor research conducted around the application of the dialogues systems in public places. One of the reasons might be that it was the company's secrecy to document it and to improve the usability of the system.

Table 7-5. **Guidelines for heuristic evaluation of speech dialogue applications**

Guideline	Primary design problems
Design prompts to elicit brief responses.	Minimizing recognition errors.
Coach a little at a time-escalate error prompts.	Recovery from repeated errors.
Offer alternative input modality, if applicable.	Recovery from repeated errors.
Prompts should guide users to stay within vocabulary.	Out-of-vocabulary words cause recognition errors.
Minimize vocabulary size and acoustic confusability of expected responses.	Confusable words lead to recognition errors.
Design prompts that encourage natural turn-taking.	Users follow the turn-taking protocol learned in human conversation.
Keep menu options simple and clearly distinguished.	Users have difficulty matching their problem with options presented.
Professional applications should follow professional protocol.	Anthropomorphism elicited by personified interfaces.
Keep prompts short, and serve the majority well.	Speech is sequential and thus slow.
N X 3 rule: no more than nine choices at any menu, no more than 3 steps to complete a task.	Limited capacity of short-term memory.

In Chapter 5, we had external discussions about different aspects related to the issue of system usability. Usability evaluation can mean rather different things to different people. To carry out any usability evaluation, we should have a clear understanding about to whom (the potential users) the usability evaluation is for and what usability means to them.

There may be a wide range of different users. According to the concept of "stockholder" as the user group, any employee or customer of the organization who will be directly or indirectly affected by the system should be considered the users (Smith, 1997). The user group can be categorized as: a) *System manager* who is responsible for the system to be operated inside an organization system. He is more interested in the efficiency and the benefit of the entire system. He identifies the need for new or revised systems and for their initial specification. b) *End user* is the one who directly uses the system. c) *Customer users* are those who are immediately affected by the input to, and output from, an information system or part of the system. d) *System user* is the person who either generates or manages the application itself rather than using the software system to support an organizational role.

If we looked at Figure 5-5, the hierarchy of the user's needs starts from the system's design that fulfills the functionality requirement. As soon as the system fulfills its functional requirement, the user's satisfaction with the system performance will take place. If the system could not fulfill the functional requirement, the satisfaction is hard to reach. The third level of usability is the pleasure the users feel, which means the users are delighted the system, and they want to use the system.

Dybkjear, et al., (2004) intend to separate the usability from user's satisfaction. In her concept the usability is more or less related only to the performance of the designed functions such as the task can be successfully performed. Thus, the usability only measures the functionality of the system. If we talk about the usability of a system, it can only come from a user's perspective. For users, without a certain degree of satisfaction, even if a system lets the users do their tasks, they would not consider it as usable unless they can be satisfied with the performance. The usability evaluation from the end users' perspective is the main interest in this chapter. There are three important components when carrying out the usability evaluation: a) who shall carry out the evaluation work? b) What are the criteria for the evaluation? c) How does one design the evaluation methods?

7.8.1 Functionality Evaluation

To evaluate the design as to whether it fulfills the functionality requirements or not, the technical evaluation is the first step of the evaluation. The technical evaluation for a spoken language dialogue system concerns different components and the entire system. Technical evaluation is usually done by the developers, and the methods are well-developed for many aspects of the system (Dybkjaer, 2004). There are a few distinctions in technical evaluation. One can distinguish language processing tasks from language processing systems (Crouch, 1995)[10]. Systems carry out tasks or functions. Tasks may be broken down into subtasks, and systems into subsystems. There is no requirement that there be an isomorphism between these decompositions. That is, task structure and system architecture need not be the same, and designers of evaluations need to take this into account. The task, specified independently of any particular system or class of

[10] http://xxx.lanl.gov/ps/9601003

systems, is the appropriate subject of evaluation. The task performance evaluation of different components includes:

a) *Automatic speech recognition*: Word and sentences error rate, vocabulary coverage, perplexity and real-time performance. For some system, it may also include the speaker identification and speaker separation.
b) *Speech synthesisers*: Speech intelligibility, pleasantness and naturalness.
c) *Natural language understanding*: Lexical and grammar coverage, conceptual error rate and real-time performance.

The technical evaluation should also consider the requirements from the system managers and system users. These specific requirements are normally set up before even the design takes place.

Nowadays, the dialogue systems are still task-orientated, mostly with a key-word spotting system for speech recognition. According to the ISO-9241 Standard, usability is the effectiveness, efficiency and satisfaction with which specified users achieve specified goals in a particular environment. Here the effectiveness and efficiency on the system are closely related to its functionality. The efficiency of a spoken dialogue systems is based of the conditions of two main factors: a) the flexibility of the dialogue strategy and b) the performance of the speech recognizer (Benitez, 2000). A flexible dialogue strategy means it can accept naturally spoken sentences, allowing a large vocabulary and a complex grammar. As the speech recognition system functions based on the statistic analysis of pattern matching, such a flexible system makes the recognition process very slow less accuracy, so there is a trade-off between these two aspects.

The objective measurement of task success is a commonly used measurement method for the functionality part of the spoken dialogue system (Dybkjaer, 2004). Normally a group of users asked to carry out a set of tasks. The task completion can be observed. The measurements include elapsed time for completing the task and the success rate, the number of utterances that are used, recognition score, proportion of repair and help requests, speech understanding and recognition rate, barge-ins and dialogue initiatives (Larsen, 2003; Walker, 1997).

7.8.2   Who Will Carry Out Usability Evaluation Work?

It is quite clear that the technical evaluation of the spoken dialogue system should be carried out by the designer of the system. He is the person who knows best about what the technical criteria and functional requirements for the design are. But who should carry out the usability evaluation of the end user? This question has not been asked by most of the papers that discuss about usability evaluation for spoken language dialogue systems. To them, it is natural that the designers or the system developers should carry out the usability evaluation. The usability we discuss now is the usability to the end user. The person who will carry out the usability evaluation should have a deep understanding of the cognitive limitation of the user, different human factors that may affect the user performance and the possible expectation from the user. The evaluators should also have knowledge on ethical questions and possible social-organizational aspects of the interaction between user and the system. At the same time, different ergonomics aspects should also be considered. Normally, the spoken dialogue designers or developers have very limited knowledge of these aspects just mentioned; therefore, the usability evaluation should be carried out by cognitive experts, or someone who has deep enough required knowledge on human before he can carry out this work.

7.8.3   The Usability Design Criteria

There are very few studies one can find in the literature regarding the usability criteria for speech interaction systems. Dybkjaer and Bernsen (Dybkjaer, 2001) discussed different criteria for spoken language dialogue systems. With their long experience with research work, especially with the performance of the European DISC project[11], they indicated certain usability qualities for spoken language dialogue systems and put them as design guidelines as well (Dybkjaer, 2004).

In Chapter 5, we discussed different versions regarding principles to support the usability design. These principles can also be used to for the usability evaluation. Jakob Nielsen's *Heuristic evaluation* (Nielsen, 1994) has been mostly used to evaluate the usability of a system. Combining these different aspects related to the usability, we could identify the following usability design principles for the spoken dialogue system design:

a) *Learnability*: The ease with which new users can begin effective interaction and achieve maximal performance (Dix, 2003). The design should always be clear about the user's experience/knowledge of the system and how quickly they can learn about the interaction. As human beings have rich experience with using spoken language to communicate with other human beings, designing the naturalness of user speech plays an important role in learnability. Speaking to a spoken dialogue system should feel as easy and natural as possible. The system's output language should be used to

[11] www.disc2.dk

control-through-priming users' input language, so that the latter is manageable for the system while still feeling natural to the user.

b) *Visibility of system status*: Always keep users informed about what is going on, through appropriate feedback within a reasonable time. For spoken dialogue systems, input recognition accuracy, output voice quality and output phrasing adequacy are important components. Good recognizer quality is a key factor in making users confident that the system will successfully get what they say. A good output voice means the system's speech is clear and intelligible, has natural intonation and prosody, uses an appropriate speaking rate and is pleasant to listen to. The user must feel confident that the system has understood the information input in the way it was intended, and the user must be told which actions the system has taken and what the system is currently doing.

c) *Aesthetic and minimalist design*: Dialogues should not contain information that is irrelevant or rarely needed. The contents of the system's output should be correct, relevant and sufficiently informative without being overly informative (Dybkjaer, 2004). Every extra unit of information in a dialogue competes with the relevant units of information and diminishes their relative visibility.

d) *Explicitness*: The system should express its understanding of the user's intention and provide the information to the user in a clear, unambiguous, correct and accurate manner and use language that is familiar to the user.

e) *Flexibility and efficiency of use*: Allow users to tailor frequent actions. Accelerators-unseen by the novice user-may often speed up the interaction for the expert user to such an extent that the system can cater to both inexperienced and experienced users (Nielsen, 1994). Adequacy of dialogue initiative is one of the important issues. A reasonable choice of dialogue initiative should consider factors such as the nature of the task, users' background knowledge, user's experience and expectations and frequency of use.

f) *Error prevention and correction*: Error handling is always important for a speech interaction system, as the error may come from the systems misrecognizing what the user said, or even users' making the error. In Chapter 6, the error handling is discussed in detail. The methods to handle the error would be different for different systems. Select good speech recognition system to reduce the misrecognition rate as much as possible.

g) *User control and freedom*: Users often choose system functions by mistake and need a clearly marked "emergency exit" to leave the unwanted state without having to go through an extended dialogue. Interaction guidance is necessary for the users to feel in control during interaction. The guidance should be designed in a flexible way that is suitable for the first-time user, as well as for some experienced users in different degrees. The system design should support undo and redo. Useful help mechanisms may be an implicit part of the dialogue, be available on request by asking for help or be automatically enabled if the user is having problems repeatedly, for instance in being recognized (Dybkjaer, 2004).

h) *User adaptation:* This is important for the user's satisfaction. Users have different domain knowledge level and system expert.

The above usability criteria can be used during the design iterative process, from the early beginning of the design plan to the final products. When it comes to the implementation of the system to the application place, organizational impact analysis, ergonomics and sociotechnical design analysis need to be incorporated.

7.8.4   Evaluation Methods

The evaluation methods generally deal with such questions as: a) what are we going to measure? b) What are the measurement equipment and process? c) Who shall be the subjects, and how many subjects shall be enough?

According to the definition of usability, the usability can only be measured for a specific combination of users, environment and tasks. In such combination, the usability is contexts-dependent and how to generalize it becomes a big question. The usability evaluation should be carried out in the application context.

To measure the functionality part of the usability, the objective measurement of task success is commonly used (Dybkjaer, 2004). Satisfaction can only be measured subjectively. The common methods used are interview and questionnaires. The users are required to respond to a number of issues related to their perception of interacting with the system. The results are obviously highly dependent on the nature of the questions (Larsen, 2003). There is no standard design of a valid psychometric instrument to measure the subjective aspects of the satisfaction toward a spoken dialogue system. One experiment involving 310 users with field trials was carried out by Larsen (Larsen, 2003). All users filled out and returned the questionnaires after two scenarios had been completed. He found that if scales especially targeted toward speech interfaces are not systematically designed and validated, but rather composed in an ad hoc manner, there will be no guarantee that what is measured actually corresponds with the real attitudes of users.

A good measurement should have the following characteristics (Sanders, 1992):

a) *Reliability*: The results should be stable across repeated administration.

b) *Validity*: The right thing is measured with the right measurement process and instruments.
c) *Sensitivity*: The technique should be capable of measuring even small variation in what it is intended to measure.
d) *Freedom from contamination*: The measurement should not be influenced by uncontrolled variables.

There are no standard design questionnaires to measure the satisfaction of any products. In the SASSI project (Hone, 2001), after reviewing literature and interviewing speech system developers, they created fifty-items questionnaires related to the user's attitude statement. All the questions are rated according to a seven-point scale labeled "strongly agree," "agree," "slightly agree," "neutral," "slightly disagree," "disagree" and "strongly disagree." The results suggest six factors in user attitude to speech systems are important: perceived system response accuracy, likeability, cognitive demand, annoyance, habitability and speed. Under each factor, over ten items are addressed.

Besides selecting the right items to ask in the questionnaires, how questions are formulated can also affect the results. The sentences should be created so that you do not bias responses. The main point to keep in mind when creating the questionnaires is to write neutral questions that do not suggest a desired response and provide a choice of the appropriate response (Barnum, 2002).

During the measurement process, normally some tasks and some scenarios are described to the test subjects. The subjects were asked to perform the tasks inside the scenario, and their performance is measured. After the task performance, some interviews and questionnaires are given to the subjects. Some general questionnaires can be distributed to the potential users without the testing of the task performance, depending on what questions are being addressed.

To evaluate the spoken language dialogue system, we need to identify the following issues when any evaluation is going to be carried out:

a) What are the characteristics of overall system performance? How does one measure it?
b) How many tasks should we ask the subjects to perform? How detailed should the scenarios be? What should their properties be? Which properties of the interface should be used?
c) What is the context of the system performance? How can context be specified?
d) What is the minimal set of applications that are needed in the evaluation?

There is no standard answer to these questions. It is not necessary that the usability evaluation we discussed here take place after the final product is released to the market. Usability evaluations should be carried out even in the early beginning of the iterative design process, then the answer to these questions would be different to different stages of the design. To encourage the usability, UCD design process is recommended.

In the usability evaluation, there are some important questions related to the test subjects:

a) Who should we choose as the test persons? Which categories in the stockholder group are more interested?
b) How should the test persons be? Should they be naïve people who are then trained extensively, or experts on existing systems who are trained to overcome their background knowledge and habits?
c) How much training is required to arrive at stable performance where we can be sure that we are evaluating the properties of the interface and not just the learning and adaptive behavior of the operators?
d) How can we consider the individual differences among different user groups? Individual factors include the culture and education background, language capacity, past experience with interactive systems and attitudes toward new technology. How can we have the right coverage of the user group? How many test persons will be good enough?

7.9   CONCLUSION

It is an attractive business for spoken dialogue system application, due to the huge economical gains and other benefits such as ease, convenience, intuitive interaction and hands/eyes freedom from the devices, etc. The usability of the system turns out to be an important issue for the success of the business.

This chapter discussed different aspects that may affect the design of the spoken dialogue systems such as cognitive aspects, human factors and social-technical factors. There are a few books on how to design the spoken dialogue system via telecommunication that provide the design guides for the dialogue design itself. This chapter does not intend to address those issues. But a few user performance problems that will obviously affect the usability of the system are shortly discussed in this chapter.

In usability evaluation, we discussed the evaluation methods, usability criteria for spoken dialogue system and usability test process. Most studies related to usability tests in the literature are focused on design functionality tests, or user task performance. Measurement of satisfaction is a tricky issue.

A user's satisfaction is an important part of the usability, and it can only be measured subjectively through interview and questionnaires. How to perform and interview and how to formulate the questions can directly affect the evaluation results.

Chapter 8

IN-VEHICLE COMMUNICATION
SYSTEM DESIGN

8.1 INTRODUCTION

With automation and technology development, there has been the
tendency to continue adding large amounts of functionality onto transport
vehicles. Drivers are beginning to experience a substantial increase in
complexity of the driver-vehicle interface. The next two decades will be no
different as a number of novel display and control operations (namely, data
receiving, data entry, menu item selection, message replay, etc.) will be
generated as a result of the widespread introduction of advanced computers
and communication-based systems (Barfield, 1998).

To let the increased communication-based systems work inside the
vehicle, one of the biggest challenges is the interface design. Since the first
priority for the driver on the road is the safety issue, any new technical
systems that will be introduced to the vehicle, should always guarantee that
they are safe, or can increase the safety aspect when the drivers use it.
Speech technology is one of the technologies that the researchers and vehicle
industry are interested in when it comes to the interface design and humans'
interaction with the in-vehicle communication systems. The reason for the
interest of speech technology is simple: it can let the driver's hands and eyes
are free from the communication device. With speech interaction, the
driver's eyes can stay on the road, and the hands can stay on the steering
wheel. What we don't know is that there are many human and environmental
factors that will affect the performance of the speech technology, and the
safety factor is still a question.

To design an effective car-driver interface, it is critical for the system to be reliable and respond with very low latency. This interface normally requires interacting with many system components. It is important that the system's architecture be flexible. In addition, it must provide the capacity for developing context-aware applications in which applications can combine various information elements to proactively aid the driver. To be able to respect the driver's attention capacity, the system should integrate with facilities to sense the driver's attention in order to develop the appropriate application. The components of the system should instantly be able to determine what the driver is focusing on and contribute information about the driver's state which is integrated in real time (Siewiorek, 2002).

In this chapter, we will discuss different factors, humans and environment, which affect the speech interaction design inside the vehicles and its safety issues.

8.1.1   Intelligent Transport System

Terms (and associated acronyms) such as Intelligent Transport Systems (ITS), Transport Information and Control Systems (TICS), Intelligent Vehicle Highway Systems (IVHS), Road Transport Informatics (RTI) and Advanced and Transport Telemetric (ATT) have been applied in the last two decades to refer to this collective group of technologies. For which, the first of the above mentioned terms, Intelligent Transport Systems (ITS), sees the largest progress in new generation automobiles and at the same time, poses the greatest challenge in ensuring if the design of such systems fully integrates safety and usability.

ITS applications seen in modern vehicles are built around the concept of information and support technology, which employ the use of advanced electronics and computer systems to fulfill an array of driving related functions. A breakdown of the various types of driving related functions relating to the use of ITS (Galer Flyte, 1995), are provided below:

a) *Driving-related functions*: Those, which "directly impinge on the primary driving task," for example, collision avoidance, adaptive cruise control and automatic lane control.
b) *Driver-Vehicle-Environment-related functions*: Those which "provide information relevant to components of the driving environment, the vehicle or the driver," for example, traffic and travel information, vision enhancement, climate control, GPS style navigation and route-guidance.
c) *Non-driving-related functions*: Those which "are unrelated to driving" – for example, cellular telephones, entertainment devices (audio/video) and office-based facilities (web browsers, e-mail, fax capabilities).

Driving the car on the road is always regarded as having potential danger, but car driving has become a part of many people's daily life. The ITS provides a different interaction between the driver and the car. The driver's attention is no longer only on the road. To manage all of these electronic systems inside the car, the in-vehicle computer system (IVCS) has been interesting to many researchers. The IVCS basically includes driver-vehicle-environment-related functions and non-driving-related functions from the categories above. The design of IVCS should provide a good support to the driver for his safe drive while still being able to carry out other activities.

Analyzing how drivers interact with themselves, external traffic and internal technology, is a central issue in the design of in-car information systems. In principle, road use is defined as a collaborative and cooperative struggle, since a majority of actors in such environments share common resources such as the road, lanes, external traffic information, etc., and, through their use, alter the conditions and possibilities of outcomes for other users (actors). In this case, drivers could be viewed as critical components of the traffic system chain, whose conduct is then reflected largely by their conduct with vehicles in a dynamic environment, in relation to their private worlds and worlds of other actors, alternating to and fro through varying situations and contexts.

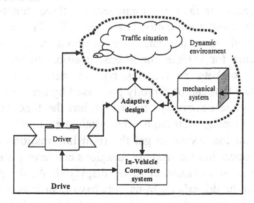

Figure 8-1. The future car facilitates with IVCS and adaptive design.

As driving is an eyes/hands busy task, speech interaction to the IVCS has naturally become a hot topics for the researchers. The research topics can be categorized into two main areas: a) interaction between driver and the IVCS and b) interaction on the road. Figure 8-1 illustrates the future car with IVCS and an adaptive system. The first research area focus on the technicalities of

IVCS design, speech control and dialogue system design and increases the intelligibility of the synthetic speech, etc. The second research area focus on how to design the interface to support the driver with enough safety margins on the road during his interaction with the in-car computer system. This includes how to design the layout of the computer screen, microphones inside the car and how to design the adaptive system to insure the drive safety also, how should one balance the driver's attention between the road and inside the car?

In this chapter, we will have a general description of the state of the art IVCS design. Our interests are the cognitive and human factor issues regarding the application of IVCS on driving behavior.

8.1.2 Design of ITS

If we look at the model in Figure 8-1, the driver is controlling the mechanical system of the car through drive performance. He controls the car speed, latency on the road, etc. The traffic situation combined with car speed and latency on the road contributes to the dynamic environment that the driver needs to cope with. This is the traditional situation of a driver with his car. When IVCS is introduced to the car, the driver will need to interact with this system, while still needing to take care of the driving situation.

The intelligent interface is organized in three functional systems: integration, control and presentation (Labiale, 1997). The *integration system* receives codes and stores the data from information of navigation, route, traffic and sensors for vehicle conditions, position on the route, distance from an obstacle or between vehicles, type of roadway, etc., and transmit them into a format that can be directly processed by the control system. The intelligent part is the *control system,* which has the functions of dialogue, strategy and learning. Through the dialogue function the drivers communicate with the system to get the right information presented in the right form. The operation for strategy functions concerns priority and timing of different items of information to be displayed. At the same time, the system will learn the driver's customs in behavior, such as reaction time and what kind of information the driver is interested in, and then link it to the information delivered by the system. The control system sends its data to the presentation system. Here it will define how to present the information with what modality (auditory or visually, speech or sound signal, text or graphics).

The proposed intelligent interface above is an ideal approach the IVCS system design. There are still a lot of cognitive problems in the interface design that need to be studied. In a way, the IVCS takes the driver's attention from the road to the inside of the car, which actually introduces a potential danger for the driver. Another problem is the extra mental workload introduced by the system (Piechulla, 2003). To insure the safety margins, the adaptive design is introduced. Adaptive design includes functions like collision avoidance, adaptive cruise control and automatic lane control, etc. One of the first experimental systems that implemented adaption to the driver is the GIDS (Generic Intelligent Driver Support) (Michon, 1993). It brought behavioral and technical scientists together in an unprecedented concerted effort to determine the feasibility of in-vehicle knowledge refinery.

The intelligent adaptive system includes many different elements, one of which is the driver's mental model of the system. It can evaluate, inference and modify the user's capacity, behavior and preferences. It can determine the driver's goal and present operation. It can take into account the dynamic dimension of interaction between the driver, the car and the driving situation (Labiale, 1997). Adaptive design can access and integrate information from the driver's behavior and the dynamic environment and provide certain control to the driver's interaction the in-car computer system. For example, if the adaptive system senses that the car's latency is in some potential danger, this system may send a warning signal to the driver or stop the driver's interaction with the in-car computer system. The IVCS and the adaptive system are regarded as the driving support system.

To support the driver's activity and improve the operation and safety of the traffic system, there has been a rapid increase in research activity devoted to the design of new ITS over the past few years. The development of such systems poses crucial questions regarding choosing suitable technologies, designing the interaction system and the consequences on a driver's future behavior and the consequences for overall safety.

Studying the integration of a new aid in a driving activity entails taking account of the essential dimensions of the road situations in which that activity takes place and the diversity and variability of the road situations that drivers may encounter during a journey. It also involves choosing functional units of analysis, making it possible to examine not only the effect of the aid on the performance of the specific task to which it is dedicated (compliance with safety margins or speed limits), but also its compatibility with the performance of other related driving tasks, such as overtaking maneuvers, interactions with other users, fitting in with traffic constraints and so on. Finally, it calls for the selection of relevant indicators likely to signal the changes that could take place in a driver's activity.

Speech interaction found in those in-vehicle computer applications built in the cars is promising. Using speech in an application interface has many advantages. It makes data entry possible when no keyboard or screen is available, and it is also an excellent choice of technology for tasks in which the user's hands and/or eyes are busy (e.g., driving). However, speech does pose substantial design challenges as well. Speech input is error-prone and speech output is often difficult to understand. In addition, speech is a time-based medium, and it is transient, taxing a user's short-term memory and making it unsuitable for delivering large amounts of data.

8.2   IN-VEHICLE SPEECH INTERACTION SYSTEMS

In Chapter 7, we discussed some general dialogue design issues. It must be remembered that the in-car environment differs quite substantially from other domains as the speech interaction is always less important than the primary task of safe driving. Therefore, existing human factors guidelines for vocabulary and syntax design cannot be directly applied to in-vehicle information systems. There are many design challenges for speech interaction inside the car (Bernsen, 2002).

The first challenge is the safety issue for the drivers. How can we design such a system that can guarantee the drivers' attention on the road? For this purpose, the speech system has to be connected with other sensors to interpret the driver/car behavior. Driving is a complex and stressful task. How the system will be designed to help the driver to cope with such task is not a simple issue. We will have external discussion on this issue in this chapter.

The second challenge is the noise inside the car. Most ASR systems are sensitive to the noise. The noise inside the car comes from many different sources such as from the car engine, from the environment like rain or wind, from traffic on the road, from the in-car entertainment system and passengers' speech. This is one of the main technical challenges for an ASR system to work inside the car. There are a few solutions for this problem. We will not deal with this problem in this chapter.

The third challenge is the possible large size of vocabulary, different speech accent and dialect and multilingual and heterogeneous users. For a possible speech interacted GPS system to be used inside the car, the drivers should be able to say the names of the city and street in any European countries. This affects the spoken dialogue system design itself (Bernsen, 2003; 2002). This is not the main interest for this chapter, however.

8.2.1   Design Spoken Input

There are a few studies for in-vehicle information system design that were carried out in the later '80s, and the results revealed that these systems employing terse phraseology, simple one-word commands such as "next," "back," or "select," seem to be more efficient and acceptable than those that use conversational-style dialogues (Halsted-Nussloch, 1989; Leiser, 1989; Ringle, 1989; Slator, 1986; Zoltan-Ford, 1991). It was found that they prefer speech inputs that are:

a) *Short and fast*: People wanted to minimize the attention and time required to produce commands, in order to maintain attention to the primary driving task.
b) *Goal-directed*: Users' choice of vocabulary tended to state specific goals e.g., "Radio on," "Air conditioning hot" rather than actions e.g., "Turn radio on," "Increase the temperature."
c) *Impersonal*; Users' interpretation of the system as an 'intelligent servant' rather than a social partner, participants in the current study used very few pleasantries e.g., "I would like . . .," ". . . please."
d) *Imperative and explicit*: Commands explicitly and directly stated the requirements e.g., "CD play Dire Straits!" rather than using implicit statements are more common in human-to-human communication e.g., "I like to hear Bon Jovi."

These short commands are mainly for controlling different functions inside the car. Again, in the later '80s and early '90s, there are a few studies concerned with the spoken input and dialogue system design. A few design "guidelines" have been created based on the understanding of ergonomic good practice and cover the many issues that need to be considered when designing and evaluating in-vehicle information systems at that time. They put safety and usability as a paramount design concern. The objective of these guidelines is to provide designers with a summary review of the factors that need to be considered in the design process of IVCS (Frankish, 1990; Green, 1993; Papazian, 1993). A brief summary of these guidelines is as follows:

a) *Provide non-threatening prompts, paraphrases, and help/error messages.* These should contain the vocabulary and phrases a user can understand.
b) *Users can interrupt these dialogues.* Cars may require input on destination, type of driving desired (highway or rural) and time constraints to arrive at a destination. A driver, who knows that these entries will be requested, should not need to wait to hear the dialogue explaining the options prior to entering the data for each interaction. So upon hearing "route type" a knowledgeable driver should be able to respond immediately instead of waiting to hear "enter one for fastest, two for fewest turns, three for shortest, four for fewest tolls, five for most scenic."

c) *Provide explicit confirmation of all commands by the user.* Depending upon the command, confirmation may be accomplished either while users are entering responses or when commands are completed. It is important that the driver knows that the input was received by the system. Provide different types of feedback for data and for commands. Feedback should be context-sensitive. To confirm to the driver that the desired action was accomplished, feedback should be unambiguous.

d) *Command names should be semantically discriminated.* Drivers are likely to make more mistakes if several commands for distinct actions have similar meanings.

The command input type of speech interaction for IVCS is more or less rejected by some studies published recently. One of the main reasons for not using commands is because there are more and more complicated tasks that need speech control to communicate. It is not likely that the drivers can remember all the commands correctly (Bernsen, 2002).

8.2.2   Multimodal Interface

There is an effort to research the development of dialogue management to make driving a car comfortable and safe. Researchers are looking into the possibility of the spoken language dialogue system enhanced with multimodal capabilities, which may considerably improve safety and user-friendliness of these interfaces.

Evaluation of different types of modalities for control and presentation of a multi-modal in-vehicle e-mail system was studied by Trbovich (Trbovich, 2003). It was found that the drivers would judge themselves to have the least cognitive load, as they would apply the most natural and convenient modality to interact with the system.

The VICO (Virtual Intelligent CO-driver) system is a research prototype multimodal spoken language dialogue system (Bernsen, 2001). This is European project that was carried out from 2001 to 2003. The prototype was designed to meet the different challenges for the in-car speech interaction system. These challenges are easy for a large population to use: spontaneous multiple languages input; multiple-task assistance such as navigation to address hotel and restaurant reservations and system information; and multimodal interface and different levels of adaptation. The adaptation design uses various methods to anticipate a user's needs, goals, interactive behavior, etc., so the system may adapt to the user, or the user can "customize" the system (Bernsen, 2003).

The prototype using the natural language dialogues on task deals with navigation, car status, and tourist information and hotel reservations. The multimodality interface includes push-to-talk-buttons, speech input/output and graphic output. Modality theory (Bernsen, 2002) has been applied to address the problems of how to combine the speech and graphics text output in the car. The modality theory cannot address which information and how much of it should be presented on the car display. Thus Wizard-of-Oz simulation is used to further investigate the solution. The findings from the simulation experiment are that the speech is acoustic and omni-directional, whereas graphics are neither eyes-free/hands-free, nor are they omni-directional. Thus it is justified to use the speech output in the car. The speech has a very high saliency, which might distract a driver's attention from traffic situations. Thus VICO addresses this problem by including an element of situation awareness, which will make its speech synthesizer stop immediately when the car brakes are being applied. There may be times when the driver might be distracted from listening to VICO in a stressful traffic situation. Since speech is temporal while graphics text is static, leaving the most important VICO information presented on the display allow the driver to get the dialogue with VICO back on track or remind him of what has already been agreed upon with VICO, once the distractions have gone away.

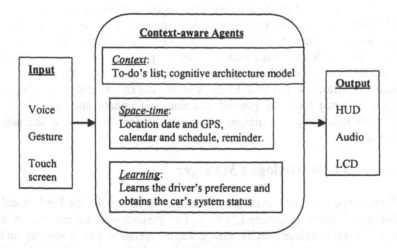

Figure 8-2. An information flow of companion system. Here the gesture recognition can sense the hand waves for giving the order of pause/stop/start the system. It also has face/fingerprint recognition for driver identification. HUD is the head-up display. LCD is the liquid crystal display.

There are many different proposals/prototypes on multimodal interface for IVCS. Here I would like specially to describe the system called

"Companion" that Daniel Siewiorek and his colleague proposed (Siewiorek, 2002). The information flow of the system is shown in Figure 8-2.

This prototype combines a wide range of input-output modalities and a display hierarchy. The intelligent agent links information from many contexts, such as location and schedule, and transparently learns from the driver, interacting with the driver when necessary. The development process of this prototype follows a User-Centered Interdisciplinary Concurrent system design methodology (Siewiorek, 2002). Based on field study and interview with users, they created the baseline scenarios. During the design process, they put the driver's cognitive and attention resources at the center of consideration. When they analyzed the input and output information hierarchy, the level of the driver's distraction was considered. Hand gesture recognition is preferred as the least distracting input modality. If the driver is listening to music or speech on the phone while an emergency occurs, a wave of the hand turns off all speech and systems at once. This works better than giving a speech command to the system. The HUD is regarded as low distraction as well, since it is located on the windshield and is where the driver is often focusing. At the same time, they are also aware that the amount of information that can be clearly displayed on the HUD is very limited. Speech input and output are regarded as a medium distraction while touch screen and LCD are highly distracting to the driver (Siewiorek, 2002).

The system tracks whether the vehicle is running, who the current driver is and what the current items are on the driver's task list. The driver's information, such as temperature, mirror position, etc., is learned, remembered and updated. Most interesting is that they introduced the cognitive architecture model into the system. It can integrate the driver's task performance with the present situation and update and provide the driver with necessary information to support his decision-making and task performance.

8.2.3   ETUDE Dialogue Manager

A prototype of a conversational system was designed for the Ford Model U Concept Vehicle (Pieraccini, 2003). The system was augmented by a graphical user interface (GUI) with a haptic (relative touch screen) and speech interface. The speech interface provided the user with additional information such as indication on the state of the dialogue, suggestions on what to say, and feedback on the recognized utterance. The UI was carefully crafted in order to allow drivers with no experience to use the system without prior training or reading a user manual. The UI also allows expert users to speed up the interaction by using the compound and natural language sentences.

The functions provided by the system are climate, communication, navigation and entertainment, which can be controlled by voice and the spoken interaction with the system and can be conducted on one of the three following levels (Pieraccini, 2003):

a) For direct dialogue, the system will prompt the user to solicit new information toward the completion of the task. A new user would start by speaking words visible on the GUI.

   For example:

   User: Climate.

   <switch to the climate control GUI and note that there is a button displayed seat temperature>

   User: Seat temperature.

   System: Driver, passenger, or both?

   User: Driver.

   System: Warmer or cooler?

b) For terse commands, the user will then be familiar with the application space and thus can speed up the execution commands by providing the necessary information in a single sentence, without being repeatedly prompted by the system.

   For example:

   User: Climate driver seat temperature.

c) For natural language commands, they are expressed using grammar at any stage of the dialogue.

   For example:

   User: Climate.

   <switch to the climate control GUI and note that there is a button displayed seat temperature>

   User: Turn the seat temperature all the way down.

Usability tests were conducted during the development process and while system was being developed. The purpose of conducting the usability test is to find macroscopic problems with the user interface during the development lifecycle of the prototype. One of the usability tests used the advanced implementation of the prototype and asked the subjects to interact with it while using a drive simulator video game. The results showed that the dialogue system reached a satisfactory accuracy.

8.2.4   DARPA Communicator Architecture

Another prototype of a dialogue system for in-vehicle navigation was developed by CAVS (Center for Advanced Vehicular Systems) located at Mississippi State University (MSU) (Baca, 2003). The system allows users to speak queries using natural unconstrained speech to obtain information needed to navigate the Mississippi State University campus and the surrounding town of Starkville, Mississippi. It employs speech recognition, natural language understanding and user-centered dialogue control to resolve queries. The architecture of the system is shown in Figure 8-3 (Baca, 2003).

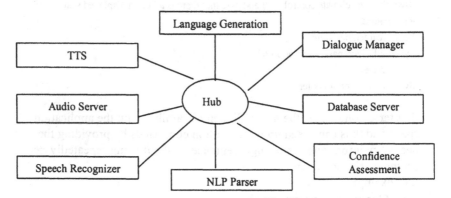

Figure 8-3: *DARPA* **Communicator Architecture.**

A hub acts as a central router through which the set of servers pass messages to communicate with each other. An audio Server accepts incoming speech signals and passes the result via hub to the Speech Recognizer. It also sends synthesized speech through the hub from TTS to the audio device. Dialogue manager (DM) initiates the dialogue, prompting the user for an initial query. The user speaks a request that is received by the speech recognizer module and parsed by the Natural Language Understanding (NLU) module. The NLU module then passes a list of possible sentence parses to the DM that will determine what data the user requires, obtaining the data from the database server and finally presenting it to the user.

The NLU module employs a semantic grammar consisting of case frames with named slots. A context-free grammar (CFG) specifies the word sequences for each slot. The grammars are compiled into Recursive Transition Networks, against which the recognizer output is matched to fill the slots. A semantic parse tree is generated for each slot with the slot name as the root. A simple example frame for a request for driving information is shown below (Baca, 2003):

```
FRAME: Drive
  [route]
  [distance]
```

The first slot in this frame, [route], allows asking for directions along a specific route, while the second slot, [distance], allows asking for the distance from one location to another. A subset of CFG rules for the route slot is shown below:

```
[route]
  (*IWANT*[go_verb] [arriveloc])
IWANT
  (I want *to) (I would *like *to) (I will) (I need *to)
[go_verb]
  (go) (drive *to) (get) (reach)
[arriveloc]
  (*to [placename] [cityname])
```

A series of usability tests were conducted to find the domain-specific data required to develop the semantic grammar and language models where users were asked to spontaneously enter requests for information about the university campus and surrounding town. Two tests were performed during the initial data gathering and system testing, followed by the final test of retesting the system on the initial data after enhancements were made to the grammar and language model. Issues identified during the testing are the level of clarification required by the user as well as the level of user initiative.

8.2.5 SENECs

Research has been carried out in the framework for the EU-project SENECs (Speech Control Modules for Entertainment, Navigation and Communication Equipment in Cars) in the 4th ESPRIT program (Minker, 2003). Its objective is to integrate and further develop the speech recognition technology together with Spoken Language Dialogue Systems for use in the car, which consists of a wide range of entertainment, navigation and communication applications. It uses an intelligent dialogue strategy that allows the system to compensate for low speech recognition confidence and word-level ambiguities by engaging in flexible clarification dialogues with the user. An example of command and control dialogue between the user and the system, created by Minker, et al., (2003) is shown below and in Figure 8-4.

> System: Yes, please?
> User: Enter address.
> System: Name of the town?
> User: Siegburg.
> System: Do you know the postal code?
> User: No.
> System: Do you know a city near the destination?
> User: No.
> System: Do you know the federal state?
> User: Yes.
> System: Name of the Federal state or region?
> User: North Rhine-Westphalia.
> System: <Displays city name> Is the name of the city correct?
> User: Yes.
> System: Navigation to destination started.

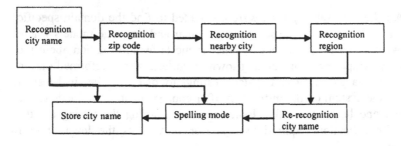

Figure 8-4: Disambiguation technique in the German version of the SENECa SLDs.

An evaluation was conducted using real driving trials. The experimental conditions were via speech mode and manual mode to perform nine tasks which were representative of different functions and included important complex operations of the driver information system: navigation, dialing and address book tasks. The data collection used are log-file recording, an audio/video recording with three cameras and the judgments on driving performance made by a professional driving instructor and the evaluation of the system by the subjects themselves. The trials took place at a middle-sized town in Germany, which involves a long course composed of express roadways, highways, entrances/exits and roads through cities/villages as well as residential streets. Sixteen subjects of different ages and experience were involved in the testing. The results showed that there was a lot of speech error due to forgetting of commands, spelling errors or using wrong vocabulary. There were more than twenty commands that the divers had to remember. But, at the same time, there were fewer driving errors with speech input compared to manual input.

8.2.6 SmartKom Mobile

Another research project is SmartKom. It is a long-term research effort funded by the German Federal Ministry for Education and Research. The project started in 1999. SmartKom Mobile is one of the applications. It is a portable personal communication assistant to be used in the office, at home and outdoors, where the user can walk around and carry it as a permanent and location-aware digital companion. It can be installed in the docking station in the car and acts as a digital assistant that provides hands-free access to trip planning and incremental route guidance. It uses different combinations of communication modalities as interaction modes, depending on the scenario as shown in Figure 8-5 (Bühler, 2002; Bühler, 2002; Bühler, 2003).

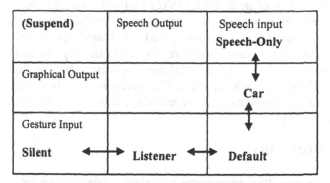

Figure 8-5: Interaction modes as combinations of communication modalities in SmartKom Mobile.

In Figure 8-5, the inter-modal transitions are:

a) *Default*: All modalities should be enabled for this mode. It is mainly used in the pedestrian environment when privacy or disturbing others is not an issue. It is from safety considerations and when the car is not moving.
b) *Listener*: This is when a user chooses spoken output but not speech input. This is used for summaries or extended background information.

c) *Silent*: this is when there is a problem with spoken language human-machine dialogue. The interface should be similar to traditional graphical user interfaces.
d) *Car*: Speech is the dominant communication modality, while displays are used only for presenting additional information such as maps.
e) *Speech-only*: This mode is mainly used when the car is moving. Gesture input or graphical output must be avoided for safety issue.
f) *Suspend*: it happens when the driving situation becomes dangerous and the driver's attention should be only on the road.

Thus, the SmartKom project adopts a common and uniform dialogue-based interaction model in these scenarios. Hence, it will adapt the dialogue strategy according to the restriction that each scenario imposes on the system.

8.3   THE COGNITIVE ASPECTS

The number and variety of multi-modal IVCS that are affordable to a wider range of users is continuously increasing in functionality and complexity due to the advanced technology, and car manufacturers are trying to offer these functions to their customers. We must carefully consider the user, such as his resource on the physical and cognitive capacity, the system and the environment, such as, the traffic situation as well as the specific capabilities of the particular sensory modality, when designing the multimodal interface for IVCS.

8.3.1   Driver Distraction

According to Verwey (1993), driver inattention plays a role in about 30% to 50% of all accidents. Most researchers agree that distraction from the main driving task is one of the most important causes of accidents. A distraction can be visual, auditory, cognitive or biomechanical (Ranney, 2000)[12]. Something that takes the driver's visual attention away from the road or changes the glance behavior is a visual distraction, e.g., looking at a map displayed on the visual display screen, listening to music while driving could be an auditory distraction to the drivers as well. An activity that demands thought can be a cognitive distraction if it takes the driver's mind off the road traffic environment, e.g., engaging in a call on a hands-free

---

[12] www-nrd.nhtsa.dot.goc/driver-distraction

cellular (mobile) phone. Something that requires the driver to lean forward, take hands off the steering wheel or stretch to operate can be a biomechanical distraction, e.g., adjusting the volume on an entertainment system. These biomechanical distractions can force drivers off their normal sight lines and impair their control of the vehicle. Many distracting activities that drivers engage in can involve more than one of these four components. For example, tuning the radio or dialing a cellular (mobile) phone may involve all four components of distraction.

The speech interface that would be used for interaction with IVCS should not distract the driver. The dialogue between a human and the system should be natural, or the driver might become distracted by the amount of concentration taken away from the driving experience.

There is an initial study done which compares different types of control for operating an in-car entertainment (ICE) system, incorporating radio, tape and CD functions with speech in traditional manual controls (Carter, 2000). The driving performance was found to be significantly better while using voice controls than the manual controls. Subjectively, the voice controls with feedback were rated easier, most likeable, and most efficient to use while driving. Similar findings were also reported by Tsimhoni, et al., (2004), who compared three address entry methods: word-based speech recognition, character-based speech recognition and typing on a touch-screen keyboard. They found that word-based speech recognition yielded the shortest total task time (15.3 seconds), followed by character-based speech recognition (41.0 seconds) and touch-screen keyboard (86.0 seconds). The measurement of latency position also showed that when performing the keyboard entry, it was 60% higher than that of all other address entry methods. So the use of speech recognition is favorable.

For a driver, the primary task is to monitor and control the vehicle's lateral and longitudinal position along a safe path. At the simplest level this involves controlling steering and speed, a task that with time and experience becomes almost automatic requiring, for example, overtaking and maneuvering appropriately at junctions and roundabouts. These tasks demand greater cognitive effort and the introduction of additional information while undertaking these tasks will create an unwanted distraction (Allen, 1971; Tijerina, 2000).

Some studies revealed that the speech-based interface is not a panacea that eliminates the potential distraction of in-vehicle computers. The study used a car-following task to evaluate how a speech-based e-mail system affects drivers' response to the periodic breaking of a lead vehicle, comparing a simple and a complex e-mail system in both simple and complex driving environments (Lee, 2001). The results of the study reveal three important points, which are speech-based interaction with in-vehicle information systems demands attention and can distract drivers and degrade safety; by increasing the complexity of a speech-based interface, it may impose a greater cognitive load; the cognitive demands of a speech-based interaction are additive with the demands of the roadway environment.

The interfaces using speech input-output in in-vehicle systems may present significant safety risks because of re-direction of attention if the concurrent tasks have a relatively high memory load or demands on the central executive function in working memory. There are additional cognitive process costs with the comprehension of synthetic speech (Paris, 2000) and dialogue grammar, understanding the menu structure and recalling the speech command. A study also revealed that the comprehension of messages in synthetic speech compared to recorded human speech was significantly lower for short navigation messages, medium length (approximately 100 words) e-mail messages and longer news stories (approximately 200 words) while driving (Lai, 2001). This study was done where twenty-four participants drove a simulator while listening to these three types of messages. There is no significance result in driving performance. But with further analyses of increasing the workload, it might reveal a negative impact on driving performance.

8.3.2   Driver's Information Processing

Different in-car systems demand additional cognitive processing and varying degrees of divided effort in order to interpret information collectively or accurately while driving. The point to realize at this juncture of evaluation is not that such a system in itself is doubtful, but when used at particular times and during certain tasks, it may have the potential for surprise, distraction and overload, all of which are simply opportunities for human error. For one, the attention demands of interacting with numerous systems could deprive attention to the relevant dimensions of the optical array (Schiff, 1995; Warren, 1990) and, in turn, degrade drivers' abilities to control their vehicles in a safe manner. Besides, distraction and inattention on the road already deemed as the major contributing factors to numerous road accidents (Wierwille, 1995).

While such symptoms could be present even while performing fundamental tasks such as manipulating controls, radios/tape players, manual windows, vents, heaters, etc., or when having a simple conversation with passengers, human information processing could be further challenged, if such operations or manipulations are expected to be performed using higher order systems and interfaces. This is probably the initial motivation to introduce the IVCS to increase the safety of driving. A typical driver information processing model of the use in such systems is provided in

Figure 8-6. When there is no IVCS, the driver's attention for perception would only attend to the traffic situation. When the IVCS is installed inside the vehicle, the driver has to divide his attention between the traffic situation and the interface of IVCS. The cognitive workload will naturally increase.

Driving inattention, an antecedent to adverse cognitive activities, occurs when attention shifts away from the driving task due to non-compelling reasons, or due to an over-reliance of information support systems. Reduced vigilance of this form gradually degrades concurrent driving performance. Driver distraction, on the other hand, refers more to the shift in attention from the primary driving task due to compelling reasons. It could be manifest in the form of a sudden in-vehicle task resulting in a visual or cognitive stall, for instance, when listening to a complex auditory message, as from a phone, electronic device, etc.; due to an inappropriate salience that captures special attention, such as images of broadcast or video playback or appealing photos accompanying navigation information, etc.; and by other unforeseen cues that elicit sudden responses, such as telephone rings, alerting beeps, etc.

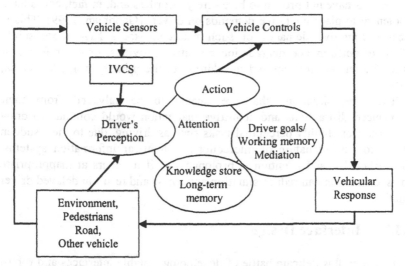

Figure 8-6. Driver Information Processing Model

There is an experiment using simulated driving tasks together with concurrent counting tasks using either visual or auditory presentation of digits by Cook and Kiekel (2001). It was observed as the memory load of the visual task increased, the level of destructive interference between tasks increased. Thus it conflicts with Wickens' multiple resource model (Wickens, 2000), which suggests that verbal information input-output is independent of visual-manual processing and can take place concurrently with little interference between the tasks.

The studies above revealed that the cognitive workload is a very important factor that determines the performance and safety of driving. Thus it is a challenge for the designers to design the IVCS so that it can maintain an appropriate level of cognitive workload. Future research should focus on how to minimize driver attention demands due to the audio distraction and visual attention deterioration caused by the in-vehicle computer system.

Earlier explorations have attempted to overcome limitations posed by inattention and distraction with ubiquitous computing concepts (Wieser, 1991); hiding attention-seeking technology within the car and its infrastructure; and touch screen interfaces relying on tactile and kinesthetic feedback. The novelty offered by speech recognition technology for IVCS approach is to enable drivers to keep their eyes on the road or their hands retained on the steering wheel.

On one hand, integrated designs envisaging haptic, tactile and kinesthetic modalities lack proof of their usability. On the other, voice-based interactions have not proven to be entirely effortless and, in fact, could have a potential to place cognitive demands on drivers (Goodman, 1999). This is because cognitive demands of such nature/interfaces have more subtle effects to performance degradation cognitive activities, or other forms of directed attention are required in addition to the primary task (Pellecchia, 2001).

Therefore, situations that are unable to be salvaged from their pronounced distractions and cognitive inattention would continue to draw resources from highly perceptual tasks such as driving, due to their sudden and/or concurrent nature of interactions with in-car information systems. Also, when focused attention is prolonged, or if it occurs at inappropriate times, distraction can reduce situation awareness and result in delayed driver reaction.

8.3.3   Interface Design

Hence, in this delicate battle of developing suitable interfaces and robust evaluation techniques to determine effective driver interaction with IVCS, research efforts should not begin by purporting if systems or technologies are effective or ineffective but rather, begin with the following:

a) To identify major human factors and cognitive issues surrounding the use of existing in-vehicle systems based on past findings, current situations and trends.

b) To identify types of tasks or circumstances that could lead to reduced performance, driver overload and accidents.

There are recommendations on visual and auditory information presentation for in-vehicle computers (Lee, 2001). For auditory information presentation, the messages should be simple, with volume adjustments and a repeat capability. It should avoid some of the potentially negative aspects of using speech presentation in light of visual presentation, for example, information provided to the driver must be timely and must be prioritized appropriately so that sufficient time is allowed for the driver to respond to any information as he needs time to hear and/or see the information, decide whether it is relevant and act upon it. When information is presented visually, it should be easy to view to avoid too much time on visual focus.

Using speech as the interactive modality for IVCS has advantages as it frees the driver's eyes/hands while driving. However, speech input is error-prone, and speech output is often difficult to understand. Additionally, speech is a time-based medium, and is transient, taxing a user's short-term memory and making it unsuitable for delivering large amounts of data. On the other hand, the information on the visual display can be retrieved whenever the driver wants. But using visual presentation will take away the driver's view of the road. Thus, the different types of modality interactions have their pros and cons. The multimodal interface of the IVCS should be carefully designed so that more information can be made accessible to the driver while not overloading his information-processing resources. There have not been many studies in assessing the effectiveness of a multi-modal interface in cars, especially in the effectiveness of the integration of the different modals in lightening the workload of the driver.

8.3.4 Human Factors

More importantly, there has been considerable research in the last two decades regarding even basic human factors issues surrounding the use of ITS devices while driving. Numerous research programs within Europe, North America and Japan have included large-scale collaborative efforts tackling a wide range of topics concerning the use of vehicular information systems. We shall outline and discuss research based on current situations and trends for two popular information systems: cellular (mobile) phones and in-vehicle navigation.

Most of the speech interactive IVCS are still at the research and prototype level. There is a long way to go before it reaches a necessary safety margin and high-level usability. As it is not in real-time use yet, there are very limited studies on the human factors and safety issues for such systems. At the same time, among the ITS, non-driving-related telemetric (referring to information and computer-based accessories in vehicles) devices such as, cellular phones and driver information support systems, namely, GPS-based route guidance/ navigation, remain as popular topics in research, due to their expanding functionality and notoriety of their inherent implications to driver behavior, performance and driving safety. There are some similarities between the use of these systems and the potential IVCS, especially in its cognitive aspects. The knowledge from these studies has certain value for the interface design of IVCS.

In a positive light, navigation systems allow drivers to travel to unknown destinations in safety, avoiding the need to ruffle through paper maps and other materials. Also, there is the likelihood that satellite navigations systems such as "SATNAV" will seem common to vehicular systems as the market matures and design/development costs fall. Estimates speculate that by the year 2005, there would be more than five million navigations systems sold annually in Europe (Rowell, 1999).

Though cost-benefit analysis fails to indicate a definite advantage for cellular phone use in driving situations (Cohen, 2003), they too have also been considered as assets to road-users, as they allow rapid access to police and emergency services (Chipman, 2000; Nunes, 2000). As a matter of fact, in-car systems such as route guidance/navigation systems and telemetry like cellular phones are intended to support safe and efficient driving, but there may be unforeseen side effects to such technology. This shall be discussed further.

8.4   USE OF CELLULAR PHONES AND DRIVING

With the wealth of new technologies and applications in vehicle telemetric systems, we notice cellular phones in particular have become increasingly popular choices for mobile and remote communication. The recent study conducted by the National Highway Traffic Safety Administration (NHTSA) (Utter, 2000), estimated that 54% of drivers carry mobile phones with them, during driving. Among which, 55% were reported to have kept their phones turned on during most of their driving episodes, 73% of drivers claimed this practice were occasional and an estimated 3% of drivers (on average one-half million passenger vehicles) are actively involved in a cellular phone conversation at any time during daylight hours.

Here I will try to summarize the studies about using cellular phones while driving; the knowledge can be very useful for the IVCS design as it relates to the safety issue.

8.4.1   Accident Study

The safety implication of using cellular phones while driving is becoming widespread, of which the most compelling evidence of human factors concerns come from accident statistics. Unfortunately, ITS device-related accidents statistics are not available for North America or Europe, but the Japanese National Police Agency Traffic Planning Department has been closely monitoring causal factors and collecting accident statistics for cellular phones and in-car navigation systems. Accident statistics seen in Table 8-1 are due to driving with cellular phones between 1997 and 1999. Data from the Japanese National Police Agency Traffic Planning Department were extracted from their website[13] and translated to English for the sake of our international readers.

As seen in Table 8-1, cellular phone related deaths are estimated to be around 25 per year based on three years of statistics.

Table 8-1. Cellular phone related accidents in Japan

Year	Cellular Phone Use	
	Injuries	Deaths
1997	2,095	20
1998	2,397	28
1999	2,418	24

In other countries such as Australia, Spain, Israel, Portugal, Italy, Brazil, Chile, Switzerland, Great Britain, Singapore, Taiwan, Sweden, Japan, Austria and several cities in North America, who share similar concerns over devices that are visually or/and cognitively demanding, they have passed laws limiting the use of phones while driving.

8.4.2   Types of Tasks and Circumstances

The types of functions that may be undertaken by drivers who carry cellular phones while driving could be classified in terms of reacting to incoming calls (beginning with retrieving phones, receiving calls, viewing caller ID, etc.), performing manual manipulation (dialing, saving/retrieving information, etc.), being involved in a phone conversation (hand-held, hands-free) and others (switching hands while speaking, inserting/removing ear-phones for hands-free mode, manipulating other ancillary in-car controls, etc.). These function characteristics could be further understood in terms of their potential for task overload-eyes-off-the-road and mind-off-the-road.

---

[13] www.npa.go.jp

As seen in the previous section, there is ample research work available on performing phone manipulations and voice conversations while driving. While fewer of studies were devoted toward receiving or reacting to sudden incoming calls, the crash statistics from Japan taken over an eleven-month span in 1999 before their ban against hand-held phone conversations were imposed are disheartening (see Table 8-2).

Table 8-2. **Driver Tasks and Crashes (January – November, 1999)**

Cellular Phone Use	
Results	Crashes
Receiving Calls	1,077
Dialing	504
Talking	350
Other	487
Total	2,418

Crashes arising from receiving calls were highest with over 1,000, making up 45% of the total crashes, followed by dialing, talking and others. This could be because tasks associated with receiving calls are often reflective of reaction to sudden changes while maintaining focus on a primary task.

Speaking of traffic conditions, Violanti, (1997) who examined Oklahoma drivers carrying cellular phones while driving, found that crashes generated by inattention, unsafe speeds or/and being on an undesirable side of the road, were more likely in the city, a location that is known for complex conditions, congestion and heightened attention demands.

8.4.3   Human Factors Study Results

There have been several studies conducted in support of such bans, largely due to devices symptomatic of visual and attention distraction or cognitive overload. They could be broken down into three categories based on research methodologies and factors specific to certain tasks.

In the first category, there are epidemiological and experimental studies conducted by Redelmeier and Tibshirani (1997) and by Violanti (1997). Work presented by Redelmeier and Tibshirani (1997) happens to be the most cited to date. It comprises of data collected from approximately 700 drivers from Toronto who were cellular phone users and were involved in motor vehicle crashes that resulted in substantial property damage. Results pointed out that risk of crashes was 4.3 times greater when a cellular phone was used than when it was not. There was also no significant difference to crash risk between using hand-held and hands-free use.

The second category of research work involved introducing a secondary task over the primary driving task. That is, drivers could be requested to perform concurrent phone conversations or cognitive/memory tasks of that equivalence in addition to driving. Alm and Nilsson (1994) measured driving performance and braking reaction time to visual stimuli, using handheld cellular phones versus hands-free cellular phone use as its variable. It was found that driving performance remained unaffected until exposed to higher workloads.

Reactions to unexpected events may be most affected only during early stages of a newly accepted phone conversation, beyond which performances decrements reduced appreciably over time. This could be true provided there are no additional human affective, cognitive and sensor motor changes experienced from changes in traffic situations and apt interruptions demanded during this recovery in performance. Otherwise, as pointed out by Pellecchia and Turvey (2001), overall coordination could still degrade if directed attention or cognitive activity occurs intermittently or in addition to the primary task. Other supports to this claim were that mobile phone users (drivers) reacted 0.5 seconds slower in detecting the decelerations of cars ahead, one second slower in time to contact and 30% less able to respond to critical traffic situations (Hancock, 2003; McKnight, 1993) such as taking a eight to fourteen meters extra when performing emergency stops from seventy mph speeds, using hand-held and hands-free phones respectively.

Hence, one has to consider other inherent relationships between cognitive activities, speech and motor coordination. Apart from reduced performance and other possible manual gestures that could arise from speech articulation (Corballis, 2002), there are more severe considerations to manual coordination and manual phone handling (such as dialing, switching hands between conversation, etc.), which then leads us into the third category of research.

McKnight and McKnight (1993) also demonstrate other detrimental effects such as lane variance and failure to respond to traffic events while attempting to dial a phone. Of course, one explanation to this could be that if one focuses attention on one's dominant (preferred) hand in a bimanual coordination task, then an increased asymmetry could result, which may be detrimental to performance (Amazeen, 1997; Riley, 1997). But the overall mechanism, which underlies the influence that, talking into or manipulating a phone has on driving, is complex and cannot be limited to one field alone. Rather, a breakdown is required to identify various types of tasks and distributed circumstances that converge toward degraded attention, unsafe behaviors and consequences.

Apart from phone conversations, dialing or phone manipulation presents other, yet unresolved issues for hands-free systems. They continue to impose cognitive, visual and manual demands, often causing the driver's eyes and attention to be diverted from the road. While some have suggested that hands-free voice dialing has the potential to eliminate risk of operating a cellular phone while driving, these data suggest the risk will be reduced but not eliminated. On another note, Matthews, et al., (2003) found that between speaker-based systems and ear-phone systems, the latter creates higher cognitive workloads and frustrations for users. However, in contrast to uttering a brief command into a voice-controlled radio, speaking into a ear-phone piece would require more thought, and the output is longer than a single phrase (as in a voice-controlled radio).

Another consideration when concerning the visual and cognitive demands of voice interaction is that it is important to provide clear distinctions between speaking to a passenger and speaking to a party over a cellular phone. The passenger lends an extra pair of eyes to monitoring the traffic environment, warning the driver of hazards and changes and even assisting in navigation when necessary. Very often passengers regulate the difficulty in their verbal speech and articulation to drivers' abilities for information processing and context. For example, very seldom do passengers discuss arithmetic problems with drivers. Instead, they observe silence and changes in driver mannerism, based on surrounding effects. In contrast, the person the driver is speaking to on the phone is visually and mentally unaware of the surrounding traffic situations and continues to converse.

In summary, there are immediate concerns for reduced perception of affordances, increased visual, manual and cognitive demands and safety risks, when using cellular phones with respect to all types of use (reacting to calls, manual manipulations and conversations) performed concurrent to driving. However, in the interest of drivers, manufacturers and economies, there is also considerable room for re-design and support of such devices, in a manner that minimizes undesirable effects such as distraction, overload and accidents.

8.5   USE OF IN-VEHICLE NAVIGATION

Similar to telemetric devices, the growing popularity of in-vehicle route guidance/navigations systems also urges the need to accord safety and usability, at least at equal importance to functionality. To begin, the intentions of such a system are aimed at providing drivers route guidance information and to enable travel to a selected destination via the most efficient route (Ashby, 1993). Apart from reduced travel time and cost, Dingus and Hulse (1993) and Warnes, et al., (1993) have suggested that in-car navigation systems could also reduce decision-making and potential stress by allowing more attention to be devoted to the primary driving task, thereby improving safety.

Fundamentally, all in-vehicle navigation systems provide drivers visual and/or auditory information in concurrence with their highly perceptual driving task. Since driving is characterized to be 90% visual in nature (Sivak, 1995), any additional workload imposed by screen-based guidance, such as by route navigation systems, could be detrimental to safety. As a countermeasure, several auditory interfaces have been developed to convey route guidance information. Verbal guidance could be easily processed or interpreted compared to written instructions or symbol notations (Verwey, 1989) and shown to be actually more preferred during driving (Burnett, 1996). However, verbal route guidance may inhibit the development of cognitive maps and prevent the normal representations of the area.

8.5.1 Accident Analysis

As mentioned previously, research concerning in-vehicular navigation systems to date has focused on cognitive and human factors issues predominantly around broader topics regarding types of information, timing of information exchange, methods and modes of how information could be presented and system performance. However, identifying the greater potential risks of a navigation system, if there are any, is extremely difficult.

Other reasons for the lack of crash information were restrictions to the design of navigation systems near the turn of the last millennium and restrictions that were not always applied to the aftermarket navigation products. With low market penetrations back then, too few cases of incidents were available to serve in statistical analysis. OEMs of the future may develop crash investigators to detect effects of in-car navigation systems but until then, we have to regard crashes arising from navigation systems as rare events.

Table 8-3. Navigation system-related accidents in Japan

Year	Cellular Phone Use	
	Injuries	Deaths
1997	117	1
1998	131	2
1999	205	2

For now, the primary source of accident and crash statistics is based on the Japanese National Policy Agency archives. Table 8-3 shows injuries and deaths related to the use of navigation systems while driving in Japan from

1997 to 1999. Although the number of deaths in most instances is small (one death per hundred injuries), there are more concerns for injuries.

One other earlier risk assessment conducted by Katz, et al., (1996) used Ali-scout and Pathmaster navigation systems over a four-segment route around Troy, Michigan. Although there were no critical incidents while driving using the Pathmaster system, there were four incidents of lane changes (without looking) using the Ali-scout system. Unfortunately, there were validity issues when comparing these two interfaces. The study suffered from the lack of baseline data as there were only five Pathmaster systems tested in comparison to over fifty for Ali-scout. Though the experiment was large and expensive, substantial safety inferences cannot be made.

8.5.2   Types of Tasks

As previously mentioned for cellular phone use, tasks and task characteristics that result in overload should be considered, so as to determine what actions to take on future systems and which opportunities of overload can be minimized. To reaffirm the concerns over eyes-off-the-road and mind-off-the-road, we shall examine tasks such as destination entry; destination retrieval and route following tasks performed in concurrence to driving and discuss concerns arising from such tasks.

Table 8-4. Driver Tasks and Crashes (January—November, 1999)

Navigation System	
Results	Crashes
Looking	151
Operating	46
Other	8
Total	205

- **Destination Entry**
  Destination entry is regarded as a challenging task, depending on the method used (Paelke, 1993). If destinations are entered through manual controls and visual displays, then they are chances for drivers to have their eyes off the road. While one alternative to lessen the shift in attention is to provide a HUD to solve problems with the direction of gaze, there still may be an issue of locus of attention. Other means of such recovery would be to allow destinations by voice, as in the case for many systems in Japan, the only ITS production that is currently in quantity. However, as seen in Table 8-4 (similar to Table 8-2 data from the National Police Agency of Japan taken between January and November 1999), these may not necessarily solve the problem. Note that operating a navigation systems (mostly,

destination entry) has lead to approximately one-third of the total number of crashes arising from the use of a navigation system during driving. Voice entry remains questionable and unresolved.

- **Destination Retrieval**

Destination retrieval typically requires two or less keystrokes to lead to a stored list, which makes information readily available either after keying the first few characters of a destination name or scrolling a list for it. Currently lists are arranged alphabetically or by order of use (last destination retrieved appears first).

- **Route Following**

During route following, drivers are prompted by arrow-like displays and a supplemented voice messages as they are guided turn by turn. If timing of voice messages are not in sync with display arrows or too authoritative or commanding, there is a possibility that drivers shift their minds off the road, put more faith in direct instructions and act without checking the surrounding traffic situation (in visual terms, eyes-off-the-road). The human factors data suggest that eye-fixation times are of moderate duration except when using map displays, owing to higher visual demands (Dingus, 1995). As seen in Table 8-4 crashes arising from the task of looking at a map are most common in Japan between January and November 1999. This may be because, unlike the maps in North America which are simpler and have less visually demanding turn-by-turn displays, maps are more critical in Japan because of their non-grid like nature of road networks, lack of street names and identification of street addresses in chronological rather than spatial order. The consequence of having to scan a map results in greater opportunity for eyes off the road.

- **Information Retrieval**

Apart from the information-rich Japanese navigation systems (also known for terms like "infomobile"), there are still few systems worldwide that support information retrieval. It is crucial to determine the types of potential tasks drivers perform, which require information retrieval, and how such information could be used when navigating to an unknown destination. It is also important to define guidelines on the types of information permitted for retrieval, during driving and when stationary.

8.5.3 Navigation System-Related Risk

Tasks that are generally considered to be acceptable, such as activating the wipers, adjusting the fan speed, and so forth, have task times more than an order of magnitude less than those for navigation data entry (Green, 1999). A study conducted by Tijerina, et al., (1998), illustrates the implications of task duration for destination entry and sustained looking onto the interface (or eyes-off-the-road). In it, drivers are made to drive on an oval test track with traffic, while using four commercial navigation systems (Alpine, Delco, Clarion and Zexel navigation systems). All sixteen subjects (eight drivers aged thirty-five or less and eight over fifty-five), were given practice using each navigation system prior to the experiment. Table 8-5 shows a summary of their performance data for the use of several navigation systems as well as cellular phone use for comparison.

Table 8-5. The comparison of task duration for destination entry with different navigation systems

	Group	Alpine Navigation	Delco Navigation	Clarion Navigation	Zexel Navigation	Cell Phone
Driver Interface Type		Hand-Held Controller	Manual Entry	Voice Operate	Scrolling List	Manual Dial
Performance measure						
Trial Time(s)	Young	78	56	74	70	22
Point of Interest (POI)	Old	158	98	76	140	30
Mean eyes-off-road	Young	60	44	24	35	18
Time(s), POI task	Old	121	78	26	108	16
Mean Glance to Device(s)	All	2.6	2.7	1.1	2.7	3.2
Mean Lane Excursions/Trial	All	0.88	0.24	0	0.99	0.07

First, the time (trial time measured in seconds) to select a point of interest (POI is a relatively easy destination-entry task) took about a minute with every navigation system available. Also, POI entry time was about three times longer than when dialing a cellular phone, despite ironically, the latter has been accused as the cause of major distractions.

Another finding was that drivers operating these devices devoted two-thirds to three-fourths of the total task time looking at their device and interfaces (or eyes-off-the-road) when executing a strictly manual (Delco) destination entry task. In addition, drivers performing manual entry using full-featured navigation systems (Alpine, Zexel) averaged, almost one line excursion per entry sequence. This is totally unacceptable per driver safety regulations. With regards to navigation systems that enabled voice entry (Clarion), drivers still tend to look towards the speakers or microphone (eyes-off-the-road) about one-third of total task time.

Task risk could be associated with task time or in other words, eyes-off-the-road. Looking from the perspective of task duration alone, information referral or other types of prolonged looking and destination entry driving/or operating (both tasks are equally complex) while driving could be deemed as our top two concerns, but when considering task frequency or others based on accidents statistics in Table 8-4, route following (visual and/or auditory turn-by-turn guidance) and destination retrieval could just as much impose higher workloads on drivers, which in turn is also equivalent to our safety concerns.

In conclusion, there is great excitement about the growing functionality of ITS/telemetric applications and the extent to which systems such as these have potential to enhance driving experience. However, if safety and usability are not accorded at least with equal importance to functionality, the law of diminishing returns could possibly take effect, and drivers could suffer the consequences of higher risk and mental overload during driving. Numerous driver interface-induced accidents have been shown to occur because of the use of cellular phone and navigation systems. Other similar and new applications entering the market could also lead to accidents because the driving environment, drivers and tasks performed using new devices are similar to those for phones and navigation systems.

In order to keep safety, human factors and cognitive considerations at pace with electronic development, research has to begin by revealing types of tasks and circumstances under which using a device could directly or indirectly increase the risk of distraction, reduce performance and expose vehicle occupants and other road users to accidents.

8.6   SYSTEM DESIGN AND EVALUATION

As we have discussed different aspects related to a driver's behavior, cognition and interaction with IVCS in the earlier part of this chapter, here I would like to propose the general principle for the system design using user-centered approach strategies.

8.6.1   System Design

According to the user-centered design principle, information about the requirement of the system, identifying the user group and their characteristics and the environmental factors should be established early in the design process. Human factors professionals should be involved throughout the design process. As with all user interface design endeavors, involving users in the design process throughout the lifecycle of a speech application is crucial.

- **System Requirements**

Detailed design issues that need to be considered vary with the type of system or systems being developed and their functionality, e.g., what do drivers need systems to do; what functionality must be provided by the system to meet these needs; and what functionality must be provided by the system to avoid misuse by the driver.

At the very early stages of the design, the users can help to define application functionality, which is critical to speech interface design. They also can provide input on how humans carry out conversations in the domain of the application. Natural dialogue pre-design studies are an effective technique for collecting vocabulary, establishing commonly used grammatical patterns and providing ideas for prompt and feedback design.

- **System users**

Information about which drivers the system is intended to help needs to be taken into account in the design process, and the system should ideally be designed with all users in mind, whether male, female, young, old, able bodied or disabled. While some systems are intended for all drivers, some will target particular groups.

It is important to appreciate that drivers differ markedly in their physical, perceptual and cognitive abilities and systems need to be designed with this in mind, e.g., the elderly. In addition, attitudes and emotional states will vary between drivers and these may critically influence their behavior.

Drivers do not perform consistently while driving; they experience lapse of concentration, suffer from fatigue and stress (whether social, work or journey-related) and sometimes consume alcohol. In contrast, drivers or developers undertaking an on-road assessment of a new system may be highly motivated and focused and consequently unrepresentative of the wider user group. It should also be recognized that driver behavior during initial use e.g., testing may be very different from that adopted after the driver has habitually used the system for a number of months.

- **Road and traffic conditions**

The designer should recognize that the system would not always be used in the conditions for which it was originally designed and that conditions on the road and traffic can vary markedly.

- **Environmental conditions**

Designers need to consider the range of environmental conditions in which their systems are likely to be used, e.g., noise from the passenger.

8.6.2   System Evaluation/Assessment

Designers should recognize that evaluation is important at all stages of design, both for improving design and reducing potential liability problems in the future. A formal evaluation methodology or procedure should ideally be applied to ensure a continuous and consistent assessment schedule.

It is advisable that user assessments (ideally with inexperienced users) should be conducted at an early stage in order to highlight unexpected circumstances of system user and misuse. It is advisable to conduct usability trials using different categories of the users, both "naïve" drivers as well as "experts," who might have experience with similar systems.

- **When does one evaluate?**

If the system uses elements that have not been tested in vehicles before, it is advisable to evaluate the system's safety and usability early in the design process and again when a prototype system is available.

In all cases, it is advisable to plan and undertake assessment trials throughout the design process. Tests should be conducted as soon as the first prototype systems are available. Final tests must also be done to ensure the systems are safe for use while driving. In some circumstances it may be possible to test pre-prototypes using mathematical or structural models, mock-ups and simulated driving techniques.

- **Who should be involved?**

Designers should involve suitably qualified and experienced ergonomic/human factors personnel at all stages of any assessment conducted, either utilizing appropriate internal resources or engaging the services of external consultants where necessary. Potential IVCS "end users" should also be involved. The amount of training and/or experience that these drivers have with the system will depend on what exactly is being tested. If it is the comprehension of the system instructions provided that is being assessed, then the users should be inexperienced. However, if the trials demand driving in busy traffic, then a certain amount of familiarity with the system would be advisable.

Both male and female drivers, and the young and elderly should participate. The required number of drivers participating in the evaluation depends on the salience of problems being investigated and their probability of occurrence (Lewis, 1994). It is unlikely that a system evaluation would be effective using anything less than ten to twelve end users (Nielsen, 1993).

- **Evaluation method**

    In human-computer interaction, there is a vast array of methods developed to assist in evaluation activities within the design process. It is common for methods to be classified primarily by the "who" factor, that is, who is providing the evaluation data e.g., users, experts, theories or best practice. Methods can also be classified according to the "why" factor, that is their objectives (diagnostic, summative or certification) and the "when" factor (their applicability within the product design cycle). A further factor may be added which is "what," as in what type of data are provided by the evaluation (quantitative, qualitative or both).

    A few methods will be discussed in this section. The first method considers the ergonomic requirements of the specific IVCS design. A second method assesses the IVCS "in-situ" (but not necessarily on the road) according to human and system performance criteria. Other methods involve the use of focus groups, mathematical models and user trials. The most appropriate method to use will depend on when it is being applied in the design process and what aspect of the system is being considered.

    The first method is essentially a test of how well the system meets a set of design principles, research outcomes and recommendations, which may be performance criteria set by the producer. It may be based on such requirements as the anthropometric and ergonomic standards for physical sizing, locations of controls and the labeling and display of information and is usually conducted by human factors experts against some pre-specified objective criteria. There are different systematic approaches to these expert evaluations such as TRL/DETR IVIS Checklist (Stevens, 1999) and the Heuristic Evaluations method (Nielsen, 1993). These expert methods can be used very early in the development and can be task-based or holistic.

    In a heuristic evaluation, an interface design is critiqued based on a checklist of design guidelines or heuristics. Such lists of guidelines enable designers to apply discount usability engineering methods to SUIs such as obtaining guidance in early stages of design. Nielsen's ten usability heuristics are too general to be useful for avoiding the specific pitfalls of speech interface design. Few published guidelines for speech interface design are not compiled in a format that can be used in a heuristic evaluation. They do not translate well to evaluations of systems for drivers as these traditional HCI methods were conceived with desktop computers in mind, and often within an office context. The driving context (e.g., users, tasks and environments) provides a range of complexities that necessitate new methods or at least adaptations of existing methods.

The second method evaluates the system more quantitatively with respect to driver and system performance. Although there is no single indicator for acceptable usability or safety performance, this method will need to take account of issues such as reliability, validity and sensitivity (ISO/DIS 17287, 2000) (Stevens, 2000). Reliability can be indicated, for example, by two types of system errors; errors of omission where a stimulus is not detected, when the "hit" rate for stimulus detection should be high and errors where a response occurs when no stimulus is present (false alarms) when the false alarm rate should be low. Some examples of reliable and validated measures are provided in Table 8-6.

Table 8-6. Measures of IVIS Safety and Usability Performance

IVCS performance	Driver and vehicle performance
Efficiency	Driver
• Number of button presses	• Eye movement behavior (e.g.,
• Number of errors	mean and maximum glance
• Task success rate	duration, glance frequency, eyes-
• Task completion time	off-road time)
	• Situation awareness
	• Reaction time to events (e.g.,
	peripheral detection task)
Driver workload	Vehicle
• Subjective rating/attitudes (e.g.,	• Lane position variance
usability and usefulness ratings)	• Unplanned lane departures
• Psychophysiological measures	• Steering reversals
(e.g., heart rate and heart rate	• Steering and speed entropy (ie.
variability)	unpredictable patterns)
• Secondary task performance	• Mean speed, speed variance
	• Minimum headway and headway
	variance
	• Minimum time to collision
	• Number of critical incidents and
	crashes, speed on impact

Designers need to recognize that system assessment should never jeopardize the safety of the participants (subjects), the evaluator or the general public. With this aim, desktop, laboratory or driving simulator assessments, e.g., Wizard-of-Oz, may be used to conduct preliminary assessments and identify more serious safety and usability concerns before conducting trials in real road situations. In Wizard-of-Oz studies, a human "wizard" (sometimes using software tools) simulates the speech interface. Major usability problems are often uncovered with these types of simulations. Usability tests of speech applications tend to be different from usability studies involving graphical interfaces. For example, it is difficult to have a facilitator in the room with the study participant because any human-to-human conversation can interfere with the human-computer conversation, causing recognition errors. It is also not possible to use the popular "speak aloud" protocols, as the voice channel in many speech applications is needed exclusively for speech input.

Ideally, final safety assessments should be performed during road tests. It is also necessary to tailor any assessment with respect to the existing situation.

8.7   FUTURE WORKS

The technical industry is rapidly developing new communication devices for in-vehicle use, for example, Internet browsers with a decreased visual and increased auditory output. Taking these developments into account strategies for coping with possible negative effects of in-vehicle communication have to be developed. Driver assistance systems may be a promising means to this end as they are designed to relieve the driver from some workload and supervise and warn the driver if a dangerous situation arises that requires additional effort.

Various types of driver assistance systems have been implemented. One of them is the collision or safety warning systems, using sensors to detect hazardous driving conditions and then processing the sensor outputs to determine when the driver needs to be warned. The warnings could be auditory (tones, buzzers, synthesized speech), haptic (vibration or torque applied to steering wheel, vibration or pressure to gas pedal or seat cushion), kinesthetic (application of brake pulse) or visual (lights on instrument panel, in mirrors or head-up display). The auditory, haptic and kinesthetic warnings could be very effective at catching the attention of a distracted driver if they are well-designed to elicit the "correct" emergency response from the driver. The visual warnings are less likely to help since the distracted driver is not necessarily going to notice them.

Another type of these systems provides control assistance to the drivers, which presents a more complicated picture relative to driver distraction. The most prominent of these systems is adaptive cruise control (ACC), which uses a forward ranging sensor such as radar to measure the distance and closing rate to the leading vehicle and then uses that information to adjust the speed of the equipped vehicle so that it maintains an "appropriate" separation behind the leading vehicle.

An experimental study is done concerning an evaluation of advanced driving-assistance systems using methods for estimating workload levels. The effects of such systems on drivers' mental workloads and driving performances were measured experimentally using the driving simulator. Six subjects were instructed to drive the simulator in a highway environment with and without Adaptive Cruise Control (ACC) and/or the collision-warning system (CWS). To assess the effectiveness of these systems on drivers' performances, the subjects were asked to calculate sums of single- or double-digit figures displayed. The results show that higher accuracy was obtained under a condition with ACC than without it. To estimate the subjects' mental workload levels, their electrocardiograms and respiration data were recorded during the sessions and the RRI, heart rate variance and respiration frequency were calculated. The results indicate that the provision of the CWS and ACC reduced the subjects' mental workloads compared with the situation without the systems (Takada, 2001).

The ACC systems may be able to improve safety by encouraging drivers to follow at somewhat longer separations from other vehicles than they do today, and they may be able to reduce rear-end crashes caused by inattentive drivers overtaking slower vehicles. However, if drivers become overly reliant on the ACC and do not really understand its limitations (inability to sense stopped vehicles, road debris and animal intrusions and inability to respond to aggressive cut-ins or abrupt stops of preceding vehicles), it has the potential to exacerbate the driver distraction problem. This could even encourage drivers to engage in more non-driving tasks than they do now while driving, which would be most unfortunate.

Further research is looking into more natural human-machine interaction, such as dialogue managers assisting in preventing the presentation of information at inappropriate times to reduce driver distraction and overload. For example, a hard-braking situation would be an inappropriate time for the system to tell drivers they have an e-mail message. The dialogue manager should block information when it detects that the driver is too busy or will be occupied with more important tasks. These same systems can also learn patterns of driver behavior and preload remote information so that it can be instantly presented whenever the driver requests.

8.8   CONCLUSION

To utilize these technologies in the field, understanding the states and behavioral characteristics of drivers will become more important and applying user-centered design concepts will be required to design an effective speech user interface.

Therefore, though the use of speech while driving may have adverse implications for driving safety, it was found that the drivers adopted a few approaches to reduce the cognitive workload caused by additional conversation by slowing down and increasing the headway. Therefore human beings may deal with such higher task demands by investing more effort, performing behavioral adaptation, changing working strategies or neglecting subsidiary information (Cnossen, 2000) to adapt to the different working environment. It is also important to be aware that there are external factors that may influence the implications of speech while driving, which are the real world constraints, such as other traffic participants and different lighting conditions.

With the car navigation systems becoming popular, much more information will be made available to drivers, which may overload them. Thus, in order to develop safe and user-friendly in-vehicle information systems, the in-vehicle systems should be carefully designed so that, although with more information being made accessible to the driver, they should not overload his information-processing resources and, most importantly, the driver should be doing his job—driving safely on the road.

Chapter 9

SPEECH TECHNOLOGY IN MILITARY APPLICATION

9.1   INTRODUCTION

In modern military activities, information overflow is a common problem. It is extremely important for the operators to have the right information presented in the right form, at the right time and leave the right space for the operator to react to it. So the human-information system interface design has turned out to be the most critical part for the successful activities of the operators.

For any system control, the interaction between human and machine means that the human has to engage part of his body (arms, fingers, legs, ears and eyes) in the performance. Sometimes such interaction is impaired due to illness, stress or tiredness, especially in the military operation situation. Speech may help with some specific conditions such as control of the system, reducing workload and training acceleration, or even increase the situation awareness.

Nowadays, most of the control functions are performed manually and displayed visually. At the same time, there is a strong interest in implementation of effective non-conventional "Alternative Control Technology" (ACT), such as voice or gaze control, which would ideally let the operator intuitively use his own communication strategies, then generate considerable benefits (Leger, 1998). Among all different types of ACT are touch screen, touch pad, head tracker, eye tracker, direct voice inputting, gesture, EMG and EEG, direct voice input, or in a larger scope of speech technology, that has constant interest and is commonly presented as the most mature ACT.

Speech technology has been of special interest to the military. The arguments in favor of using speech technology are as follows:

a) Speech is a natural way of information processing.
b) Speech technology allows maximum hands-and eyes-free operation.
c) Use of additional unused processing capacity.
d) Separation of information and information sources.
e) Increasing power and sophistication of speech recognition technology.
f) Reduction and redistribution of the workload for well-defined tasks.

In general, speech and language technologies are useful to all military application areas. The differences between the military application and civil applications have the following characteristics (Steeneken, 1996):

a) Adverse military conditions such as poor radio channels, noise, vibration, G-force and stress and high workload with multiple task performance.
b) Multi-lingual speech and language technology is particularly important because of the multi-national, joint or coalition military operations.
c) Military application requires special security, robustness against noise and jamming and limited bandwidth channels.
d) Speech communication systems must operate with high performance for native and non-native speakers.

Military people have a long history of interest in speech technology. The NATO research study group on speech processing (AC243 Panel 3 RSG.10) was already established as early as 1978. A lot of experiments and surveys focused on military applications of language processing have been carried out in this organization (Steeneken, 1996). The application of speech technology into different military areas has been proposed, and some of them have been tested time after time in the real-life working environment. Compared to civil applications, tactical communication has a restricted vocabulary, normally consists of short utterances and limited to the task at hand with well-trained users.

In general situations, the application in the military sets higher demands on the interaction design and usability the system. Central questions when considering of application of different speech systems or products can be summarized as follows:

a) What accuracy will the user expect?
b) Is the speech recognizer robust enough to meet the expectation of the user in different application domains? Does the user trust the system?
c) Where and when shall one use the speech input/output system to enhance the system performance?
d) How does one design the interface so that the speech technology system can be performed in a natural way?
e) How does one correct the error made by voice so it will not irritate the user?
f) Does the benefit of using speech recognition in this application outweigh its cost compared to alternative technology?

Speech technology used in military environments must be robust enough to handle adverse conditions. This is because many military situations involve difficult acoustic and mechanical environments, such as high and variable noise, vibration and G-force, while the operators are under high stress levels. Normally under such conditions, human beings are easily make speech errors, and ASR recognition accuracy is decreasing dramatically. The distance between expected accuracy and the possible accuracy of ASR performance in military environments has become the key factor that limited the applications. Besides, the six questions related to the implementation of speech technology for military application are all interaction design and human factors issues. A better understanding of the task, the constraints from the domain task and the environment and the user's characteristics and behaviors are important to lead the research work and application work.

9.2 THE CATEGORIES IN MILITARY APPLICATIONS

Speech technologies are constantly improving and adapting to new requirements. Many technologies have been developed or enhanced in recent years. Some of them can now be integrated into a wide range of military applications and systems:

a) Speech input and output systems can be used in control and command environments to substantially reduce the workload of operators. In many situations, operators have busy eyes and hands and must use other media such as speech to control functions and receive feedback messages.
b) Large vocabulary speech recognition and speech understanding systems are useful as training aids and to prepare for missions.

c) Speech processing techniques are available to identify talkers, languages, and keywords and can be integrated into military intelligence systems.

d) Automatic training systems combining automatic speech recognition and synthesis technologies can be utilized to train personnel with minimum or no instructor participation (e.g., air traffic controller).

The following sections outline military applications and associated specialization and design requirements in six areas (Steeneken, 1996): a) Command and control; b) Communication; c) Computers and information access; d) Intelligence; e) Training; f) Joint forces.

9.2.1   Command and Control

Speech technology can help Command and Control systems work more effectively by introducing multimodal interaction systems and ASR control. But this application requires a high performance of speech and language technology in real time, under adverse conditions, including motion and noise, with multi-lingual input and output.

The required speech technologies for command and control are speech coding, speech enhancement, speech synthesis, speech recognition, language understanding, interactive dialogue system, multi-model communication and 3-D sound display (Steeneken, 1996). Human factors issues for the implementation of speech technology into command and control application should be very critical.

To design the interface, the commonly asked questions can be how does one design the commands for different functions? How many commands can the operators remember during stress situations? How does one design the feedback so the user knows that he has issued the right commands and the system understand him correctly? As both human speech and ASR can make errors, when errors happen, how can they be corrected in an easy way? When we consider the multimodal interface, then the first question is which modality should be used for what functions? How are different modalities combined to give the most natural and robust performance? All of these questions have to be answered under each design and application context. The requirements of the speech interface design for the air traffic control system and for the devices that the foot soldiers carry in the battlefield are very different.

9.2.2   Computers and Information Access

Computers and information access are a crucial part of modern military operations. In all levels of military operation, the quantity and range of all types of available digital information have increased dramatically. They include not only the text and numerical numbers, but also sound, graphics, images and video. It is highly necessary for the operator for entry and retrieval of information. Nowadays, manual input and visual output are the most common designs for the interface between human and computer. This type of input/output requires the use of the hands and eyes, which in military operations are often busy with other tasks. For example, it would be difficult for a forward observer to watch the objective and at the same time enter data via keyboard. The design of the human/computer interface can be a bottleneck for the capability to access and process information. For such military applications, speech recognition can provide a useful input mechanism while speech synthesis can provide a useful output mechanism (Steeneken, 1996). The technologies required for this application are multi-lingual speech recognition and synthesis.

Military-related spoken dialogue system design research is not as popular compared to different civil spoken dialogue system studies that we discussed in Chapter 7. One of the problems may be due to the security reasons. Military-related information is hardly open to anyone.

9.2.3 Training

Well-trained personnel are imperative for the success of military missions. The training of forces for military operations can be significantly aided by applying speech technology to allow people to interact with advanced simulation systems by voice. New applications of speech and language processing are able to support training by using computer-supported learning systems; production of computer-aided test facilities; and production of didactic material. In addition, for joint (coalition) operations, training in foreign languages is essential; such training can be aided by utilizing speech and language technology to provide machine-aided foreign language exercise and tutoring for military personnel (Steeneken, 1996). Speech and language technologies can be integrated at all levels of simulated training. In particular, verbal man-machine dialogue, man-machine interfaces, voice-activated system command and feedback will play a growing part in training activities. The potential benefits of introducing speech and language technologies are to reduce training time and costs; increase the applicable number of languages; improve proficiency levels; make the telelearning available; and maintain language knowledge.

9.2.4   Joint Forces at Multinational Level

The change in the international military landscape results in a frequent use of multinational joint task forces. One of the main problems is foreign language understanding. Joint forces operations require the coordination of forces speaking multiple languages. Here speech and language understanding and translation technology have great potential to increase the efficiency of operations. The major benefits of introducing speech and language technologies in a multinational military environment are:

a) Enhancing the mutual understanding process.
b) Increasing the speed of operational exchange of multilingual information.
c) Reducing the reaction time of operational units in critical situations.
d) Speeding-up of foreign language acquisition and learning.

However, the demands on the technology are high, and initial applications probably need to focus on limited domains for translation and multilingual information exchange.

9.3   APPLICATION ANALYSIS

Integrating speech into a system is not trivial. It has to be taken into account at the very beginning of the design of the global interaction between operator and system. Each potential application has to be properly analyzed in order to know if and to what extent speech can help, given the potentialities of speech technology. A thorough human factors analysis is therefore required (Steeneken, 1996). The design considerations to be taken into account may include:

a) Why use speech? What are the benefits of speech for the operator: improvements in accuracy, response time or workload decrease? In particular, the benefit of speech versus any other means of interaction has to be assessed.
b) For what functions should speech be used? Which interactions are to be mediated through speech, bearing in mind the impact of security, workload and usefulness since it is not necessary to build a function which is 100% speech-based if it is seldom used?
c) For which speakers should speech be used? This implies knowing whether one or several persons will use the system. Consequently, recognition in either speaker-dependent or speaker-independent mode has to be chosen. It is also necessary to know the intellectual level of the

users, since it has an influence on the choice of the means of interaction between the user and the system.

d) Benefit versus costs: What will be the total cost for the system? This cost should include not only design and development costs but also operator training and maintenance costs.

There are several potential positive aspects regarding using a speech recognition system in task performance in a military area: It is a natural way of presenting information; it allows hands-and eyes-to be free and does not require typing skills. It may reduce the mental workload and save time in different situations. At the same time, there were two weak points regarding applying speech recognition technology: unreliability and lower error tolerance for the user. These two weak points have strongly limited the application of speech technology. The unreliability of the ASR is caused by stress and workload and can influence the accuracy of recognition.

However, speech can be very useful to control discrete and non-critical functions, e.g., radio channel selection, check list control, data input and management. In these non-critical tasks, a recognition performance of 95% is in many situation satisfactory and better than human performance (Steeneken, 1996).

Automatic speech recognition used in military environments must be robust to adverse conditions. This is because many military situations involve difficult acoustic and mechanical environments, such as high and variable noise, vibration and G-force levels. Together with the adverse environmental factors, other psychological factors such as stress, fatigue, multiple-task performance and divided attention also need to be seriously studied.

In military situation, there are such situations as the operator's hands and eyes being occupied with control and monitoring functions, the operator having difficulties typing on the keyboard because of vibration, or having different type of gloves on, or the lighting is not bright enough, or the integration of different information is carrying from many sensors, or input data while observing at the same time, etc. The introduction of ASR provides an additional channel for control, data input and allowing the operator to have better performance, at the same time reducing the workload on hands and eyes. At the same time, the awareness of the situation is increased.

Although ASR has a number of operational applications in a military environment, many potential applications are still under development. The environmental factors, stress and technical limitations play a major role in influencing the successful use of ASR technology.

There is very little study on the integration and interface design regarding the application. A few basic questions have to be answered for the interface designers:

a) What accuracy will the operator expect? One has to accept the fact that the ASR system can never reach 100% recognition accuracy under certain conditions.

b) What is the operator's expectation to this speech system? Does the operator trust the system? Or will it make the operator over-trust the system?

c) Is the speech recognizer accurate enough for the task? Can error rate in speech input be comparable with manual input?

d) How much is the operator's workload reduced and the efficiency increased through the use of the ASR system in the tasks?

e) Does the benefit of using speech recognition in this application outweigh its cost compared to alternative technology?

There can be the problem of integrating speech technology into any complex control system. New types of equipment within the physical constraints of space and weight and the psychological constraints of maintaining maximum consistency in the mode of interaction with different subsystems must be considered.

In military applications, highly dynamic environments, noise, high workload and stress will have negative effects on the speech recognition systems, since either the operators could not talk properly, or the machine was affected by the environment, and even highly robust ASR systems can hardly reach a very high recognition rate. It is relatively hard to find the regulation between workload/stress and the speech changes. Designers have to find out other solutions to compromise the problem.

There are solutions that can be introduced the speech understanding system (Anderson, 1998). Many speech understanding systems attempt to make use of several different areas of knowledge about the speech and the situation in which it is being used. The system recognizes some key words instead of the whole commands, so the operators do not need to remember the commands exactly as they were trained. The second method is to introduce the intelligent and adaptation system. By integrating the possible information about the activity situation of the system and the environment, the system can provide certain suggestions about the right commands or stop some illegal or unsuitable commands.

Another way is to apply the alternative control, or set up the back-up systems. Speech control works in parallel with manual control; when the speech recognition rate is decreasing due to stress or a hazardous environment, the operator can have the opportunity to switch into manual control as he did before.

9.4 COMPARISON BETWEEN SPEECH INPUT AND MANUAL INPUT

The comparison of using a keyboard or using speech for data inputting is interesting in military applications. Quite a few studies were carried out in the '80s and early '90s. The results were sometimes contradictory to each other. This is due to what they compared under what kind of condition. Was the comparison carried out in the situation of hands/eyes busy, or in an office environment? Do they compare the task performance in total, or typing or saying exactly the same commands? Do they only consider the error rate or even the error correction process? It is not so simple to say speech input is better than manual input, or vice-versa, but it is important to understand this issue in military applications.

9.4.1 The Argumentation between Poock and Damper

Poock's (1980; 1982) study is notable for the degree of superiority which it purported to show for speech over keying. As a primary task, his subjects entered simulated military command and control instructions on the ARPANET. Speech was found to be 17.5% faster than keyboard entry for the primary task, while there were 183.2% more errors for keying than for speech. Also, speech input allowed subjects to transcribe 20% more weather report information as a secondary concurrent task than was possible during manual entry. This work has been widely quoted in support of the view that command and control is one of the specific tasks in which ASR holds clear advantages.

This interpretation was questioned by Damper and Wood (1995), who argued that Poock's experimental methodology embodied an implicit bias toward speech and against keying. The putative flaw is that while Poock's verbose commands had to be entered character-by-character when keyed, they were spoken as single (whole-phrase) utterances. Damper and Wood believe that there is no sensible reason why verbose commands need be keyed in full, on the grounds that acronyms, key assignment or abbreviation completion constitutes the "natural" language for a key-press interface. They compared speech and keying in experiments modeled on those of Poock but using terse commands. Since Poock used a military, classified system, some details of his work (e.g., the precise command vocabulary) are not reported, so that a faithful replication of his experimental conditions is not possible.

Where there were differences, Damper and Wood attempted to make these favor speech to provide a maximally stringent test of their hypothesis that speech input was shown in an unduly favorable light in Poock's work. Contradicting Poock's results, Damper and Wood found that speech was 10.6% slower (not statistically significant) and 360.4% more error-prone than keying. This was interpreted as strong support for the hypothesis that Poock's methodology was flawed. They conclude that a fair comparison of input media requires that the experimenter explicitly attempts to minimize for each medium the number of user actions necessary to elicit a system response, just as one would do in a practical interface design.

The alternative explanations of the differences between Poock's results and Damper and Wood's are that Poock used a dual-task paradigm. As well as the primary (command and control) task, his subjects were given a secondary manual task (transcribing weather report information) to perform in system-idle time. Since Damper and Wood (1995) used a stand-alone PC as host, there was no system-idle time due to network delays, and so they omitted the secondary task. However, speech input is generally considered to be advantageous in such situations (Damper, 1984; Martin, 1989; Mountford, 1980; North, 1976; Wickens, 1983) by allowing a classical separation of modalities. Thus, it is possible that the absence of concurrent task played a part in the observed differences.

Damper and Tranchant (1996) reported another study to determine if the major difference between the omission of concurrent, secondary tasking from their study could explain Damper and Woods observed superiority of keying over speech. In their study, simulated command and control experiments are described in which speech input, abbreviated command keying and full command keying are compared under dual-task conditions. They find that speech input is no faster (an insignificant 1.23% difference) and enormously more error-prone (10.38%, highly significant) than abbreviated keying for the primary data entry task, but allows somewhat more (11.32%, not significant) of a secondary information-transcription task to be completed. Full keying has no advantages whatsoever. They believe that this confirms the methodological flaw in Poock's work. If recognizer errors (as opposed to speaker errors) are discounted, however, speech shows a clear superiority over keying. This indicates that speech input has potential for the future—especially for high workload situations involving concurrent tasks—if the technology can be developed to the point where most errors are attributable to the speaker rather than to the recognizer.

9.4.2   Effects of Concurrent Tasks on Direct Voice Input

Speech input is likely to bring most benefits to data entry tasks where the hands and eyes are busy performing other operations. When performing low-flying operations and maneuvering with manual controls, pilots need to be able to obtain visual feedback from outside the aircraft, and speech has the advantage of allowing data to be entered in parallel (Hapeshi, 1988). The advantages conferred by parallel processing can also be very useful in a wide range of industrial applications such as material handling (sorting parcels, mail or luggage) and quality inspection (e.g., cars and electronic components), as well as shipping and receiving applications.

A number of trials have compared the effectiveness of speech and manual input in different phases of flight (Jones, 1992). Speech was most advantageous during critical periods involving "HOTAS" (hands on throttle and stick), or maps of the earth flying as close to the ground as possible using features like trees to conceal the aircraft. Jones did not refer to any data indication and how such a comparison was conducted. In other situations, manual data entry was often quicker. No systematic investigation appears to have been conducted into the problem of switching between these two input modes.

Because the vast majority of ASR applications will involve speech input while another task is being carried out, the interaction between the concurrent task and speech should be of major interest to system designers. Usually, manual performance is the concurrent activity to ASR. Manual performance can vary in complexity from a simple button press to the control of a fighter aircraft.

Welch (1980) found that button presses were too trivial an activity to affect performance on an ASR device. A more complex task, such as interpreting a sentence and converting it into a series of data fields, did have a detrimental effect on voice entry. Welch (1980) also found that this effect was worse if the primary entry mode was via a manual keyboard than by voice. However, it is not clear if the experimental task used by Welch (button pressing while entering data manually) has a meaningful parallel in real applications. If manual entry is to be used, it is unlikely that operators would need to hold a button down with one hand while keying in data with the other.

Operators have great difficulty using voice and manual input simultaneously when the manual task is complex, as when they are transcribing TV dialogue for subtitling (Damper, 1984). Voice input was introduced to allow the simultaneous entry of subtitle text manually and style specifications by voice. Voice input had a twenty-five words vocabulary for selecting color, position and size of each subtitle frame. The users were not instructed initially to enter both types of information concurrently, and they showed no natural inclination to do so. This could be because they were all trained to work with the previous all-manual system, where the text was typed on the keyboard before style selection was made on a special function keyboard. When instructed to perform the voice and manual input simultaneously, only one of the five subtitles could be done, entering approximately a third of the style parameters in parallel with the text. According to Damper, et al., this reduced the extra time spent on style entry on its own, but increased the total input time by around 10%, owing to a large increase in text entry time attributed to dual task interference. Performance deteriorated on the keyboard task, with many errors occurring during speech input, and the recognition error rate of the ASR device increased by 10%. This was attributed to the effects of stress on speech, but interference from keyboard noise also played a role. Damper, et al., (1984) concluded that concurrent entry of "style" information by voice during manual text entry might be faster for some users but only with significant practice.

9.4.3   Voice Input and Concurrent Tracking Tasks

In general, it can be shown that manual data entry is faster than voice data entry when operators are skilled at using the manual device, but in a dual task application, the difference disappears (Aretz, 1983; Dennison, 1985). There have been a number of studies intended to simulate the dual task conditions of the in-flight aircraft cockpit in order to determine potential difficulties in the use of speech input in aviation. In an experiment reported by Aretz (1983) pilot subjects were required to perform a horizontal and vertical tracking task on a simulator while entering data for communication, navigation and weapon systems. The tracking task was made more difficult by the need to "fly" the simulator within certain speed limits. It was found that pilots made significantly greater errors on airspeed maintenance when data was being entered, compared to occasions when no data was being entered. Aretz argued that voice input was generally at least as good as keyboard input during the flight task—a conclusion supported by Damos (1985). However, Aretz (1983) suggested that voice entry might be slightly better than manual entry if the difference in the time to initiate input was ignored. If the data was complex, manual input was initiated faster than for voice input, but this could be attributed to more practice that pilots had with the keyboard entry relative to speech entry.

For applications, such as avionics, it is important to consider the effects of different types of data input on the primary task, that of flying the aircraft safely. In a simulation carried out by Mountford and Narth (1980), using students for whom both voice and keyboard data entry were novel, it was found that voice data entry had an insignificant effect on tracking performance, compared to manual data entry. In addition, they found that the tracking task disrupted manual input to a greater extent than voice input. Similar findings were reported by Laycock and Peckham (1980). In this last case, the flying task was simulated by asking subjects to track a moving target on a screen by operating a joystick with the right hand while carrying out keyboard entry or recognizer activation with the left hand. Subjects were clearly instructed to regard the tracking task as the primary task. Even so, performance on the tracking task was found to suffer more than performance on the secondary task involving data input. When comparing the effects of voice versus manual input on tracking, it was found that manual input produced a greater decrement in performance than did voice input.

Dual task interference can be particularly severe in the helicopter cockpit during many in-flight maneuvers, when the pilots must keep both hands on the controls (Coler, 1984). Control of the aircraft and weapon operations in the military cockpit is always the primary task, and the operation of communications and navigation systems is secondary. It is these dual task requirements that have led to the argument for the introduction of voice input to transfer some of the pilot's workload from the visual and manual channels to the audio channel (Roy, 1983; Simpson, 1982). This has been supported by empirical observations made during in-flight testing in three different helicopters (Coler, 1984). Coler reports that pilots suffered no distractions due to the ASR task when they were monitoring instruments or looking for other air traffic during flight. Moreover, Coler argued that pilots were able to maintain reasonably high ASR accuracy under high manual workload conditions, such as difficult wind conditions, that almost certainly would have prevented keyboard entry of the same information. Also, pilots stated that they had no difficulty with the ASR device in all the flight modes in which it was used (level cruise or sustained hover).

9.5   AVIATION APPLICATION

As we can see, the speech technology is very attractive in the aviation application. In aircraft, the operator's hands and eyes are occupied with control and monitoring functions. The introduction of ASR provides an additional channel for control, allowing the operator to reduce the workload on his hands and eyes. At the same time, he can increase the situation awareness. Modern aircraft are required to fly under extremely severe conditions (low-altitude, high-speed, night attack, bad weather, etc.). In addition, crews are often reduced to one pilot because of economic constraints. In aviation application, ASR technology is constrained by the less-than-perfect reliability of speech recognition accuracy. Most application studies were restricted to simulations, although some in-flight trials have also been considered for the following functions:

a) *Navigation*: Entering waypoints, latitude and longitude data, and updating routes.
b) *Communication*: Selecting radio frequencies.
c) *Information retrieval*: Calling up displays and controlling speech format.
d) *Weapons selection*: Calling up and activating weapons (with weapon firing remaining under manual control).

9.5.1   The Effects from Stress

Military avionics is probably the most stressful situation in which individuals will be use the speech recognizer. The ever-increasing complexity of aircraft systems coupled with the requirement to operate at very low altitude level and in all types of weather creates a high workload in military cockpit, especially in single seat aircraft. The aircraft cockpit in flight represents a very hostile environment for the ASR device. The effects of stress and fatigue on voice will be particularly difficult to accommodate, and may require a system that can adapt to changes in the voice during the onset of such phenomena.

When considering the application of ASR in aviation, many environmental factors can course the difficulties, such as a) high levels of acoustic noise; b) high vibration; and c) high levels of "G" in maneuvering flight. Also, pilots need to wear protective helmets and mask microphones, which will introduce noise from breathing. The principal sources for noise in the aircraft cockpit will be engines, airborne noise and air-conditioning. Jones investigated the effects of environmental stressors, such as noise, vibration and G-forces (Jones, 1992) at several of the sites visited. In general, flight trials with trained aircrew indicate that physical stressors are less likely to affect recognition accuracy than cognitive and emotional stress. Although G-forces were found to have significant effect on speech, the level at which recognition was seriously impaired (in the region of 7 g) is above the level at which manual input would be affected. There are studies that demonstrated that (Morrison, 1994) the G-force can decrease the subject manual and cognitive performance and the effects can be cumulated.

Although there have been few studies on the effect of stress on speech patterns and automatic speech recognition accuracy, it is generally considered that both physical and task-induced stress represent a major obstacle to the success of speech recognition applications (Simpson, 1985a;

1985b; 1985c; 1986; 1982). Because of these difficulties, the applications of ASR in the aircraft cockpit can only be in non-emergency situations to alleviate the workload or provide a more user-friendly man-machine interface at the present ASR technical level.

To be able to implement the ASR system into a cockpit, one has to answer these questions in the right way. There are three research steps that can be taken:

a) Set up critical aspects for application of ASR in the cockpit. These criteria will be based on the limitation of the ASR technology and the special condition of cockpit tasks.
b) Studying the factors that influence the accuracy of the recognition, and it may give a better understanding of unstable performance of the ASR system, which will assist in overcoming the problems.
c) Integrate ASR in the cockpit. To implement ASR in the cockpit, the syntax design, error correction system and feedback design can be very special compared to other applications. To combine ASR technology with other speech technologies may make the system work more reliably and more effectively.

9.5.2   Compare Pictorial and Speech Display

Two types of emergency information displays were considered for implementation into aircraft cockpits: computer-generated speech and computer-generated pictorial displays. Some applications are clearly better suited for certain display methods. As an example, a map of a strategic air strike area is clearly more effectively portrayed to the pilot via a graphical display than via digitized speech. On the other hand, whether an on-board system failure should be described to the pilot via graphics or via speech, or even both, is less clear. Some guidelines suggest that warning messages should be displayed by speech because the auditory sense is omnidirectional (Simpson, 1985). Other guidelines, especially in the process control area, have emphasized that warning messages should be pictorial so that the display can graphically "zoom in" on the trouble area (Goodstein., 1980; Rasmussen, 1980).

Synthesized speech was being considered in a number of cockpit displays in the '80s (Robinson, 1987) for a variety of reasons. First, it can increase the time to view the visual information outside the cockpit (Butler, 1981). A second advantage has come from basic research on information processing and dual-task performance. In particular, Wickens' theory on multiple resources (Wickens, 2000) predicts that using another input modality for secondary information will incur less mental workload than using the same modality as for the primary flying task. Finally, experimental data have indicated that speech can be responded to faster than pictorial displays in some case. Wickens, et al., (1983; 2000) confirmed the predictions of the multiple-resource theory in showing that responses to speech displays could be time-shared with a visual/manual tracking task with no decrement in performance from the single to the dual task condition for the speech display.

One possible reason that performance with speech displays is better than or equal to performance of graphic displays, e.g., Hawkins, et al., (1983) is that the speech displays use the context inherent in the grammar and the temporal separation of information slots to reduce the uncertainty of later items. The study of comparing synthesized speech and pictorial displays in the flight simulation showed that (Byblow, 1990) speech messages with low-redundancy levels were effective in minimizing message length and ensuring that messages did not overload the short-term memory required to process and maintain speech in memory. System response times were quicker when synthesized speech warnings were used.

Pictorial displays have been hypothesized to be beneficial because many tasks in which the operator must process information about physical systems have a code of representation that is spatially based (Bainbridge, 1974; Mozeico, 1982; Rasmussen, 1980; Wickens, 1983; 2000). The graphics, therefore, provide a means of establishing the compatibility between the display and the underlying code of representation, or mental model, that the operator has of the task (Wickens, 1983). Wickens, et al., (1983) showed that certain kinds of information lend themselves to a spatial display format and can elicit better performance than the same information displayed verbally. They attribute these results to the stimulus/central-processing/response (S-C-R) compatibility. This compatibility occurs for spatial displays when there is a relationship between the spatial configuration of the display, the operator's code of representation of the task and the physical location of the manual response buttons.

Robison and Eberts (1987) compared response performance over single and dual tasks when information was presented pictorially and in speech in two experiments. Pictorial subjects responded more quickly than did speech subjects. The addition of a visual tracking task in the dual-task condition had a differential effect on performance, depending on the modality of the primary task and the rate at which information was presented. The dual task impeded performance more in the fast and medium presentation rates for the speech condition but had little differential effect across rates for the pictorial condition. An analysis of the error data indicated that subjects in the pictorial condition were better able to maintain the context of the emergency than those in the speech condition.

Robison and Eberts' (1987) experiments showed advantages, in terms of response times, of pictorial displays over speech displays for both single task performance and when displayed concurrently with a visual tracking task. This result would not be predicted by a multiple-resource theory (Wickens, 2000), because more resources should be available to perform concurrent tasks if they draw upon separate resource pools. Both visual tracking and the visual processing of the emergency displays in the pictorial condition drew on the same pool of resources in the encoding stage. The processing of the speech display and the visual tracking drew on separate resource pools, yet little or no advantage was found for this format.

The multiple-resource theory has been expanded (Wickens, 2000) to take into account the important role of S-C-R compatibility in the processing of information. Although Wickens, et al., (1983) admit that a) some tasks are more compatible with speech information, b) some are more compatible with pictorial information, and c) the relationship of task type to compatibility is as yet poorly defined, an argument could be made that the code of representation for emergency information is pictorial or spatial in nature because it refers to a physical system. For example, when a system emergency occurs, a practiced operator will most likely isolate it in terms of a spatial reference as occurring at a certain location in the aircraft rather than as a verbal reference.

9.5.3 Eye/Voice Mission Planning Interface (EVMPI) Model

Since the cockpit is a hands-busy task environment, replacing the number of tasks requiring hand manipulation by more efficient and primarily cognitive tasks through a speech generation should improve pilot performance. While there is typically more information in an utterance than other forms of user input (e.g., pointing), speech is difficult to handle in a noisy environment (like a cockpit) and when the operator is performing in stressful conditions. While speech recognition systems have achieved dramatic improvements over the past several years, they still cannot resolve all interpretation ambiguity. This residual ambiguity must be resolved somehow and Hatfield, et al., (1996) investigated a concept for integrating eye-tracking and voice recognition in an aviation interface. An eye-gaze-tracking device, voice recognition system and the traditional mouse/keyboard are integrated into the interaction interpreter. The system sampled all information from these three data input modalities and integrated information to interpret the user's intention by a different single message, and provided the feedback information to the user via visual cues, synthetic speech or even non-speech audio signals. The Eye/Voice Mission Planning

Interface (EVMPI) concept is intended to increase the quality and efficiency of aviation mission planning task performance. The concept and approach recognize the limitations and constraints imposed by the individual input modalities, eye point-of-gaze and voice, and are intended to compensate for them in the best possible way.

In an EVMPI (Hatfield, 1996) model, the user monitors a visual display and can issue verbal commands and requests while visually attending to the display. In some cases, only the verbal portion of the interaction will be used to interpret the user's intent, e.g., latitude and longitude entry; in other cases, both eye movement and parsed voice will be used to establish context and operator intent. System feedback will be provided in the form of visual events (e.g., changes in object color, intensity or other properties) and audio events (e.g., synthesized speech and non-speech audio such as button clicks). Button pressing can also be accommodated, for example, stick and throttle button presses can be integrated with other modalities.

Input from one or more modalities, such as mouse (representing throttle and stick buttons), eye or voice, are time-stamped according to their time of occurrence and passed to a fusion component which produces a single message of user intent. As speech is parsed, the timestamp associated with the utterance (or an individual word) is associated with the operator's point-of-gaze (POG) at the time the utterance was made. An eye-tracker data smoothing process examines the data stream and estimates the eye POG at a particular time. This estimation is used as the basis for identifying the user interface object or specific pixels that were fixated at the time the utterance was made. Information in the utterance may be further used to disambiguate the objects of interest.

Introducing eye/voice technology (Hatfield, 1996; 1995) offers the potential for further reducing cockpit workload by eliminating many of the physical action requirements altogether and replacing them with coordinated spoken-word and eye-pointing input.   For example, with eye/voice interaction, a pilot could specify a general mission plan using a relatively small number of commands, and using eye point-of gaze to indicate specific geographical locations and points of interest.

The multi-function displays (MFDs) in a typical high performance combat aircraft implement about sixty to hundred functions, requiring substantial hand manipulation possibly in connection with the stick. Stokes and Wicken (1988) point out that the problem with MFDs is that the pilot must remember what is not being displayed and how to obtain it. If hand-manipulation MFDs were simply replaced by voice commands alone, the pilot would be forced to memorize in the order of one hundred command utterances. Hatfield, et al., (Hatfield, 1996; 1995) selected a tactical air strike scenario, and pilots manipulated with MFDs in their experiments. They found that eye/voice input could significantly reduce pilot manual workload in this hands-busy environment without incurring the additional cognitive (memory) load that would arise through use of voice alone.

Hatfield, et al., (Hatfield, 1996) presented some principles/guidelines for conceptualizing and designing eye/voice interaction dialogues:

a) The user should be able to interact with the system (via the interface) in as natural a manner as possible. More specifically, this means that normal visual scanning and verbalization should be accommodated as much as possible.
b) Minimize the requirement for training.
c) Use eye point-of-gaze measurement as deistic to either establish general interpretive context or to resolve verbal referents to objects with precise spatial coordinates.
d) Provide feedback on user verbal commands, e.g. voice feedback to conform commands.
e) Provide feedback on user object selection, e.g. visual feedback.
f) Provide context-dependent memory aids, e.g. text representations of valid utterances.

Table 9-1. Cockpit function tasks supported in the current EVMPI

Mission Planning Task	Dialogue Type	Protocol/Modalities
Selection of ground target or waypoint	Specific deictic reference	Eye to select geographic location of target, voice to designate
Designation of ground target as a navigation waypoint	Specific deictic reference	Eye to select target, voice to instantiate as waypoint
Designation of ground target as a navigation waypoint	Scene navigation: Instantaneous teletransport	Eye to designate position about which to zoom in, voice to initiate zoom
Target hand-off to FLIR	Approximate deictic reference	Eye to select MFD, voice to hand-off target to FLIR
Control of FLIR camera	Scene navigation: smooth teletransport	Voice to start, stop and lock camera; eye to indicate panning direction
Entry of waypoints (lat. and long.)	Simple numerical entry	Voice to enter numerical data

It is necessary to design interaction dialogues that generate a single message of user intent from potentially multiple input modalities. Hatfield, et al., (1996) has given some examples (see Table 9.1)

9.5.4   Application in Cockpit Fast Jet

The technology has developed to such a level that some ASR systems can achieve near 100% accuracy at 0 dB speech-to-noise ratio. The ASR is also more robust to variations in speech. The cognitive aspects and human factors issue in the implementation of an ASR system in cockpits requires careful study. Not all tasks in the cockpit are suited to voice input. Once the appropriate tasks have been chosen, the vocabulary and syntax must be designed, and suitable feedback methods implemented.

There are critical requirements for the implementation of the ASR in the fast-jet-cockpit (Steeneken, 1996). A very high level of accuracy is required. The pilot must be confident that the aircraft system will respond as he desires. Voice input must be regarded as an integral part of the cockpit design, and not as an optional extra.

The ASR systems can enhance data entry operation for the communication, navigation and identification (CNI) systems, and display management and control functions in the cockpit. The voice feedback can be used for Interactive "crew assistant" applications (Steeneken, 1996).

There are a series of flight trials of ASR for different times and functions in the cockpit. Most of the testing purposes concerned the accuracy of the recognition rate and pilot's overall opinion of the system. Table 9-2 gives a short summary of the flight trials. Anderson (1998) has given a good description of these tests and the respective articles are listed in Table 9-2. Some general conclusions one can draw from these tests are as follows:

a) Noise and G-force can influence the accuracy of ASR
b) Different parts of the syntax do not lead to the same results.
c) Speakers' habituation, vocal characteristics and training affect the results.
d) ASR technology can be very useful to reduce the workload, but the system's ability to control its own recognition and manage erroneous recognition is important.

There is no detailed report published about each flight trial, especially detailed description of the interface design. There was no measurement of the total effectiveness of the cockpit tasks performance with and without speech interface. Harvey (1988) reported, "When voice commands were used, average airspeed stayed at 441 knots and height at 547 feet. But when manual control was used, things deteriorated. The AFTI's average height for the manual run was 603 feet AGL and its speed 435 knots, additionally, not all the tasks were done in the required time. Some were done out of order."

Table 9-2. Flight trials of ASR system

Flight type (year)	Testing functions	Results
BAC 111 civil airline (1982–1985)	Use ASR to control the display, radios and the experimental flight management system	Average recognition accuracy 95%, 249 words vocabulary. Less noise and stress, pilots found ASR interface very useful.
F-16 jet aircraft (mid-1980s)	Control Multi-Function Display in the cockpit Two-word vocabulary, Programmable switch.	Performance approximately 90%
19 flights were made during 1989-1993 by U.S. Air Force, NASA and U.S. Navy	Navigator's main interface, via the television tabular display a simple operations may require many key presses	95% accuracy, 99 word vocabulary
NASA OV-10 aircraft	Controlling communications and navigation functions 1g to 3 g 95 dB to 115 dB	97% performance level, 53 words or phrases
Mirage IIIB aircraft Over 80 flights (?)	10,000 vocal commands	

9.5.5 Battle Management System

A preliminary test was carried out in the Rome Air Development Center for command, control, communications and intelligent (C3I) research by implementation of speech recognition to the weapons control station in 1987 (Hale, 1987). In the weapons control station, the weapons control operator controls aircraft, monitors the progress of the battle, notes targets of interest and situations of concern and alerts the battle commander when critical situations develop. The operator must simultaneously monitor the screen, manipulate the keyboard and thumbwheel to zoom in and out of the display and pay attention to the commander's orders and to comments from other station operators, as well as cope with a variety of distractions. The test used some short voice commands instead of some keyboard functions and for data entry. The results showed that operators' efficiency improved and data entry errors were reduced. It reduced 15 to 30% of time needed to execute a short series of commands, and the operator's eyes never needed to leave the display.

9.5.6   UAV Control Stations

Unmanned Aerial Vehicles (UAVs) are powered, aerial vehicles, which do not carry a human operator, but can fly autonomously or be piloted remotely. UAVs can use commercial off-the-shelf components and can be used for both military and civilian applications. Besides the flight control, another important task for the GCS (ground control system) operators is to control different sensors in the UAVs, observing, integrating the information and acting as the command and control central to coordinate with other activities in the entire operation system. According to the NATO standard, the design of the ground station should allow five levels of control (Wall, 2002):

a) *Level 1*: Allows receipt and transmission of secondary images or data.
b) *Level 2*: Provides for receipt of images or data directly from the UAV.
c) *Level 3*: Enables control of the UAV payload.
d) *Level 4*: Authorizes control of the UAV, except for takeoff and landing.
e) *Level 5*: Represents full UAV control.

New requirements brought new challenges to the design of the GCS interfaces for communication of the information. Most studies about the human factors aspects on GCS interface design are still at the research stage. Very few relevant reports on this issue can be found in the literature (Gugerty, 1999).

The mission control system attributes can be categorized according to following items:

a) Task allocation by phases of mission.
b) On board/off board common operating picture.
c) Dynamic mission planning & replanning.
d) Knowledge based functions to aid in decision-making and mission-execution.
e) Dynamic distributed control.

In designing an optimal human-machine interface for UAV and CUAV operations, an essential issue is information display. The operating of future UAVs will be far different from flying manned aircraft. The emphasis is on multimodal or multi-sensor technology (Wilson, 2002). A wide range of possibilities, such as synthetic vision technology, tactile alert systems, spatial (3D) audio, speech recognition and voice control and head mounted displays, etc. An advanced interface should increase operator situational awareness, manage operator workload and measure improved overall UAV system performance.

The most frequently asked questions for the design of the multimodal GCS interface are:

a) What is the proper information the operator needs for situation awareness?
b) How can it be presented in an intuitive way, so it does not have to be deciphered?
c) What input/output modality is optimal for supporting the control-display?
d) Is unimodal or multimodal interaction with the control-display preferable?
e) What should be done about the interaction between the modalities in the task context?

Controlling an UAV or a cluster of UAVs is connected to an augmented role as an information provider in a larger command/control system. Thus, the operator has to keep control of the communication and sensor platform, co-ordinate his/her activities with other operators and decision makers. The aims of such communication/coordination are:

a). *Develop and distribute the battlefield picture*: The accurate and concise picture of the battlefield can be achieved by using all available information sources such as the sensors in multiple UAVs. It also requires having interfaces with existing army battle command systems. At the same time, the battlefield information disseminates to joint and combined arms team and processes from team sensors and command assets. This communication may lead to the development of dynamic mission/battle plans.

b). *Monitor the battle*: It includes monitoring team status and unfolding tactical situations, initiating new plans as battle conditions change and integrate tactical situation displays.

Multimodal interaction might be used for different purposes, for instance:

a) Increasing performance.
b) Reducing the impact of high mental workload.
c) Drawing attention to certain information.
d) Enhancing the "senses."
e) Augmentation of "critical" information.
f) Prevention of errors by increasing redundancy.

Some common human cognitive aspects related to perceiving remembering and understanding information displays can be referenced to display design.

9.6   ARMY APPLICATION

9.6.1   Command and Control on Move (C2OTM)

C2OTM is an Army program aimed at ensuring the mobility of command and control for potential future needs. It allows the battle information from foot soldiers, tanks, transportation vehicles and other units on the field, from UAV or other fight aircraft to be exchanged with command and control centers via satellite system. Speech technology can be especially useful for such systems. A foot soldier may use it to forward observer report and use it as a translation for allies. He can use speech recognition to enter a stylized report that would be transmitted to command and control headquarters over a very low-rate, jam-resistant channel. Different vehicles may use it, and repair and maintenance in the field can be facilitated by voice access to repair information and helmet-mounted displays to show the information. In a mobile command and control vehicle, the commander can use voice to enter and update plans, access to battlefield information by multi-modal input/out such as combining voice, text, pen, pointing and graphics together. The available voice input and output may also help to increase the tank driver better situation awareness and weapons system selection. Normally typing is often a very poor input medium for the mobile user whose eyes and hands are busy with pressing tasks (Weinstein, 1995).

9.6.2   ASR Application in AFVs

AFVs (Armored Fight Vehicles) are usually operated by three to five individuals (the minimum number would be a commander and a driver). The potential advantages in introducing ASR can be (Noyes, 2000):

a) Alleviate workload (hands-eyes-busy environment).
b) Current tasks carried out on a keyboard may provide a suitable choice for speech recognition.
c) Users are dedicated, speaker-dependent. Users could be highly trained to use ASR.

The first drawback factor that may have a strong effect on the application is the life-threatening and safety-critical nature of the military environment. It implies that the soldiers are facing high stress and heavy workload. High workload may influence the speech products and needs to be considered. The safety-critical nature requires the system to be highly reliable and robust. The second drawback factor is the noise (110 dB) and vibration. Two common techniques include the application of sophisticated noise canceling and adaptation algorithms. In recent years, researchers have paid strong attention to the noise effect, since many ASR systems actually can tolerant quite high noise. Vibration can be difficult, however, and less research was carried out in the respective areas.

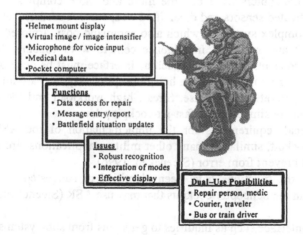

Figure 9-1. The soldier's computer: system concept; functions that would be assisted by human-machine communication by voice; technical issues and possible dual-use application.

Noyes has carried out a feasibility study (Noyes, 2000) on sixteen male soldiers from variety of AFV backgrounds (e.g., infantry fighting vehicles, reconnaissance and main battle tanks). They worked in 8 two-people teams on a sophisticated, fixed-base research AFV simulator. A speaker-independent, isolated word, speech recognizer was compared with push buttons for the data entry task that involved entering reports into the battlefield database. It was found that there was no significant difference in the reporting time between the two input modes; therefore, it was tentatively concluded that ASR technology is feasible in this environment.

9.6.3   The Soldier's Computer

In a modern war, foot soldiers may carry several different electronic instruments as shown in Figure 9-1. The soldier's computer is an Army Communications and Electronic Command (CECOM) program that responds to the information needs of the modern soldier. The function is similar to C2OTM, but it is not feasible for a foot soldier to carry a keyboard with him, so the voice input, or integration with other input/output modes, can be very crucial for them (Figure 9-1) (Weinstein, 1995).

9.6.4   Applications in Helicopter

Military helicopters have become more and more complex. There are more sophisticated sensors and diversified weaponry. A modern helicopter is a part of a complex system in which also command post, armored vehicles, other aircraft and real-time intelligence centers work together (Howells, 1982). New concepts of the man-machine interface must be developed. The hazard environmental conditions a helicopter pilot may face are night flight, bad weather, vibration and noise. Stress, high workload and the eye/hand busy situation are similar to the fast-jet-cockpit pilots.

The critical requirements for the implementation of the ASR in the helicopter cockpit, similar as many other military applications, are simple to perform and prevent from error (Steeneken, 1996).

Some of the functions have been identified or suggested by different researchers in the helicopter cockpits that may use ASR (Steeneken, 1996):

a) System interface such as inquiries to get status from sub-systems of the aircraft (engine status, fuel management).
b) Sensor suite management.
c) Non-decisive actions concerning the weaponry.

ASR was tested on a tactical management system. The system includes an on-board station and a ground station. Both stations share a common mapping of the local area. This implies the management of a cartographic database and of additional objects that are related to the military units in the area. It simplifies the interface for all operators. They only need to deal with visualization and configuration for the output and using voice for input. The test found that the performance for the voice input system was 95% correct responses to the spoken commands even under adverse environment conditions (Howells, 1982).

It was also pointed out that the reliance on helmet-mounted displays can create a problem for the aircrew in operating switches and controls inside the aircraft, so voice input is an important adjunct to the visually coupled system (Anderson, 1998).

Some on-flight tests were carried out (Anderson, 1998) using speech to control the functions such as underlying map presentation, the overlaying symbol presentation, the loading and saving of the mission data and the aircraft navigation. The results showed that it can reach 98% accuracy and the subject belief that speech interface can provide a tremendous amount of increased ability while decreasing the workload (Anderson, 1998).

In 1987, there were three flight tests on voice control of different systems in helicopters. In test one, voice commands were used only to operate three on-board radios. In test two, a voice control was extended to the copilot, and navigation functions and operation of the target hands-off system were put under voice control. In test three, voice control even included controlling of digital map display used for route planning, target location and navigation. The results showed that the performance time was shorter from 12,5% to 34.4% (Henderson, 1989).

In 1997, the ASR system was tested for changing radios and inserting radio frequency in helicopters by the National Research Council of Canada (North, 1997; Swail, 1997). This study made careful interface design (Swail, 1997) to produce an intuitive voice control interface which was unobtrusive to use and allowed hands-free, eyes-out operation. The information required for each pilot task associated with radio control was determined through a review of the Avionics Management System (AMS) pages. This information was studied, along with the subject matter expert (SME) comments, to identify AMS functions that were prime candidates for optional control by voice. The software design engineers took also human factors into consideration. The template was trained by the user together with the background noise. Only forty-seven words were used. The pilot pressed the "talk" key to activate the microphone and said "computer" to activate the ASR system, "enter" to confirm the frequency insertion. A female voice repeated the recognized frequency as feedback (Swail, 1997). The recognition accuracy was about 94% accuracy. The test pilots did not have training to get used to the ASR system. The ASR system was very helpful. The interface design still retained some problems as it was said "human factors on pilot workload in helicopters, and the establishment of a priority of which function to be incorporated in a DVI (direct voice input) system, needed to be done."

9.7   AIR TRAFFIC CONTROL APPLICATION

The application of speech technology in ATC (air traffic control) can be categorized as follows (Weinstein, 1995):

a) Training of air traffic controllers.
b) On-line recognition of flight identification information from a controller's speech to quickly access information on that flight.
c) Processing and fusion of multimodal data to evaluate the effectiveness of new automation aids for air traffic control and gisting of pilot/controller communications to detect potential air space conflicts.

The application of speech recognition and synthesis to emulate pseudo-pilots in the training of air traffic controllers was studied at the Federal Aviation Administration (FAA) Academy in Oklahoma City and at a Naval Air Technical Training Center in Orlando (Weinstein, 1995). Figure 9-2 illustrates the application. Advances in speech and language technology will extend the range and effectiveness of these training applications. Speech recognition technology and data fusion could be used to automate training session analysis and to provide rapid feedback to trainees.

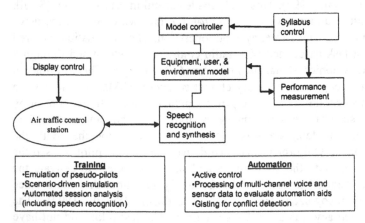

Figure 9-2. The possible application of speech technology in air traffic control systems.

In the ATC application, it is to have high speech recognition accuracy. As we have discussed in Chapter 6, the speech recognition error may not only because the ASR system performance, but also because human speak error. When it is in a high stress dynamic environment, it is easier for the users to make errors. The effort is to introduce the user's cognitive model and model of situations based on detail task analysis. Here we give two examples, one is to make the ATC controller training system and one is the real time gisting in the ATC application.

9.7.1 Training of Air Traffic Controllers

Most ATC simulation facilities use the pseudo pilot concept to simulate the communication with aircraft pilots. Each controller working position is equipped with a radio communication link to pseudo pilots in an adjacent room. The pseudo pilots listen to the clearances (permission for an aircraft or other vehicle to proceed during certain traffic controls) and enter the relevant parameters via a terminal that is connected to the simulation computer. As the pseudo pilots have to enter the control parameters and read back the instructions at the same time, their workloads often become unacceptably high. A possible solution to this problem is the use of automatic speech recognition (ASR) to replace the pseudo pilots.

One of the most remarkable studies on introducing speech technology to ATC training systems was made by Schaefer (2001). In the proposed system, the speech technology was used to replace the functions of the pilot's role. The ASR is used to recognize the controller's comments, synthesized voice is used for responding and the corresponding changes will show on the radar screen. Schaefer made a detailed task analysis of the controllers in the ATC simulator and developed a cognitive model of the air traffic controller. Then he integrated this cognitive model into the ASR system. Figure 9-3 gives an overview of the processes involved in the ATC simulation and ASR environment (Schaefer, 2001). He introduced the cognitive controller's model, which would use the aircraft's flight parameters (altitude, heading, airspeed, etc), from simulation to generate the most probably and possible syntaxes, integrated with a list of the aircraft presently in the control sector together with their flight parameters to limit the speech recognition error and generate the synthesized voice.

By such a system design, Schaefer (2001) successfully increased the ASR's accuracy (50%) by using a dynamic, context-sensitive syntax generated by this cognitive model. In other words, he can reach 95% of recognition rate under operational condition if additional enhancements in other system components are envisaged, such a recognition rate is achieved,

ASR could widely be used in the ATC simulation domain, supporting research and development as well as controller training (Schaefer, 2001).

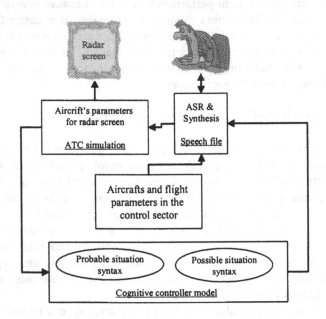

Figure 9-3, The ATC simulation and speech recognition environment.

9.7.2   Real Time Speech Gisting for ATC Application

Dunkelberger (1995) developed a system concept for real time gisting in the ATC application. It provides runway situation and threat assessment. This concept includes real time channel conditioning, speech recognition, speech data analysis, situation assessment, threat recognition, alerting elements and voice commands. In this concept system, all ATC clearance communication information spoken by or to the controller passes over his or her communication channel, and the speech recognizer element recognizes and translates the controller's verbal clearance to aircraft, ground vehicles and other controllers. Situation and threat assessment are performed by relating speech data to proper control sequences and positioning data for controlled aircraft and ground elements. Audio and/or visual alert will be provided to the controller using existing media.

The speech recognizer utilized in the system is based on single sentences, it is not answer, nor does it capitalize on, inter-sentence knowledge. Gisting fills this void, taking over where speech recognition leaves off. Gisting applies reasoning under uncertainty techniques using application-specific knowledge bases to choose the best phrase, sentence and paragraph hypotheses. The dataflow architecture for the gisting subsystem is proposed by Dunkelberger (1995) as shown in Figure 9-4.

The sentences from the ASR system are submitted to the phrase spotter, which proposes a rank-ordered set of phrase hypotheses based on inexact syntactic pattern recognition techniques. Each phrase hypothesis in the rank-ordered set provided by the phrase spotter is validated in isolation by the phrase qualifier. The qualifier has a deep understanding of the semantics pertaining to real-world parameters contained in the hypotheses. Its qualifier, utilizing appropriate mathematical metrics, correlates these implied values with those available from ATC data sources. The situation-based qualified phrases remaining after the qualifier sifting are passed to the phrase splicer for sentence integration. The phrase hypotheses carry delimiters defining their span within the sentences together with the classification confidence provided by the spotter. A subset of the rank-ordered set of sentences, provided by the splicer can then be asserted into the hypo world. The preliminary testing results indicate over 95% key command phrase recognition, possibly approaching 98% (Dunkelberger, 1995).

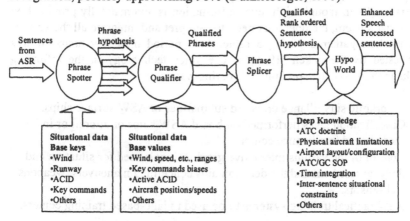

Figure 9-4, dataflow architecture for gisting subsystem.

9.8   NAVY APPLICATION

9.8.1   Aircraft Carrier Flight Deck Control

Not many speech application researches were found in the Navy application. In the mid 1990s, both Weistein and Steeneken made some predictions about possible applications in the Navy.

In the carrier flight deck control application, speech recognition can be used to provide updates to aircraft launch, recovery weapon statues, and maintenance information (Weinstein, 1995). Navy Oceans System undertook demonstrating using a commercially available recognizer for Admiral Tuttle in 1991 and 1992 (Weinstein, 1995). The results indicated the potential advantages but needed enhancement in the overall human-machine interface systems.

Currently, on board submarines, surface ships or maritime patrol aircraft, the sonar operators are overloaded. Today more and more often the complexity level and the amount of information available at a sonar suite output are exceeding the capacity limits of the classical man-machine interface. Even though high level information is automatically provided by the sonar system, the operators need to interpret and integrate all the various information gathered during a rather long period of a few hours. It was suggested by Steeneken (1996) to use voice technology in the following tasks in the ASW domain (anti-submarine warfare):

a) Panoramic surveillance on board submarines or ASW surface ships.
b) Classification tasks performed on board ASW ships or aircraft or in shore-based intelligence centers.
c) Control of the various interactive graphical tools used for situation and threat assessment and the decision aids concerning maneuvers, weapons and sensors.
d) ASW tactical training systems to be used in land-based training centers.

In all these sonar-related tasks, voice input based on connected word recognition provides the following benefits:

a) The user can look continuously at the object without looking at menus or the keyboard.
b) The cursor remains available for pointing out objects on the screen rather than pointing out menu items.
c) The screen surface allocated to menus is reduced.
d) Consequently the interaction can be faster.

These advantages become more important as the control information and Command room generally has a low light level and as the use of the cursor (controlled by the trackball) and of the keyboard can be difficult in this situation especially when the ship is rolling and pitching.

The sonar supervisor on board a surface ship needs to control displays, direct resources and send messages while moving about the command center and looking at command and control displays.

It was proposed by Navy Personnel Research and Development Center that speech-and language-based technology could help with Navy combat team tactical training. It includes a mix of real forces and force elements simulated by using advanced simulation technology. Personnel in the combat information center (either at sea or in a land-based test environment) must respond to developing combat scenarios using voice, typing, trackballs, and other modes and must communicate both with machines and with each other. As suggested in the figure, speech-based and language-based technology and fusion of language with multiple data sources can be used to correlate and analyze the data from a combat training exercise to allow rapid feedback (e.g., what went wrong?) for debriefing, training and re-planning. These language-based technologies first developed and applied in training applications where risk is not a major issue can later be extended to operational applications, including detection of problems and alerting users, and also to development of improved human-machine interfaces in the combat information center (Weinstein, 1995).

9.9   SPACE APPLICATION

It has been quite a long history for NASA to finance the research work on speech technology and its possible application in space. It is mandatory for crewmembers in the space environment to be restrained when working at a workstation for long periods of time in order to acquire some stability. At any time the crewmember might be utilizing the hand controllers, which support robotic operations' and also needing to do another task requiring the activation/deactivation of any number of switches such as video camera control and speech recognition. When a crewmember is working within a glove box, they could have their procedures displayed on a monitor and page through them by voice control rather than stop the task, remove their arms from the gloves and turn the piece of paper to their next step.

One of the earliest investigations about using voice input/output in the space was carried out in the mid 1980s (Hoskins, 1984). In the space shuttle, there is a central control system called MCDS (Multifunction Cathode Ray

Tube Display System). Astronauts used this system to interact with five flight computers. The MCDS has a 32-key oversized keyboard designed for use with the bulky gloves of a space suit. An ASR system was used as an alternative to the keyboard. The similar applications of voice I/O are being considered for the space station as well (Castigione, 1984). It indicated some positive results.

A couple of tests were performed using voice I/O to control different functions in the space shuttle, such as control TV cameras and manipulator arms (Anderson, 1998), using limited vocabulary, and astronauts were pleased with this interface.

The main problem with application is the microgravity environment, and it may influence the voice of speech and the placement of microphone. Some laboratory tests showed the effects (Morris, 1993). Since the ASR technology has been in continuous development, the recognition problem ten years ago may not be the same as today.

9.10   OTHER APPLICATIONS

There are many other applications that are on the way to being discovered. Here we give some examples.

9.10.1   Computer Aid Training

Almost in every application area—aviation, Army, Navy, air traffic control and space—one of the common applications is for training. For a few years now, experiments have been conducted in various countries on the effectiveness and cost/effectiveness of Computer Assisted Instruction (CAI) versus conventional instruction. It is significant that CAI permits each student to proceed at his own rate of learning, to correct errors or misunderstandings as soon as they occur and to reward him with knowledge of success when that occurs. This is difficult or impossible to achieve in conventional instruction where such knowledge is not immediately available to the instructor and where all students, regardless of their ability or style of learning, must proceed at the same pace (Noja, 1993).

The computer technology development in its present stage promises a large memory capacity, with reliability, low price and flexibility to put optimal media, such as graphics, video, pictures and sounds, together in class. Most cases are the development of PC-based virtual technology at different levels of the fidelity to the real world.

One of the main costs for using CAI for training is to program the training lesson, which depends on the contents. Besides, it is almost impossible to have a perfect program to take care of almost all possible problems. So the teacher should always be there. The integration and interaction of different technologies, teachers and students was studied by Noja (Noja, 1993). Speech technology can be very helpful for this system.

9.10.2 Aviation Weather Information

Weather is a major contributing factor in accidents. Some data showed that about 33% commercial carrier and 27% general aviation accidents were due to lack of weather situation awareness and poor decisions (Stough, 2001). The current information systems require significant attention for heads-down time. The research group proposed that voice recognition may significantly reduce heads-down time and workload. The research objects will address issue of using voice command and control of FISDL-type weather information system (accuracy in a high noise environment, vocabulary, menu structure automation, hybrid interface system, effect on workload and improvements in weather situation awareness). The expected results should result in: improved human-machine interface technology that reduces workload and human error and guidelines for operational use of voice interface with graphical weather information system (Stough, 2001).

9.10.3 Interface Design for Military Datasets

Military operations today depend heavily on the C4ISR (Command, Control, Communications, Computing, Intelligence, Surveillance and Reconnaissance) framework. To date, unfortunately, many military systems make it difficult for users to develop a useful understanding of the information to immediate requirements, even though it may be contained within the massive amount of data that flows from the various intelligent resources. The useful may be buried in the flood of irrelevant data. The users may not be able to use the systems to extract the information from the data, or they may not be able to create displays that allow them to see what they need. Potential information sources may be ignored, or not well used, because techniques for extracting information are deficient. As a consequence, users of many current systems discard much data un-accessed (Taylor, 2001).

The architecture of new systems must support a flexible, responsive and mobile approach to military processes. A component-based approach must be adopted so that the system can adapt to changes. It is recognized that for future military visualization systems to be operational, they will have to be oriented specifically to the task, application and user's expertise. Furthermore, there is a need to assess the performance of any visualization system both subjectively and objectively to determine their effects on user performance, beneficial or otherwise (Taylor, 2001). Combining with the visualization system is the multimodal speech interaction system that can be interesting to the user. Integrating the multimodal speech interaction technology into the military system requires a careful human factors study.

9.11   INTEGRATING SPEECH TECHNOLOGY INTO MILITARY SYSTEMS

To apply the speech technology into the military system, the integration issue can be split up between two axes: human factors and system engineering. Attention should be paid to the tasks the operator has to carry out, not only those affected by the speech as new interface mode, but also in regard of possible indirect effects of this new interaction on the whole system operation. A user-centered approach to the design process should be suitable (Leger, 1998). Main steps of such an process should be as follows: identification of top level task requirements, analysis and task modeling, determining the user-system communications needs, developing recommendations for interaction requirements, developing initial requirements for interface technology, identifying the tasks where the speech technology can be used, selecting the suitable ASR and syntheses generation products, designing the suitable vocabulary and syntax, identifying the performance criteria, rapid prototyping, evaluation and iteration to obtainment of the required performance and evaluating the final performance within the system context. Within this design process, to select the right speech technology to right application is important. During the interface design, human factors and human-system interaction should be taken into main consideration.

9.11.1   The Selection of Suitable Functions for Speech Technology

There is not a systematic study on whether those proposed tasks are really suitable for the speech technology or not. The '80s studies that rely on interviews with and administering questionnaires to experienced pilots to determine their preferences for the application of speech technology find that data entry and information retrieval tasks are preferred (Moore, 1989). Since then there is not a process, or regulation, or guidelines for engineers or researchers to select which functions from the systems are suitable to use speech interface.

At the same time, there is a big gap between the state of the art of the technology and off the shelf testable ASR products. At the same time, there are many speech products that have different advantages and disadvantages. There is not a usable procedure for the user to select the right technology according to the requirement. It is essential that technologists work with the users to narrow this gap. It is important for the technologists to understand the user requirements under certain application domain. The software and hardware shall be portable and PC-based for most of the applications. It shall also be easy to adapt to new domains or to unforeseen variations in the user's needs. Eventually, the user should be able to take over and continue adapting the technology to the changing needs.

The principle of how to select the function that was suitable for using speech input and output is by using the task analysis and comparing the task demands and speech technical specification/limitation. The choice and implementation of the speech technology into the system should be made with the understanding of the demands that the task placed upon the human operators and the alternative ways in which it is possible for the operator to do the task. When introducing the speech technology into the existing system, one should be able to observe and measure those important factors and compare it in terms of the effectiveness or performance against the existing system (Rood, 1998). Beside that, one should be very aware of the constraints when speech as input/output mode.

Speech is naturally quite a slow communication process. It becomes very difficult to organize speech in a rigid way; until the moment speech is even too slow to follow the action. Speech recognizers are themselves quite slow in processing the commands and add to this problem. This introduces a serious limitation to the use of direct voice input in muddled and time-constrained situations when a fast and intuitive communication channel would be most useful (Leger, 1998).

9.11.2   Recognition Error, Coverage and Speed

The correct recognition of the spoken sentence is essential for ASR performance. Recognition error will add extra stress to the user. Besides the study of the recognition accuracy, there is very little study about performance efficiency using speech technology compared with the traditional manual performance from an operator's point of view. Most of the real-life test concluded that speech recognition is not accurate enough with about 3% to 5% recognition error. There was not a detailed study about what of this error came from. It may come from human error due to stress or other factors. It is not possible to solve the last 3% to 5% recognition error only from phonetic and acoustic analysis. There must be a better understanding of human speech behavior.

Misrecognition cannot be avoided by any speech recognition system, and it can strongly influence the performance of the system. Even if the user generates an appropriate command (i.e. one that the system should understand), the interaction can still fail due to inaccurate speech recognition. A perfectly well-formed command can fail completely, leaving the user wondering what went wrong and what to do next.

Two key issues exist when considering accuracy rates: first, the precision rates must be within acceptable levels for the particular system. Additionally, error-handling measures are necessary as these systems are not perfect, and some errors will occur. The error levels vary with each system's purpose. Voice dictation programs for word processors allow users to correct their mistakes. But military application normally requires the highest degree of accuracy possible.

There are many different ways to handle the recognition error, especially in the military applications, when normally the users are well-trained experts of the task domain knowledge and the vocabularies are normally specified toward the specific task. Accuracy can be greatly improved by limiting coverage, i.e. by limiting the number of different commands that the system understands. The fewer commands that the system is required to understand (and the less that these commands sound alike), the more likely it is that a well-constructed command will be recognized correctly by the system. Coverage can be limited either by reducing the number of different ways to perform the same action or simply reducing the number of functions that the system can perform. This may not necessarily mean to reduce the robustness and the flexibility of the system. Here a balance between the robustness and ASR error rate is important. The feedback design is crucial, and the user needs to be able to determine what has happened when his commands are not executed: Has the recognizer failed to understand a well-formed command (accuracy problem), is the particular command phrasing that was used unavailable, as in the first coverage problem, or is the desired functionality not available in the system? These problems can appear identical to users with no other feedback when the system fails to execute (James, 2000).

An additional dimension to coverage is that of consistency. The CHI community is well aware of the advantages of maintaining interface consistency, and it is one of the usability design principles. In a speech interface, this translates to having similar commands available for dealing with similar entities or functions. For example, if temperature and pressure are both properties in the domain of a system, while the system allows the command "what is the temperature?" but not "what is the pressure?" it exhibits poor command consistency. In this case, when the command "what is the pressure" is misrecognized, the user may become confused because the command *type* is familiar. If the user assumes that the problem is accuracy, he may repeat the same command multiple times (James, 2000).

Interaction speed is also an important consideration in a speech interface. Decreasing the grammar size can increase the speed of processing spoken input. This in turn may increase accuracy but will reduce coverage. The potential problem for the user is that there are fewer ways to perform a specific function and a greater need to remember exact commands. A system with greater coverage will usually take longer to produce a result, thus decreasing speed. In this case, the user may either become frustrated at constantly having to wait for a response, or may mistake a long processing time for failure and make incorrect assumptions about the grammar coverage.

It was pointed out that the misrecognition will induce external stress to the user, but manual performances also make errors. There is very little study on the error tolerance to trained subjects. In the military operation, most operators are used to trained systems. If the operator can get used to the recognition error, by accepting the recognition errors as they accepting the manual performance error, then the comparison become meaningless. Moore (Moore, 1989) has pointed out that once the pilot/operator gains sufficient familiarity with the equipment, acceptance of the technology increases.

9.11.3 Interface Design

The interface design for military application should have the following characteristics:

a) The system has to be easy to use.
b) A vocabulary and syntax structure must be devised which are logical to the user and use familiar terms.

The selection of vocabulary and syntax has significant effects on the performance of the ASR system. For different systems and functions the vocabulary can be different.

In military applications, most of the studies use limited vocabulary to build short commands. One reason to use short commands is to increase the recognition accuracy; military operators are trained to use short commands. But the negative side of the restricted vocabulary is that it is no longer the natural language. To remember those commands gives extra memory loading to the operators. Besides, human speech is more verbose and needs to embellish communication to improve emphasis and add information.

Producing information-rich communication may be benefit for the listener as a method of self-review and monitoring one's own knowledge. A restricted vocabulary does not have the qualities of a natural language used as a native tongue since birth, which is an inherently slow but effective method of communication.

There are other things regarding the implementation to take into consideration such as the place of the button, microphone, the selection of microphone, etc. Layout design and ergonomics factors of the whole system may affect the usability of the system as well. For example, if we apply the speech technology into a foot soldier's hand-computer system, they need to have a microphone and headphone. The selection of the microphone and headphone would not only be considered for the quality of speech system performance, but also whether they can fit into the rest of the equipment they carry with them and not add extra weight, or add uncomfortable factors.

9.11.4   Alternative and Parallel Control Interface

An important aim of multi-modal interfaces is to maximize human information uptake. Effective visual and auditory information presentation means that information is prepared using the capacities of the human's sensory and mental capabilities such that humans can easily process the information. New interface technologies are being developed that seek to optimize the distribution of information over different modalities such as (Essens, 2001):

a) Image fusion: Real-time sensor images (CCD, thermal, II, SAR, Ladar, etc.); stored images; geographic images (maps).
b) New display technologies: 3D stereoscopic displays; head-mounted displays other "body-mounted" displays; 3D auditory "displays;" high quality virtual environments.
c) Automated pre-selection and decision support and automated target detection.

A problem in this context is the control of information overload. Or, how can overall workload are used to redirect the distribution and presentation of information? If workload is high, information presentation should not add to this. When speech technology is applied to the new interface design of military system, the workload and stress effects to the speech may not be the same as before.

Some studies indicated the interference between the concurrent speech tasks and the visual tasks (Cresswell-Starr, 1993). At the same time, the visual memory and acoustic memory have different characteristics. Some studies showed some typical problems with acoustic display (Cook, 1997):

a) Operators would often forget the mode the system was currently in.
b) It was difficult to remember the acoustic information when it was too long.
c) Forgetting important information or failing to process important signals.

A clear feedback and avoiding extra memory loading should be important issues in the interface design. It should make clear distinctions between multimodal interfaces and parallel or alternative control system.

9.11.5 Innovative Spoken Dialogue Interface

The speech technology application in military domain normally concern in the situation where hands and eyes are busy or it is a difficult to press keyboard- There may also be certain limitations for visual display, or safety demands that the operators maintain visual contact with and attend to tracking instruments. Speech will be considered the alternative technology for effective input/output modality.

Military spoken dialogue interfaces should be developed so they can contribute to increased efficiency, robustness and safety of these missions. An important part of creating a performance-enhancing system is to have interface that do not in them creates additional work or cognitive burden. The best situation would be to use a computer system just the same way as you would delegates to a human assistant or as you would collaborate with a colleague.

The key part of the work is the question of understanding and modeling how people communicate and of exploring what representations can facilitate that communication either with other humans or with computer systems.

Commercial dialogue systems are not motivated to address issues of portability to multiple domains. Besides, dialogue systems have strong context orientation. Every dialogue system should be designed based on its application area. The difference between military dialogue systems and civil systems is that military application requires more natural speech, robustness and portability to multiple information sources. There is not any military dialogue system in this application yet.

There is no systematic evaluation methodology regarding the effects after the speech technology was implemented. Most of the studies just used some simple subjective evaluation by asking a few questions. The evaluation should be the combination of performance measurement, physiological measurement and subjective measurement together. It is important to analyze the whole set of concurrent and subsequent tasks that the operator has to carry out and not just focus on the local context.

9.12   CONCLUSION

The research work on speech technology in the military has a long history. In the '80s and '90s of last century, there were quite a lot of cognitive aspects that related to speech interface design that were studied. A lot of potential advantages are explored if speech technologies are applied for certain task performances. As in most of the application situations, military works are often performed in a hazardous environment with high stress and the ASR's recognition accuracy decreased. This limited the real-time application of speech technology, especially the ASR technology in the application.

References

Abe, T., and Suzuki, J. (1979). A distribution center increases efficiency by use of a voice input unit and a laser scanner. *79' International Physical Distribution Conference,* April, Tokyo.

Ainsworth, W. A. (1988). Optimization of string length for spoken digit input with error correction. *International Journal of Man-Machine Studies,* 28, 537—581.

Ainsworth, W. A. and Pratt, S. R. (1992). Feedback strategies for error correction in speech recognition systems. *International Journal of Man-Machine Studies,* 36, 833—842.

Akoumianakis, D., and Stephanidis, C. (2001). Re-thinking HCI in terms of universal design. In *Universal Access in HCI: Towards an Information Society for All,* ed. C. Stephanidis, NJ: Lawrence Erlbaum & Associates, 8—12.

Allen, T. M., Lunenfield, H., and Alexander, G. J. (1971). Driver Information Needs. in *Highway Research Board 366,* Technical report of "Highway Research Record Number 366."

Allwood, J. (2001). Cooperation and flexibility in multimodal communication. *Cooperative Multimodal Coomunication, Second International Conference, CMC'98,* Tilburg, The netherlands, January, 1998, Selected papers, eds. B. H. Bunt and R. J. Beun, Berline: Springer-Verlag, 113—124.

Alm, H., and Nilsson, L. (1994). Changes in driver behavior as a function of hands-free mobile telephone. *Accidents Analysis and Prevention,* 26, 441—451.

Altmann, E. M. and Gray, W. D. (2000). Managing attention by preparing to forget. *Proceedings of the IEA 2000/HFES 2000 congress*. July 29 - August 4, 2000, San Diego, California, USA, 152—155.

Alty, J. L. (2002). Multimedia Interface design: issues and solutions. *Tutorial Presentation of OZCHI- the Annual Conference of the Computer Human Interaction Special Interest Group of the Ergonomics Society of Australia*. November 26-28, 2003, Brisbane, Australia

Amalberti, R., Carbonell, N., and Falzon, P. (1993). User representations of computer systems in human-computer speech interaction. *Internatioanl Journal of Man-Machine Studies*, 38, 547—566.

Amazeen, E., Amazeen, P., Treffner, P. J., and Turvey, M. T. (1997). Attention and handedness in bimanual coordination dynamics. *Journal of Experimental Psychology, Human Perception and Performance*, 23, 1552—1560.

Andersen, T. S., Tiippana, K., and Sams, M. (2004). Factors influencing audiovisual fission and fusion illusions. *Cognitive Brain Research*, 21, 301—308.

Anderson, T. R. (1998). Applications of speech-based control. RTO EN-3, Lecture Series on *Alternative Control Technolgies: Human Factors Issues*, Brétigny, France. 5-1 to 5-9.

André, E., Finkler, W., Graf, W., Rist, T., Schauder, A., and Wahlster, W. (1993). WIP: The automatic synthesis of multimodal presentations. In *Intelligent Multimedia Interfaces*, ed. M. Maybury, AAAI Press, 75—93.

Arens, Y., Hovy, E., and Vossers, M. (1993). On the knowledge underlying multimedia presentations. In *Intelligent Multimedia Interfaces*, ed. M. Maybury, AAAI Press, 280—306.

Aretz, A. J. (1983). A comparison of manual and vocal response for the control of aircraft systems. *Human Factors Society 27ᵗʰ Annual Meeting*. October, Santa Monica.

Ashby, M. C., and Parkes, A. M. (1993). Interface design for navigation and guidance. In *Driving Future Vehicles*, eds. A. M. Parkes and S. Franzen, London: Taylor & Francis. 295—310.

Baber, C., and Stammers, R. B. (1989). Is it natural to talk to computers? An experiment using the "Wizard of Oz" technique. *Contemporary Ergonomics: Proceedings of the Ergonomic Society's Annual Conference*. London: Taylor & Francis. 234—239.

Baber, C., Stammers, R. B., and Taylor, R. G. (1991). An experimental assessment of automatic speech recognition in high cognitive workload situations in control room systems. In *Design for Everyone*, eds. Y, Quinnec, and F. Daniellou, London: Taylor & Francis, 833-835.

Baber, C., Usher, D. M., Stammers, R. B., and Taylor, R. G. (1992). Feedback requirements for automatic speech recognition in the process control room. *International Journal of Man-Machine Studies*, 37, 703—719.

Baber, C., and Hone, K.S. (1993). Modelling error recovery and repair in automatic speech recognition. *International Journal of Man-Machine Studies*, 39, 495—515.

Baber, C., and Noyes, J. (1995). Automatic speech recognition systems: effects of environmental stress. *ESCA-NATO Workshop on Speech Under Stress,* September, 1995, Lisbon, Portugal, 37—40.

Baber, C. (1996a). Automatic speech recognition in adverse environments. *Human factors*, 38, 142—155.

Baber, C., and Mellor, B. (1996b). The effects of workload on speaking: Implications for the design of speech recognition systems. In *Proceedings of the Annual Conference of the Ergonomics Society,* April 10-12, 1996. University of Leicester, 513—517.

Baber, C., Mellor, B., Graham, R., Noyes, J.M. and Tunley, C. (1996c). Workload and the use of automatic speech recognition: The effects of time and resource demands. *Speech Communication*, 20, 37—53.

Baber, C., and Siddell, T. (1999). The effects of combined stressors on the production of speech. In *Engineering Physiology and Cognitive Ergonomics Volume Four - Job Design, Product Design and Human-Computer Interation*, ed. D. Harris, Aldershot: Ashgate. 263—268.

Baber, C. (2001). Task analysis for error identification. In *International Encyclopedia of Ergonomics and Human Factors*, 1908—1910.

Baber, C., and Noyes, J. (2001). Speech control. In *User Interface Design for Electronic Appliances*, eds. K. Baumann and B. Thomas, London: Taylor & Francis. 190—208.

Baca, J. A., Zheng, F., Gao, H., and Picone, J. (2003). Dialog systems for automotive environments. *Proceedings of the 8thEuropean Conference on Speech Communication and Technology—Eurospeech 2003,* September 1-4, 2003, Geneva, Switzerland, 2003.

Baddeley, A. D. (1994). *Human Memory: Theory and Practice.*
Hillsdale, NJ: Lwarence Erlbaum Associates.

Baddeley, A. D., and Hitch, G.J. (1974). Working memory. In *The Psychology of Learning and Motivation*, ed. G. H. Bower, Vol. 8, London: Academic Press. 47—89.

Bainbridge, L. (1974). Anatysis of verbal protocols from a control task. In *The human operator in process control*, eds. E. Edwards and F. P. Lees, New York: Halsted Press, 146—158.

Balentine, B., and Morgan, D. P. (2001). *How to Build a Speech Recognition Application: A Style Guide for Telephony Dialogues.* 2nd ed. San Ramon: EIG Press.

Ball, L. J., and Ormerod, T.C. (2000). Putting ethnography to work: the case for a cognitive ethnography of design. *International Journal of Human-Computer Studies*, 53, 147—168.

Banbury, S. P., Macken, W. J., Tremblay, S., and Jones, D. M. (2001). Auditory distraction and short-term memory: Phenomena and practical implications. *Human Factors*, 43, 12—29.

Barfield, A. A., and Dingus, T. A. (1998). *Human Factors in Intelligent Transportation Systems.* London: Lawrence Erlbaum & Associates.

Barnum, C. M. (2002). *Usability Testing and Research.* New York, San Francisco, Boston: Longman.

Bates, M. (1994). Models of national language understanding. In *Voice Communication between Human and Machine*, eds. D. B. Roe, and J. G. Wilpon, Washington D.C.: National Academy Press, 238—253.

Bauchage, C., Fritsch, J., Rohlfing, K. J., Wachsmuth, S., and Sagerer, G. (2002). Evaluating integrated speech- and image understanding. In *ICMI Fourth IEEE International Conference on Multimodal Interfaces.* October14- 16, 2002, Pittsburgh, Pennsylvania, USA, 9—14.

Benitez, M. C., Rubio, A., Garcia, A., and de la Torre. (2000). Different confidence measures for word verification in speech recognition. *Speech Communication*, 32, 79-94.

Benoit, C., Martin, J. C., Pelachaud, C., Schomaker, L., and Suhm, B. (2000). Audio-visual and multimodal speech system. In *Handbook of Standards and Resources for Spoken Language Systems, Supplement Volume*, ed. D. Gibbon, Dordrecht: Kluwer Academic.

Benyon, D. (1993). Accommodating individual differences through an adaptive user interface. In *Adaptive User Interface*, eds. M. Schneider-

Hufschmidt, T. Kühme and U. Malinowski, North-Holland: Elsevier Science, 149—165.

Benyon, D., Turner, P., and Turner, S. (2005). *Designing Interactive Systems*. Mateu Cromo, Spain: Pearson Education Limited.

Bergan, M., and Alty, J. L. (1992). Multimedia interface design in process control. In *IEE Colloquium on Interactive Multimedia: A Review and Update for Potential Users*, 9/1—9/6.

Bernsen, N. O., Dybkjar, H., and Dybkjar, L. (1998). *Designing Interactive Speech Systems—from first ideas to user testing*. Berline: Springer-Verlag.

Bernsen, N. O. (2000). Speech- related technologies. Where will the field go in 10 years? *Position Statement for the Elsnet Brainstorming Workshop on Speech and Language Research 2000-2010*, November 23-24, 2000, Katwijk aan Zee, The Netherlands.

Bernsen, N. O. (2001a). Natural human-human-system interaction. In *Frontiers of Human-Centred Computing, On-line Communities and Virtual Environments*, eds. R. Earnshaw, R. Guedj, A. van Dam, J. Vince, Berline: Springer-Verlag. 347—363.

Bernsen, N. O. (2001b). Natural Interactivity. Natural Interactive Systems Laboratory, University of Southern Denmark. HLT Project Class, deliverable D1.7.

Bernsen, N. O. (2001c). Natural interactivity—Draft report on emerging market opportunities. HLT project Class, diverable D1.7, NISLab.

Bernsen, N. O., and Dybkjar, L. (2001). Exploring natural interaction in the car. *Proceedings of the CLASS Workshop on Natural Interactivity and Intelligent Interactive Information Representation*, December, Verona, Italy.

Bernsen, N. O. (2002). Multimodality in language and speech systems - from theory to design support tool. In *Multimodality in Language and Speech System*, ed. B. Granström, D. House, I. Karlsson, Boston: Kluwer Academic, 93—148.

Bernsen, N. O., and Dybkjaer, L. (2002). A multimodal virtual co-driver's problems with the driver. *ISCA Tutorial and Research Workshop on Multi-Modal Dialogue in Mobile Environments Proceedings*, June 17 - 19, 2002, Kloster Irsee, Germany.

Bernsen, N. O. (2003). On-line user modelling in a mobile spoken dialogue system. *Proceedings of the 8th European Conference on Speech*

Communication and Technology—Eurospeech 2003, September 1-4, 2003, Geneva, Switzerland, 737—740.

Bevan, N. (2001). International standards for HCI and usability. *International Journal Human-Computer Studies,* 55, 553—552.

Billi, R., and Lamel, L.F. (1997). RaiTel: Railway telephone services. *Speech Communication,* 23, 63—65.

Billi, R., Castagneri, G., and Danieli, M. (1997). Field trial evaluations of two different information inquiry systems. *Speech Communication,* 23, 83—93.

Bond, Z. S., Moore, T. J., and Anderson, T. R. (1987). The effects of high sustained acceleration on the acoustic-phonetic structure of speech: A preliminary investgation. *Journal of the American Voice I/O Society,* 4, 1—19.

Bond, Z. S., Moore, T. J., and Gable, B. (1989). Acoustic-phonetic characteristics of speech produced in noise and while wearing an oxygen mask. *Journal of Acoustical Society of America,* 85, 907—912.

Bordegoni, M., Faconti, G., Maybury, M., Rist, T., Ruggieri, S., Tahanias, P., and Wilson, M. (1997). A standard reference model for intelligent multimedia presentation systems. *Computer syandards & Interfaces.*18, 477—496

Bosch, L. T. (2003). Emotions, speech and the ASR framework. *Speech Communication,* 40, 213—225.

Bower, G. H. (1981). Mood and memory. *American Psychologist,* 36, 129—148.

Bradford, J. H. (1995). The human factors of speech-based interfaces: A research agenda. *Proceeding in SIGCHI Bulletin,* 27(2).

Brajnik, G., Guida, G., and Tasso, C. (1990). User modeling expert man-machine interface: a case study in intelligent information retrieval. *IEEE Transactions on Systems, Man and Cybernetics SMC,* 20, 166—185.

Brenner, M., and Cash, J. R. (1991). Speech analysis as an index of achohol intoxication - the exxon valdez accident. *Aviation, Space, and Environmental Medicine Journal,* 62, 893—898.

Brenner, M., Doherty, T., and Shipp, T. (1994). Speech Measures indicating workload demand. *Aviation, Space, and Environmental Medicine Journal,* 65, 21—26.

Broadbent, D. E. (1958). *Perception and communication*. Oxford: Pergamon.

Brookings, J. B., and Damos, D. L. (1991). Individual differences in multiple-task performance. In *Multiple-Task Performance*, ed. D. Demos, Taylor and Francis, 363—386.

Browne, D. P. (1993). Experiences from the AID project. In *Adaptive User Interface*, eds. M. Schneider-Hufschmidt, T. Kühme and U. Malinowski, North-Holland: Elsevier Science, 69—78.

Bullinger, H. J., Ziegler, J., and Bauer, W. (2002). Intuitive human-computer interaction—toward a user-friendly information society. *International Journal of Human-Computer Interaction*, 14(1), 1—23.

Burger, J., and Marshall, R. (1993). The application of natural language models to intelligent multimedia. In *Intelligent Multimedia Interfaces*, ed. M. Maybury, AAAI Press, 174—196.

Burgoon, M., and Klingle, R. S. (1998). Gender differences in being influential and/or influenced: a challenge to prior explanations. In *Sex Differences and Similarities in Communication*. eds. D. J. Canary and K. Dindia. Mahwah, NJ: Lawrence Erlbaum & Associates. 257—285.

Burnett, G. E., and Joyner, S. M. (1996). Vehicle Navigation Systems: getting it right from the driver's perspective. *Journal of Navigation*, 49, 174—177.

Burns, C. M. (2000). Navigation strategies with ecological displays. *Internatioanl Journal of Human-computer Studies*, 52, 111—129.

Butler, R., Manaker, E., and Obert-Thorn, W. (1981). *Investigation of a voice synthesis system for the F-14 aircraft (ACT-81-001)*. Warminster, PA: Naval Air Development Center.

Byblow, W. D. (1990). Effects of redundancy in the comparison of speech and pictorial display in the cockpit environment. *Applied Ergonomics*, 21, 121—128.

Bühler, D., Minker, W., Häubler, J., and Krüger, S. (2002). The smartkom mobile multi-modal dialogue system. In *ISCA Tutorial and Research Workshop on Multi-Modal Dialogue in Mobile Environments Proceedings*, June 17-19, 2002, Kloster Irsee, Germany.

Bühler, D., Minker, W., Häubler, J., and Kruger, S. (2002). Flexible multimodal human-machine interaction in mobile environments. *7th International Conference on Spoken language processing, ICSLP-2002*, September 16-20, 2002, Denver, Colorado, USA. 169—172.

Bühler, D., Vignier, S., Heisterkamp, P., and Minker., W. (2003). Safety and operating Issues for mobile human-machine Interfaces. *IUI'03, the 8th International Conference on Intelligent User Interfaces*, Jan. 12-15, Miami, Florida, USA.

Cameron, H. (2000). Speech at the interface. *Proceedings of the COST 249 workshop: Voice Operated telecom Services—Do they have a bright future*, Ghent, May 2000.

Card, S. K., Moran, T. P., and Newell, A. (1983). *The Psychology of Human Computer Interaction*. Hillsdale, NJ: Lawrence Erlbaum & Associates.

Carli, L. L. (1990). Gender, language, and influence. *Journal of Personality and Social Psychology*, 59, 941—951.

Carli, L. L., LaFleur, S. J., and Loeber, C. C. (1995). Nonverbal behavior, gender, and influence. *Journal of Personality and Social Psychology*, 68, 1030—1041.

Carlson, R., and Granstrom, B. (1989). Evaluation and development of the KTH Text-To Speech system on the segmental level. *Proceedings of ESCA Workshop*, September 1989, Noordwijkerhout, The Netherlands. 1.3.1—1.3.4.

Carroll, J. M. (2000). *Making use: Scenario-Based Design of Human-Computer Interactions*. Cambridge: MA MIT Press.

Carter, C., and Graham, R. (2000). Experimental Comparison of Manual and Voice Controls for the Operation of the In-Vehicle Systems. *Proceedings of the IEA 2000/HFES Congress*. July 29 - August 4, 2003, San Diago, California, USA.

Castigione, D., and Goldman, J. (1984). Speech and the space station. *Speech Technology Magazine*, 1984, 19—27.

Castiliano, U., and Umiltà, C. (1990). Size of attentional focus and efficiency of processing. *Acta Psychologica*, 73, 195—209.

Chang, H. M. (2000). Is ASR ready for wireless primetime: measuring the core technology for selected applications. *Speech Communication*, 31, 293—307.

Chapanis, A. (1975). Interactive human communication. *Scientific American* (March), 36—42.

Chen, F. (1999). Effect of noise and workload on Automatic speech recognition system. *Proceedings of the 10th Year Anniversary of M.Sc.*

Ergonomics International Conference, Oct 29-30, 1999, Luleå, Sweden., 286—292.

Chen, F., and Masi, C. (2000). Effect of noise on automatic speech recognition system error rate. *Proceedings of the IEA 2000/HFES 2000 Congress*, July 29 to August 4, 2000. San Diego, USA, 606—609.

Chen, F., and Sääv, J. (2001). The effect of time stress on automatic speech recognition accuracy when using second language. *Proceedings of Eurospeech 2001—Scandinavia*, Aalborg, Sept. 3 - 7, Danmark.

Chen, F. (2003). Time-stress and accented voice input can affect subject's second language speaking. *12th International Symposium on Aviation Psychology*, April 14-17, 2003, Dayton, OH.

Chiles, W. D., and Alluisi, E. A. (1979). On the specification of operator or occupational workload with performance-measurement methods. *Human Factors*, 21, 623—641.

Chin, D. (1984). An analysis of scripts generated in writing between users and computer consultants. *AFIPS conference proceedings*, 53, 637—642.

Chipman, S. E., Schraagen, J. M. C., and Shalin, V. L. (2000). Introduction to cognitive task analysis. *NATO report*. RTO-TR-24.

Christianson, S. Å. (1992). Emotional stress and eyewitness memory: a critical review. *Psychological Bulletin*, 112, 284—309.

Christiansen, M. H., and Chater, N. (1999). Connectionist natural language processing: The state of the art. *Cognitive Science*, 23, 417—437.

Christoffersen, K., Hunter, C. N., and Vicente, K. J. (1998). A longitudinal study of the effects of ecological interface design on deep knoweldge. *International Journal of Human-computer Studies*, 48, 729-762.

Church, K. W. (2003). Speech and language processing: Where have we been and where are we going? *Proceedings of the 8th European Conference on Speech Communication and Technology—Eurospeech 2003*, September 1-4, 2003, Geneva, Switzerland.

Clark, A. (1999). Embodied, situated and distributed cognition. In *A Companion to Cognitive Science*, eds. W. Bechtel, and G. Graham. Oxford: Blackwell, 506—517.

Clark, H. H., and Clark, E. V. (1977). *Psychology and Language: An Introduction to Psycholinguistics*. New York: Harcourt Brace and Jovanovich.

Cnossen, F. (2000). *Adaptive Strategies and Goal Management in Car Driving*, Doctoral Dissertation, University of Groningen. the Netherlands.

Coe, M. (1996). *Human Factors for Technical Communicators*. New York: John Wiley & Sons.

Cohen, J. T., and Graham, J. D. (2003). A revised economic analysis of restriction on the use of cell phones while driving. *Risk Analysis*, 23, 5-17.

Cohen, M. H., Giangola, J. P., and Balogh, J. (2004). *Voice User Interface Design*. Boston: Addison-Wesley.

Cohen, P. R. (1991). The role of natural language in a multimodal interface. *Proceedings of the ACM Symposium on User Interface Software and Technology*, November 11-13, 1991, South Carolina, USA, 143—149.

Coleman, R. F., and Williams, R. (1979). Identification of emotional states using perceptual and acoustic analysis. *Transcript of the 8th Symposium: Care of the professional voice*, New York: Voice Foundation. 7—13.

Coler, C. (1984). Helicopter speech communication systems: Recent noise tests are encouraging. *Speech Technology*, 1, 76—81.

Colle, H. A., and Reid, G. B. (1998). Context effects in subjective mental workload rating. *Human Factors*, 40, 591—600.

Cook, M. J., Crammer, C., Finan, R., Sapeluk, A., and Milton, C. (1997). Memory load and task interference: hidden usability issues in speech interface. In *Engineering Psychology and Cognitive Ergonomics* Vol 1, ed. D. Harris, Aldershot: Ashgate, 141—150.

Cook, M. J., Wilson, K., and Proctor, L. (2001). Disruptive effects between multi-modal tasks. In *Engineering Psychology and Cognitive Ergonomics: Industrial Ergonomics, HCI and Applied Cognitive Psychology*, Aldershot: Ashgate, 263—269.

Cook, T. D., and Campbell, D. T. (1979). *Quasi-Experimentation - design & analysis issues for field settings*. Boston: Houghton Mifflin Company.

Corballis, M. (2002). *From hand to mouth: The origins of language*. Princeton: Princeton University Press.

Coutaz, J. (1992). Multimedia and multimodal user interfaces: A software engineering perspective. *St. Petersburg International Workshop on Human Computer Interaction*.August 2-8, 1992, St. Petersburg.

Cowie, E. D., Campbell, N., Cowie, R., and Roach, P. (2003). Emotional speech: Towards a new generation of databases. *Speech Communication*, 40, 33—60.

Cox, T., and Griffiths, A. (1995). The Nature and measurement of work stress: theory and practice. In *Evaluation of human work*, eds. J, R. Wilson and E. N. Corlett. London: Taylor & Francis Inc, 783—802.

Cresswell-Starr, A. F. (1993). Is control by voice the right answer for the avionics environment? In *Interactive Speech Technology: Human factors Issues in the Application of Speech Input/Output to Computers*, eds. C. Baber and J. M. Noyes. London: Taylor Francis.

Crouch, R., Gaizauskas, R. J., and Netter, K. (1995). Interim report of the study group on assessment and evaluation, Technical report, EAGLES project, Language Engineering Programme, European Commission.

Damos, D. L. (1985). The effect of asymmetric transfer and speech technology on dual task performance. *Human Factors*, 27, 409—421.

Damper, R. I., Lambourne, A. D. and Guy, D. P. (1984). Speech input as an adjunct to keyboard entry in television subtitling. In *Human-computer Interaction*, ed. B. Shakel, Amsterdam: Elsevier.

Damper, R. I., and Woods, S. D. (1995). Speech versus keying in command and control application. *International Journal of Human-Computer Studies*, 42, 289—305.

Damper, R. I., and Tranchant, M. A. (1996). Speech versus keying in command and control: Effect of concurrent tasking. *International Journal of Human-Computer Studies*, 45, 337—348.

Danis, C., Comerford, L., Janke, E., Davies, K., and DeVries, J. (1994). A. storywriter: A speech oriented editor. *Proceedings of CHI'94: Human Factors in Computing Systems: Conference Companion*, Boston, Apr. 24-28. New York: ACM Press, 277—278.

Davis, A., and Rose, D. (2000). The experimental method in psychology. In *Research Methods in Psychology*, eds. G. M. Breakwell, S. Hammond, C., Fife-Schaw, London: SAGE, pp, 42—58.

Delogu, C., Conte, S., and Sementina, C. (1998). Cognitive factors in the evaluation of synthetic speech. *Speech Communication*, 24, 153—168.

Dennison, T. (1985). The effect of simulated helicopter vibration on the accuracy of a voice recognition system. In Unpublished manuscript - cited by Malkin and Dennison, 1986 (in this reference list).

Deutsch, J. A., and Deutsch, D. (1967). Attention: some theoretical considerations. *Psychological Review*, 70, 80—90.

Diaper, D. (2002). Scenarios and task analysis. *Interacting with Computer*, 14, 379—395.

Diaper, D. (1990). *Task Analysis for Human-Computer Interaction*. Chichester: Ellis Horwood.

Dieterich, H., Malinowski, U., Kühme, T., and Schneider-Hufschmidt, M. (1993). State of the art in adaptive user interfaces. In *Adaptive User Interface*, eds. M. Schneider-Hufschmidt, T. Kühme, and U. Malinowski. North-Holland: Elsevier Science, 13—48.

Dillon, T. W., and Norcio, A. F., (1997). User performance and acceptance of a speech-input interface in a health assessment task, *International Journal of Human-Computer Studies,* 47, 591-602.

Dingus, T. A., and Hulse, M. C. (1993). Some human factors design issues and recommendations for automobile navigation information systems. *Transportation Research Part F: Traffic Psychology and Behavior*, IC 2, 119—131.

Dingus, T., McGehee, D., Hulse, M., Jahns, S., Manakkal, N., Mollenhauer, M., and Fleischman, R. (1995). TravTek Evaluation Task C3-Camera Car Study. McLean, *VA: U.S. Department of Transportation, Federal Highway Administration.* Technical Report FHWA-RD-94-076.

Dix, A., Finlay, J., Abowd, G. D.,and Beale, R. (2003). *Human-Computer Interaction*. 3rd ed. Harlow: Pearson and Prentice Hall.

Dunkelberger, K. A. (1995). Real time speech gisting for ATC applications. *SPIE*, 2464, 149—157.

Dybkjaer, L., and Bernsen, N.O. (2001). Usability evaluation in spoken language dialogue system. *Proceedings of the Workshop on Evaluation for Language and Dialogue Systems, Association for Computational Linguistics 39th Annual Meeting and 10th Conference of the European Chapter (ACL/EACL) 2001*, July, Toulouse, France. 9—18.

Dybkjaer, L., Bernsen, N. O., and Minker, W. (2004). Evaluation and usability of multimodal spoken language dialogue systems. *Speech Communication*, 43, 33—54.

Eason, K. D. (1995). User-centered design: For users or by users? *Ergonomics*, 38, 1667—1673.

Ebukuro, R. (1984). Discussion on application of voice input/output systems. *NEC Research and Development.* 1, 99—108.

Ebukuro, R. (1984). Discussion on application of voice input/output systems in industrial field. *Journal of the Robotics Society of Japan*, 2, 28—34.

Ebukuro, R. (2000). Development of online phone call voice input output system and its application for campus guide. March 30, 2000. Ashikaga Institute of Technology.

Egeth, H. (1977). Attention and preattention. In *The Psychology of learning and motivation*, ed. G. H. Bower, Vol. II, 277—320.

Eimer, M., and Driver, J. (2000). An event-related brain potential study of cross-modal links in spatial attention between vision and touch. *Psychophysiology*, 37, 697—705.

Ellis, H. C., Thomas, R. L., and Rodriguez, I.A. (1984). Emotional mood states and memory: Elaborative encoding, semantic processing, and cognitive effort. *Journal of Experimental Psychology*, 10, 470—482.

Ericksen, C. W., St. James, J. D. (1986). Visual attention within and around the field of focal attention: A zoom lens model. *Perception and Psychophysics*, 40, 225—240.

Essens, P. (2001). *Human factors issues for future command*. NATO report. RTO MP-077.

Eysenck, M. W. (1982). *Attention and arousal: Cognition and performance*. Berlin, Germany: Springer.

Eysenck, M. W., and Keane, M. T. (1995). *Cognitive Psychology - a student's handbook*. East Sussex, UK: Psychology Press,

Falcone, R., and Castelfranchi, C. (2002). Issues of trust and control on agent autonomy. *Connection Science*, 14, 249—263.

Felleman, D. J., and van Essen, D. C. (1991). Distributed hierarchical processing in the primate cerebral cortex. *Cerebral Cortex*, 1, 1—47.

Fernandez, R., and Picard, R. W. (2003). Modeling drivers' speech under stress. *Speech Communication*, 40, 145-159.

Finan, R., Cook, M. J., and Sapeluk, A. (1996). Speech-based dialogue design for usability assessment. In *C5 (Human-Computer Interaction) Digest No.: 96/126*, London: IEE Publications.

Flach, J. M. (1998). Research on information form in human-machine interface. Head, Human Factors Research Laboratory, Department of Reactor Safety Research, Japan Atomic Energy Research Institute, Tokai-mura, Naka-gun, Ibaraki-ken, Japan.

Flach, J. M., Tanabe, F., Monta, K., Vicente, K. J., and Rasmussen, J. (1998). An ecological approach to interface design. *Proceedings of Human Factors and Ergonomics Society 42nd Annual Meeting*, October, Chicago, Illinoice, USA. 295—299.

Flach, J. M. (2000). Discovering situated meaning: an ecological approach to task analysis. In *Cognitive Task Analysis*, eds. J. Schraagen, S. Chipman, V. Shalin. NJ: Erlbaum. Mahwah, 87—100.

Flach, J. M. (2000). Ecological interface design: some premises. In *Human Error and System Design and Management - (lecture notes in control and information sciences*, London: Springer-Verlag. 125—135.

Flach, J. M. (2001). A meaning processing approach to analysis and design. In *Usability Evaluation and Interface Design: Cognitive Engineering, Intelligent Agents and Virtual Reality*, eds. M. J. Smith, G. Salvendy, D. Harris, R. J. Koubek, New Jersey: Lawrence Erlbaum & Associates, 1405-1409.

Fonagy, I. (1978). A new method of investigating the perception of prosodic features. *Language and Speech*, 21, 34—49.

Francis, A. L., and Nusbaum, H.C. (1999). Evaluating the quality of Synthetic Speech. In *Human Factors and Voice Interactive Systems*, ed. D. Gardner-Bonneau, Boston : Kluwer Academic, 63—97.

Frankish, C., and Noyes, J. (1990). Sources of Human Error in Data Entry Tasks Using Speech Input. *Human Factors*, 32, 697—716.

Frankish, C. R., Jones, D. M., and Hapeshi, K. C. (1990). Maintaining recognition accuracy during data entry tasks using speech input. In *Contemporary ergonomics*, ed. E. J. Lovesey, London: Taylor & Francis, 445—453.

Fukuto, J. (1998). An advanced navigation support system for a coastal tanker aiming at one-man bridge operation. *The International Conference on Control Applications in Marine Systems (CAMs '98)*. October, Fukuoka Japan.

Fukuto, J. (1998). Use of speech communication as interface of a navigation support system for coastal ships. *7th IFAC/IFIP/IFORS/IEA Symposium on Analysis, Design and Evaluation on Man Machine Systems (MMS '98)*, October, Kyoto Japan.

Furnas, G. W., and Landauer, T. K., Gomez, L. M., and Dumais, S. T. (1987). The vocabulary problem in human-system communication. *Communications of the ACM*, 30, 964—971.

Furukawa, H., Inagaki, T. (1999). Situation-adaptive interface based on abstration hierarchies with an updating mechanism for maintaining situation awareness of plant operators. *IEEE SMC'99 Conference Procedings Systems, Man, and Cybernetics*. 693—698.

Gage, N., Poeppel, D., Robert, T. P. L., and Hickok, G. (1998). Auditory evoked M100 reflects onset acoustics of speech sounds. *Brain Research*, 814, 236—239.

Gaillard, A. W. K., and Wientjes, C. J. E. (1994). Mental workload and work stress as two types of energy mobilization. *Work & Stress*, 8, 141—152.

Gaizauskas, R. (1998). Evaluation in language and speech technology. *Computer Speech and Language*, 12, 249—262.

Galer Flyte, M. D. (1995). The safe design of in-vehicle information and support systems: the human factors issues. *International Journal of Vehicle Design*, 16, 158—169.

Gallivan, M. J. (2001). Striking a balance between trust and control in a virtual organization: A content analysis of open source software case studies. *Information Systems Journal*, 11, 277—304.

Garner, W. R. (1974). *The processing of information and structure*. Hillsdale, NJ: Erlbaum Lawrence Associates.

Gawron, V. J., Schiflett, S. G., and Miller, J. C. (1989). Measures of in-flight workload. In *Aviation Psychology*, ed. R. S. Jensen, Aldershot: Brookfield, 240—287.

Gevins, A., Smith, M.E., Leong, H., McEvoy, L, Whitfield, S., Du, R., and Rush, G. (1998). Monitoring working memory load during computer-based tasks with EEG pattern recognition methods. *Human Factors*, 40, 79—91.

Gevins, A., and Smith, M. E. (1999). Detecting transient cognitive impairment with EEG pattern recognition methods. *Aviation, Space, and Environmental Medicine Journal*, 70, 1018—1024.

Gibbon, D., Moor, R., and Winski, R. (1997). *Handbook of Standards and Resources for Spoken Language System*. Berlin: Mouton de Gruyter.

Gibbon, D. (2000). Handbook of multimodal and spoken dialogue systems: resources, terminology and product evaluation. In *The Kluwer international series in engineering and computer science*, Boston: Kluwer Academic.

Gibson, J. J., Olum, P., and Rosenblatt, F. (1953). Parallax and perspective during aircraft landings. *American Journal of Psychology*, 68, 372—385.

Gillan, D. J., and Bias, R. G. (2001). Usability science. I: Foundations. *International Journal of Human-Computer Interaction*, 13, 351—372.

Goldstein, R., Bauer, L. O., and Stern, J. A. (1992). Effect of task difficulty and interstimulus interval on blink parameters. *International Journal of Psychophysiology*, 13, 111—117.

Goodman, M. J., Tijerina, L., Bents, F. D., and Wierwille, W. W. (1999). Using cellular phones in vehicles: safe of unsafe? *Transportation Human Factors*, 1, 3—42.

Goodstein., L. P. (1980). Discriminative displays support for process operators. In *Human detection and diagnosis of system failures*, eds. J. Rasmussen and W. B. Rouse, New York: Plenum, 433—449.

Gopher, D., and Donchin, E. (1986). Workload: An examination of the concept. In *Handbook of Human Perception and Performance*, eds. K. Boff and L. KauFman, New York: John Wiley, 1—49.

Gould, J. D., Conti, J. and Houanyecz, T. (1983). Composing letters with a simulated listening typewriter. *Communications of the ACM*, 26, 295—308.

Granström, B., House, D., and Karlsson, I. (2002). Introduction. In *Multimodality in Language and Speech System*, eds. B. Granström, House, D., I. Karlsson, Boston: Kluwer Academic, 1—6.

Green, P., Levison, W., Paelke, G., and Serafin, C. (1993). *Suggested Human Factors Design Guidelines for Driver Information Systems*. Technical Report UMTRI-93-21.

Grice, H. P. (1975). Logic and Conversation. In *Syntax and Semantics: Speech Acts*, ed. P. Cole, and J. Morgan, Vol. 3, New York: Academic Press.

Griffin, G. G., Williams, C.E. (1987). The effects of different levels of task complexity on three vocal measures. *Aviation, and Space Environmental Medicine*, 58, 1165—1170.

Griffin, M. J. (1990). *Handbook of human vibration*. London: Academic.

Gröschel, J., Philipp, F., Skonetzki, S., Genzwurker, H., Wetter, T., and Ellinger, K. (2004). Automated speech recognition for time recording in out-of-hospital emergency medicine-an experimental approach. *Resuscitation*, 60, 205—212.

Gugerty, L., DeBoom, D., Walker, R., and Burns, J. (1999). Developing a simulated uninhabited aerial vehicle (UAV) task based on cognitive task analysis: Task analysis results and preliminary simulator performance data. *Proceedings of the Human Factors and Ergonomics Society 43rd Annual Meeting*, September 27-October 1, Houston, Texas, USA. 86—90.

Guha, A., Pavan, A., Liu, J. C. L., and Roberts, B. A. (1995). Controlling the process with distributed multimedia. *IEEE MultiMedia: Summer*. 20—29.

Hackos, J. T., and Redish, J. C. (1998). *User and Task Analysis for Interface Design*. New York: Wiley.

Hale, M., and Nordman, O. D. (1987). An application of voice recognition to battle management. In *Speech Technology*, 5, Oct/Nov, 24—28.

Halsted-Nussloch, R. (1989). The Design of Phone-Based Interfaces for Consumers. Proceedings of the ACM CHI 89 Human Factors in Computing Systems Conference, April 30 - June 4, Austin, Texas, 347—352.

Hamilton, V., and Warburton, D. M. (1979). *Human Stress and Cognition -An information processing approach*. John Wiley & Sons.

Hammersley, M., and Atkinson, P. (1995). *Ethnography: Principles in Practice*. London: Routledge.

Hammond, S. (2000). Using psychometric tests. In *Research methods in Psychology*, eds. G. M. Breakwell, S. Hammond, C., Fife-Schaw. London: SAGE, 175—193.

Hancock, P. A., Lesch, M., and Simmons, L. (2003). The distraction effects of phone use during a crucial driving maneuver. *Accident Analysis and Prevention*, 35, 501—514.

Hansen, J. H. L. (1996). Analysis and compensation of speech under stress and noise for envrioenmental robustness in speech recognition. *Speech Communication*, 20, 151—173.

Hansen, J. H. L., and Arslan, L. M. (1995). Foreign Accent Classification using source generator based prosodic features. *International Conference on Acoustics, Speech, and Signal Processing, ICASSP-95*, May 9-12, 1995, Detroit, USA. 836—839.

Hapeshi, K., and Jones, D. M. (1989). Concurrent manual tracking and speaking implications for automatic speech recognition. *Proceedings of HCI International 89*, September 18-22, 1989, Boston, Massachusetts. 412—418.

Hapeshi, K., and Jones, D. M. (1988). The ergonomics of automatic speech recognition interfaces. In *nternational Reviews of Ergonomics*, ed. D. J. Oborne, Vol. 2, London: Taylor & Francis, 251—290.

Harrington, S. J., and Ruppel, C. P. (1999). Telecommuting: A test of trust, competing values, and relative advantage. *IEEE Transactions on Professional Communication*, 42, 223—239.

Hart, S. G., and Staveland, L.E. (1988). Development of NASA-TLX (task Load Index): Results of empirical and theoretical research. In *Human Mental Workload*, eds. P.A. Hancock and N. Meshkati, North-Holland: Elsevier Science. 139—183.

Harvey, D. S. (1988). Talking with Airplanes. In *Air Force Magazine*, January, 88—98.

Haslegrave, C. M. (1991). Auditory environment and noise assessment. In *Evaluation of Human Work*, eds. J.R. Wilson & E.N. Corlett, London: Taylor & Francis, 406—439.

Hatfield, F., Jenkins, E. A. and Jennings, M. W. (1995). *Eye/Voice Mission Planning Interface. Armstrong Laboratory*, Wright-Patterson AFB, OH,. Final Report, AL/CF-TR-1995-0204.

Hatfield, F., and Jenkins, E.A. (1996). Principlea and guidelines for the design of eye/voice interaction dialogs. *Proceedings of the 3rd Annual Symposium on Human Interaction with Complex System, HICS. IEEE*, August 25-28, 1996, Dayton, Ohio, USA.

Hawkins, J. S., Reising, J. M., Lizza. G. D., and Beachy, K. A. (1983). Is a picture worth a 1000 words-Written or spoken? *Proceedings of the Human Factors Society 27th Annual Meeting*. October 10-14, Norfolk, Virginia, 970—972.

Hebb, D. O. (1949). *Organization of behavior*. New York: John Wiley.

Hecker, M. H. L., Stevens, K.N., von Bismarck, G., and Williams, C.E. (1968). Manifestations of task-induced stress in the acoustic speech signal. *Journal of Acoustical Society of America*, 44, 993—1001.

Henderson, B. W., and Francisco, S. (1989). Army pursues voice-controlled avionics to improve helicopter pilot performance. In *Aviation Week & Space Technology*, May 22, 43—46.

Hickok, G., and Poeppel, D. (2000). Towards a functional neuroanatomy of speech perception. *Trends in Cognitive Science*, 4, 131—138.

Hockey, G. R. J. (1979). Stress and cognitive components of skilled performance. In *Human stress and cognition: An information processing approach*, ed. V. Hamilton and D.M. Warburton, England: Wiley, 141—178.

Hohnsbein, J., Falkensten, M., and Hoormann, J. (1998). Performance differences in reaction tasks are reflected in event-related brain potentials (ERP). *Ergonomics*, 41, 622—633.

Hollan, J., Hutchins, E., and Kirsh, D. (2000). Distributed cognition: toward a new foundation for human-computer interaction research. *ACM Transaction on Computer-Human Interaction*, 7, 174—196.

Hollnagel, E., and Woods, D. D. (1983). Cognitive systems engineering: new wine in new bottle. *International Journal of Man-Machine Studies*, 18, 583—600.

Hollnagel, E. (2001). Time and control in joint human-machine systems. *Second International Conference on Human Interfaces in Control Rooms, Cockpits and Command Centres, 'People in Control'.* IEEE Press. 19-21 June, 2001, Manchester, UK, 246—253.

Hollnagel, E. (2003). Cognition as control: a pragmatic approach to the modelling of joint cognitive systems. Submitted to *IEEE Transaction and Systems, Man, Cybernet.*

Holzman, T. G. (2001). Speech-audio interface for medical information management in field environment. *International Journal of Speech Technology*, 4, 209—226.

Hone, K. S., and Graham, R. (2001). Subjective assessment of speech-system interface usability. *Proceedings of the 7th European Conference on Speech Communication and Technology—Eurospeech 2001.* September 3-7, 2001, Aalborg, Denmark.

Hone, K. S., Graham, R., maguire, M. C., Baber, C., and Johnson, G. I. (1998). Speech technology for automatic teller machines: An investigation of user attitude and performce. *Ergonomics*, 41, 962—981.

Hoskins, J. W. (1984). Voice I/O in the space shuttle. *Speech Technology*, 3, 13-18.

House, A. S., Williams, M. H. L., Hecker, M. H. L., and Kryter, K. D. (1965). Articulation-testing methods: consonantal differentiation with a closed-response set. *Journal of Acoustical Society of America*, 37, 158—166.

Howells, H. (1982). *Verbal Protocols and Indices of Task Loading.* Technical report of Royal Aircraft Establishment, Farnborough, Hants: HM50.

Howie, D., and Vicente, K. J. (1998). Making the most of ecological interface design: The role of self-explanation. *International Journal of Human-computer Studies*, 49, 651—674.

Huber, R., Batliner, A., Buckow, J., Noth, E., Warnke, V., and Niemann, H. (2000). Recognition of emotion in a realistic dialogue scenario. *Proceedings of 6ᵗʰ International Conference on Spoken language Processing(ICSLP 2000)*, Oct. 16-20, Beijing, China, 665—668.

Huckvale, M. (1997). 10 things engineers have discovered about speech recognition. *NATO ASI Workshop Speech Pattern Processing*, Aug, Jersey.

Isaacowitz, D. M., Charles, S.T. (2000). Emotion and cognition. In *The Handbook of Aging and Cognition*, eds. F. I. M. Craik, and T. A. Salthouse, New Jersy: Lawren Erlbaum Associates, 593—631.

James, F., Rayner, M., and Hockey, B.A. (2000). Accuracy, coverage, and speed: What do they mean to user? *CHI 2000 Workshp on Natural-Language Interaction*, June, The Hague, Netherlands.

Jekosch, U. (1992). The cluster-identification test. *Proceeding 1992 Intenational Conference on Spoken Language*, October 1992. Banff, Canada. 205—209.

Jekosch, U. (1994). Speech intelligibility testing: On the interpretation of results. *Journal of the American Voice Input/Output Society*, 15, 63—80.

Jennifer, P. (2002). *Interaction Design*. New York: John Wiley & Sons, Inc.

Jensen, R. J. (1981). Prediction and quickening in prospective flight displays for curved landing and approaches. *Human Factors*, 23, 233—264.

John, B. E., and Gray, W. D. (1995). GOMS analyses for parallel activities. *CHI '95 Conference Proceedings: Conference on Human Factors in Computing Systems*, May 7-11 1995, Denver, Colorado.

John, B. E., and Kieras, D. E. (1996). The GOMS family of user interface analysis techniques: comparison and contrast. *ACM Transaction on Computer-Human Interaction*, 3, 320—351.

John, B. E., and Kieras, D. E. (1996). Using GOMS for user interface design and evaluation: Which technique? *ACM Transaction on Computer-Human Interaction*, 3, 287—310.

Johnson, K., Pisoni, D. B., and Bernacki, R. H. (1990). Do voice recordings reveal whether a person is intoxicated? A case study. *Phonetica*, 47, 215—237.

Jones, D., Hapeshi, K., and Frankish, C. (1989). Design guidelines for speech recognition interfaces. *Applied Ergonomics*, 20, 47—52.

Jones, D. M. (1992). Automatic speech recognition in practice. *Behavior and Information Technology*, 11, 109—122.

Jonides, J. (1983). Further towards a model of the mind's eye-movements. *Bulletin of The Psychonomic Society*, 21, 247—250.

Jordan, P. W. (1998). Human factors for pleasure in product use. *Applied Ergonomics*, 29, 25—33.

Jordan, P. W. (2000). *Designing Pleasurable Products*. London: Taylor & Francis.

Jordan, P. W. (2001). Pleasure with products - new human factors. In *User Interface Design for Electronic Appliances*, eds. K. Baumann, B. Thomas, London: Taylor & Francis. 303—328.

Jurafsky, D., and Martin, J. H. (2000). *Speech and Language Processing*. Upper saddle River: Prentice Hall.

Kahneman, D. (1973). *Attention and Effort*. Englewood Cliffs, New Jersey: Prentice-Hall.

Kahneman, D., Ben Ishai, R., and Lotan, M. (1973). Relation of a test of attention to road accidents. *Journal of Applied Psychology*, 58, 113—115.

Kamm, C. (1994). User interfaces for voice applications. In *Voice Communication between Humans and Machines*, ed. R. D. B. Wilpon, Washington DC: National Academy Press, 422—442.

Karat, C., Halverson, C., karat, J. (1999). Patterns of entry and correction in large vocabulary continuous speech recognition systems. *CHI'99 Proceedings of the CHI 99 conference on Human Factors in Computing Systems: The CHI is the limit*, May 18-20, 1999, Pittsburg. 568—575.

Katz, S., Fleming, J., Hunter, D. R., Green, P., and Damouth, D. (1996). *On-the-Road Human Factors Evaluation of the Ali-Scout Navigation System*. Ann Arbor, MI: University of Michigan Transportation Research Institute. Technical Report UMTRI-96-32.

Kellner, A., Rueber, B., Seide, F. Tran, B. H. (1997). PADIS - An automatic telephone switchboard and directory information system. *Speech Communication*, 23, 95—111.

Kieras, D. (1997). A guide to GOMS model usability evaluation using NGOMSL. In *Handbook of Human-Computer Interaction*, eds. M.

Helander, T. K. Landauer, P. Prabhu, North-Holland: Elsevier Science, 733—766.

Kieras, D. (1988). Towards a practical GOMS model methodology for user interface design. In *Handbook of Human-Computer Interaction*, ed. M. Helander, North-Holland: Elsevier Science, 135-156.

Kirwan, B., and Ainsworth, L. K. (eds) (1992). *A Guide to Task Analysis*. London: Taylor & Francis.

Kirwan, B. (1994). *A Guide to Practical Human Reliablility Assessment*. London: Taylor & Francis.

Kitai, M., Hakoda, K., Sagayama, S., Yamada, T., Tsukada, H., Takahashi, S., Noda, Y., Takahashi, J., Yoshida, Y., Arai, K., Imoto, T., and Hirokawa, T. (1997). ASR and TTS telecommunications applications in Japan. *Speech Communication*, 23, 17—30.

Kolb, B., and Whishaw, I.Q. (2000). *Fundamentals of Human Neuropsychology*. 4th ed. New York: W. H. Freeman and Company.

Kolb, B., and Whishaw, I.Q. (2001). *An Introduction to Brain and Behavior*. New York: Worth.

Kotelly, B. (2003). *The Art and Business of Speech Recognition*. Boston: Addison-Wesley.

Krahmer, E., Swerts, M., Theune, M., and Weegels, M. (2001). Error detection in spoken human-machine interaction. *International Journal of Speech Technology*, 4, 19—30.

Kumashiro, M. (1995). Chapter 29, Practical measurement of psychological functions for determining workload. In *Evaluation of human work*, eds. J. R. Wilson, and E. N. Corlett, London: Taylor & Francis Inc, 865—888.

Kühme, T., Schneider-Hufschmidt, M. (1993). Introduction. In *Adaptive User Interface*, eds. M. Schneider-Hufschmidt, T. Kühme, U. Malinowski, North-Holland: Elsevier Science, 1—9.

La Rue, A., Swan, G. E. and Carmelli, D. (1995). Cognition and depression in a cohort of aging men: results from the Western Collaborative group Study. *Psychology and Aging*, 10, 30—33.

Labiale, G. (1997). Cognitive ergonomics and intelligent systems in the automobile. In *Erogonomics and Safety of Intelligent Driver Interfaces*, ed. Y. I. Noy, New Jersey: Lawrence Erlbaum Associates, 169—184.

Lai, J., Cheng, K., Green, P., and Tsimhoni, O. (2001). On the road and on the web? Comprehension of synthetic and human speech while driving. *Conference on Human Factors and Computing Systems, CHI 2001,*March 31-April 15, Seattle, Washington, USA.

Lamel, L. F., Bennacef, S. K., Rosset, S., Devillers, L., Foukia, S., Gangolf, J. J., and Gauvain, J.L. (1997). The LIMSI RailTel System: Field trial of a telephone service for rail travel information. *Speech Communication*, 23, 67—82.

Lane, H. L., Tranel, B., and Sisson, C. (1970). Regulation of voice communication by sensory dynamics. *Journal of the Acoustical Society of America*, 47, 618—624.

Larsen, L. B. (2003). Assessment of spoken dialogue system usability-what are we really measuring? *8ᵗʰEuropean Conference on Speech Communication and Technology— Eurospeech 2003*, September 1-4, 2003, Geneva, Switzerland, 1945—1948.

Larson, K., and Mowatt, D. (2003). Speech error correction: The story of the alternates list. *International Journal of Speech Technology*, 6, 183—194.

Laycock, J., and Peckham, J. B. (1980). *Improved Piloting Perfornzance Whilst Using Direct Voice Input.* RAE Technical Report 80019, RAE Farnborough, UK.

Lazarus, R. S. (1991). *Emotion and Adaptation*. New York: Oxford University Press.

Lee, J. D., Caven, B., Haake, S., and Brown, T.L. (2001). Speech-based interaction with in-vehicle computers: The effect of speech-based e-mail on drivers' attention to the roadway. *Human Factors*, 43, 631-640.

Lee, C. H. (2004). From konwledge-ignorant to knowledge-rich modeling: A new speech research paradigm for next generation automatic speech recognition. *Proceedings of 8ᵗʰ International Conference on Spoken Language Processing (ICSLP2004)*, October 4-8, 2004, Jeju Island, Korea. 109—112.

Leger, A. (1998). *Synthesi and Expected Benefits Analysis*. NATO report. RTO EN-3.

Leggett, J. W., G. (1984). An empirical investigation of voice as an input modality for computer programming. *International Journal of Man-Machine Studies*, 21, 493—520.

Leiser, R. G. (1989). Exploiting convergence to improve natural-language understanding. *Interacting with Computers*, 3, 284—298.

Levelt, W. J. M. (1993). Speaking: From intention to articulation. Cambridge: MIT Press.

Lewis, C. H., and Griffin, M. J. (1978). Predicting the effects of dual-frequency vertical vibration on continuous manual control performance. *Ergonomics*, 21, 637—650.

Lewis, C. H., and Griffin, M. J. (1978). Predicting the effects of vibration frequency and axis, and seating condition on the reading of numeric displays. *Ergonomics*, 23, 485—501.

Lewis, J. (1994). Sample sizes for usability studies: Additional considerations. *Human Factors*, 36, 368—378.

Li, Y., and Zhao, Y. (1998). Recognizing emotions in speech using short-term and long-term features. *Proceeding ICSLP 1998: The 5th International Conference on Spoken Language Processing*. November 30 - December 4, 1998, Sydney, Australia. 2255—2258.

Lieberman, A. M., Cooper, F. S., Shankweiler, D. P., and Studdert-Kennedy, M. (1967). Perception of the speech code. *Psychological Review*, 74, 431—461.

Lieberman, P. (1961). Perturbation in vocal pitch. *Journal of Acoustical Society of America*, 33, 597—603.

Lieberman, P., and Blumstein, S. E. (1988). *Speech physiology, speech perception, and acoustic phonetics*. Cambridge: Cambridge University Press.

Life, M. A., and Long, J. B. (1994). Providing human factors knowledge to non-specialists: A structured method for the evaluation of future speech interfaces. *Ergonomics*, 37, 1801—1842.

Lind, A. J. (1986). Voice recognition - an alternative to keypad. *Proceedings of Speech Tech' 86*. April 28-30, 1986, New York. 66—67.

Lindegaard, G. (1994). *Usability Testing and System Evaluation*. London: Chapman & Hall.

Lo, S., and Helander, M.G. (2004). Decoupling usability heuristics. *Proceedings of the 7th International Conference on WWCS (Work with Computing Systems)*, June 2 - July 2, 2004, Kuala Lumpur, Malaysia. 32—37.

Logan, G. D. (1985). Executive control of thought and action. *Acta Psychologica*, 60, 193—210.

Logan, J. S., Greene, B. G., and Pisoni, D. B. (1989). Segmental intelligibility of synthetic speech produced by rule. *Journal of Acoustical Society of America*, 86, 566—581.

Luce, P. A., Feustel, T. C., and Pisoni, D. B. (1983). Capacity demands in short-term memory for synthetic and natural speech. *Human Factors*, 25, 17—32.

Luczak, H. (1988). Task analysis. In *Handbook of Human-Computer Interaction*, ed. M. Helander, North-Holland: Elsevier Science, 340—416.

Lynch, S. (1984). Glass quality inspection system. *Proceedings of Speech Tech 84*, April 2-4, 1984. New York, 32—35.

Macaulay, C., Benyon, D., and Crerar, A. (2000). Ethnography, theory and system design: from intuition to insight. *International Journal of Human-Computer Studies*, 53, 35—60.

Macleod, M. (1996). Performance measurement and ecological validity. In *Usability Evaluation in Industry*, eds. P.W. Jordan, B. Thomas, B. A. Weerdmeester, I. L. McClelland, London: Taylor & Francis, 227—235.

Maguire, M. (2001a). Context of Use within usability activities. *International Journal of Human-Computer Studies*, 55, 453—483.

Maguire, M. (2001b). Methods to support human-centred design. *International Journal of Human-Computer Studies*, 55, 587—634.

Malkin, F. J., and Dennison, T.W. (1986). The effect of helicopter vibration on the accuracy of a voice recognition system. *IEEE, O547-3578/86/0000-0813*.

Mantovani, G. (1996). Social context in HCI: A new framework for mental models, co-operation and communication. *Cognitive Science*, 20, 237—270.

Martin, G. L. (1989). The utility of speech input in user-computer interfaces. *International Journal of Man-Machine Studies*, 30, 355—375.

Martin, J. C., Veldman, R., and Béroule, D. (1998). Developing multimodal interfaces: a theoretical framework and guided propagation networks. In *Multimodal Human-Computer Communication - Systems, Techniques and Experiments*, eds. H. Burt, R-J. Beun, T. Borghuis, Berline: Springer-Verlag, 158—187.

Martin, T. B. (1976). Practical applications of voice input to machines. *Proceedings of The IEEE*, 64, 487—501.

Martin, T. B., and Welch, J.R. (1980). Practical speech recognizers & some performance effectiveness parameters. In *Trends in Speech Recognition*, ed. W. A. Lea, New Jersey: Prentic Hall, 24—38.

Maslow, A. (1970). *Motivation and Personality*. New York: Harper and Row.

Massaro, D. (1997). *Perceiving Talking Faces: From Speech Perception to A Behavioral Principle*. Cambridge: MIT Press.

Matthews, R., Legg, S., and Charlton, S. (2003). The effect of cell phone type on drivers subjective workload during concurrent driving and conversing. *Accident Analysis and Prevention*, 35, 451—457.

Maybury, M. T. (1997). *Intelligent Multimedia Information Retrieval*. AAAI / Cambridge: MIT Press.

McCauley, M. E. (1984). Human factors in voice technology. In *Human Factors Review*, ed. F. A. Muckler, Santa Monica: Human Factors Society, 131—166.

McClelland, J. L., and Elman, J. L. (1986). The TRACE model of speech perception. *Cognitive Psychology*, 18, 1—86.

McDowd, J., Vercruyssen, M., and Birren, J. E. (1991). Aging, divided attention, and dual-task performance. In *Multiple-Task Performance*, ed. D. Demos, London: Taylor & Francis, 387—414.

McGurk, H., and MacDonald, J. (1976). Hearing lips and seeing voices. *Nature*, 264, 746—748.

McKnight, J., and McKnight, A. S. (1993). The effects of cellular phone use upon driving attention. *Accident Analysis and Prevention*, 25, 259—265.

Meshkati, N. (1988). Heart rate variability and mental workload assessment. In *Human Mental Workload*, eds. P. A. Hancock, N. Meshkati, Amsterdam: Elsevier Science, 101—115.

Meshkati, N., Hancock P. A. Rahimi, M., Dawes, S. M. (1995). Chapter 29, Techniques in mental workload assessment. In *Evaluation of human work*, eds. J. R. Wilson, E. N. Corlett, London: Taylor & Francis, 749—785.

Michon, J. A. (1993). *Generic intelligent driver support. A comprehensive report on GIDS*. London: Taylor & Francis.

Miller, G. A. (1956). The magical number seven, plus or minus two: Some limits on our capacity to process information. *Psychological Review*, 63, 81—97.

Minker, W., Haiber, U., Heisterkamp, P., and Scheible, S. (2003). Intelligent dialog overcomes speech technology limitations: The SENECa example. *IUI'03, The International Conference of Intelligent User Interfaces*, January 12-15, 2003, Miami, Florida, USA.

Mirenda, P., Eicher, D., & Beukelman, D. R. (1989). Synthetic and natural speech preferences of male and female listeners in four age groups. *Journal of Speech and Hearing Research*, 32, 175–183.

Moon, B. S., Lee, H.C., Lee, Y.H., Park, J.C., Oh, I.S., and Lee, J.W. (2002). Fuzzy systems to process ECG and EEG signals for quantification of the mental workload. *Information Sciences*, 142, 23—35.

Moore, R. K. (1986). The NATO research study group on speech processing: R5GI0. *Proceedings of Speech Tech 86*, April 28-30 1986, New York, 201—203.

Moore, R. (1994). Intergation of speech with national language understanding. In *Voice Communication between Human and Machine*, eds. D. B. Roe, and J. G. Wilpon, Washington D.C.: National Academy Press, 254—271.

Moore, T. J. (1989). Speech technology in the cockpit. In *Aviation Psychology*, ed. R. S. Jensen, Gower Technical, 50-65.

Moore, T. J., and Bond, Z. S. (1987). Acoustic-phonetic changes in speech due to environmental stressors: Implication for speech recognition in the cockpit. *Proceedings of the 4th International Symposium on Aviation Psychology*, Columbus, OH: Aviation Psycholgy Laboratory. 77—83.

Moray, N. (1999). The psychodynamics of human-machine interaction. In *Engineering Psychology and Cognitive Ergonomics*, ed. D. Harris, Vol. 4, Ashgate, 225—235.

Morris, R. B., Whitmore, and M. Adam, S. (1993). How well does voice interaction work in space. In *IEEE AES Systems Magazine*, August, 26—30.

Morrison, J. G., Forster, E., and Hitchcock, E. M. (1994). Cumulative effects of +Gz on cognitive performance. *Proceedings of the Human Factors and Ergonomics Society 38th Annual Meeting*, October 24-28, 1994, Nashville, Tennessee, USA. 46-50.

Mountford, S. J., and North, R. A. (1980). Voice entry for reduced pilot workload. *Proceedings of the Human Factors Society*, 24, 185—189.

Mozeico, H. (1982). A human/computer interface to accommodate user learning stages. *Communications of the ACM*, 25, 100—104.

Muir, B. M., and Moray, N. (1996). Trust in automation. Part II. Experimental studies of trust and human intervention in a process control simulation. *Ergonomics*, 39, 429—460.

Mulder, L. J. M., and Mulder, G. (1987). Cardiovascular reactivity and mental workload. In *The beat-by-beat investigation of cardiovascular function*, eds. R.I. Kitney and O. Rompelman, Oxford: Clarendon press, 216—253.

Mullennix, J. W., Stern, S. E., Wilson, S. J., and Dyson, C.L. (2003). Social perception of male and female computer synthesized speech. *Computer in Human Behavior*, 19, 407—424.

Murray, I. R., Baber, C., and South, A. (1996). Towards a definition and working model of stress and its effects. *Speech Communication*, 20, 2—12.

Möller, S., Jekosch, U., Mersdorf, J., and Kraft, V. (2001). Auditory assessment of synthesized speech in application scenarios: Two case studies. *Speech Communication*, 34, 229—246.

Nass, C., Moon, Y., and Green, N. (1997). Are machines gender neutral? Gender-stereotypic responses to computers with voices. *Journal of Applied Social Psychology*, 27, 864—876.

Nass, C., and Gong, L. (2000). speech interfaces - from an evolutionary perspective. *Communications of the ACM*, 43(9), 36—43.

Navon, D., Gopher, D., Chillag, M., and Spitz, G. (1984). On separability of and interference tracking dimensions in dual-axis tracking. *Journal of Motor Behavior*, 16, 364—392.

Nelson, D. L. (1986). User acceptance of voice recognition in a product inspection environment. *Proceedings of Speech Tech '86*. April 28-30 1986, New York, 62.

Nielsen, J. (1993). *Usability engineering*. San Francisco: Morgan Kaufman.

Nielsen, J. (1994). Heuristic evaluation. In *Usability Inspection Methods*, eds. J. Nielsen, and R. L. Mack, New York: John Wiley & Sons.

Nikolaev, A. R., Ivanitskii, G. A. and Ivanitskii, A. M. (1998). Reproducible EEG alpha-patterns in psychological task solving. *Human Physiology*, 24, 261—268.

Niwa, S. (1971). Changes of voice characteristics in urgent situations. Japan: Japan Air Self Defence Force, Aeromedical Laboratory. Tech. Report 11.

Nixon, C., Anderson, T., Morris, L. J., McCavitt, A., McKinley, R., Yeager, D. G., and McDaniel, M. P. (1998). Female voice communications in high level aircraft cockpit noises -Part II: Vocoder and automatic speech recognition systems. *Aviation, Space and Environmental Medicine*, 69, 1087—1094.

Nixon, C. W., Morris, L. J., McCavitt, A. R., McKinley, R. L., Anderson, T. R., McDaniel, M. P., and Yeager, D. G. (1998). Female voice communications in high level aircraft cockpit noises -Part I: Spectra, levels and microphones. *Aviation, Space and Environmental Medicine*, 69, 675—683.

Noja, G. P. (1993). DVI and system integration: a further step in ICA/IMS technology. *Advanced technologies applied to training design*, Eds. R. J. Seidel and P. R. Chatelier, New York: Plenum Press, 161—189.

Nooteboom, S. G., Broke, J. P. L., and de Rooij, J. J., (1977). Contributions of prosody to speech perception. In *Studies in the Perception of Language*, eds. W. J. L. Levelt, and G. B. Flores D'Arcais, New York: Wiley, 75—107.

Norcio, A. F., Stanley, J. (1989). Adaptive human-computer interfaces: A literature survey and perspective. *IEEE Transactions on Systems, Man and Cybernetics*, 19, 399—408.

Norman, D. (1988). *The Design of Everyday Things*. New York: Basic Books,.

North, D. M. (1997). Voice command system improves helicopter situational awareness. *Aviation week & space technology*, June 16, 195—197.

North, R. A., and Gopher, D. (1976). Measures of attention as predictors of flight performance. *Human Factors*, 18, 1—14.

Novick, D. G., Hansen, B., Sutton, S., and Marshall, C. R. (1999). Limiting Factors of Automated Telephone Dialogues. In *Human Factors and Voice Interactive Systems*, ed. D. GardnerBonneau, Boston: Kluwer Academic, 163—186.

Noyes, J., and Baber, C. (1999). *User-Centred Design of Systems*. Berline: Springer-Verlag.

Noyes, J., Baber, C., and Leggatt, A. P. (2000). Automatic speech recognition, noise and workload. *Proceedings of the IEA 2000/HFES 2000 Congress*, July 29 to August 4, 2000. San Diego, USA, 3, 762—765.

Nunes, L., and Recarte, M. A. (2000). Cognitive demands of hands-free phone conversation while driving. *Transportation Research Part F: Traffic Psychology and Behavior*, 5, 133—144.

Nusbaum, H. C., Francis, A. L., and Henly, A. S. (1995). Measuring the naturalness of synthetic speech. *International Journal of Speech Technology*, 1(11), 7—19.

Ogden, G. D., Levine, J. M., Eisner, E. J. (1978). Measurement of workload by secondary task: A review and annotated bibliography. *Prepared under contract no. NAS2-9637 for the National Aeronautical and Space Administration*. Ames Research Center, Washington, DC: Advance Research Resources Organization.

Okawa, S., Endo, T., Kobayashi, T., and Shirai, K. (1993). Phrase recognition in conversational speech using prosodic and phonemic information. *IEICE Transactions on Information and Systems*, E76-D(1), 44—50.

Ominsky, M., Stern, K. R., and Rudd, J. R. (2002). User-centered design at IBM consulting. *International Journal of Human-Computer Interaction*, 14, 349—368.

Oviatt, S. L., Cohen, P. R., Wang, M., Gaston, J. (1993). A simulation-based research strategy for designing complex NL systems. *ARPA Human Language Technology Workshop*, NJ: Princeton.

Oviatt, S., van Gent, R. (1996). Error resolution during multimodal human-computer interaction. *Proceedings of the 4th International Conference on Spoken Language Processing (ICSLP 96)*, October. 3-6, 1996, Philadelphia, USA.

Oviatt, S. (1997). Multimodal interactive maps: designing for human performance. *Human-Computer Interaction*, 12, 93-129.

Oviatt, S., MacEachern, M., and Levow, G. (1998). Predicting hyperaticulate speech during human-computer error resolution. *Speech Communication*, 24, 87—110.

Oviatt, S. (2002). Multimodal Interface. In *Handbook of Human-Computer Interaction*, eds. J. A. Jacko, and A. Sears. New Jersey: Lawrence Erlbaum & Associates.

Ozkan, N., Paris, C. (2002). Cross-fertilization between human computer interaction and natural language processing: Why and how. *International Journal of Speech Technology*, 5, 135—146.

Paelke, G., & Green, P. (1993). *Entry of Destinations into Route Guidance Systems: A Human Factors Evaluation*. Ann Arbor, MI: The University of Michigan Transportation Research Institute. Technical Report UMTRI-93-45.

Pallett, D. S. (1985). Performance assessment of automatic speech recognitioners. *Journal of the National Bureau of Standards*, 90, 1—17.

Paouteau, X. (2001). Interpretation of guestures and speech: a practical approach to multimodal communication. *Cooperative Multimodal Coomunication, 2ⁿᵈ international conference, CMC'98, Tilburg, The Netherlands, January 1998 Selected papers*, eds. B. H. Bunt and R-J. Beun, Berline: Springer-Verlag, 159—175.

Papazian, B. (1993). *Management of errors in spoken-language system dialogues: A literature review*. Cambridge, MA: BBN Systems and Technologies. BBN Technical report 7910.

Paris, C. R., Thomas, M. H. Gilson, R. D. and Kincaid, J. P. (2000). Linguistic cues and memory for synthetic and natural speech. *Human Factors*, 42, 421—431.

Pashler, H. (1994). Dual-task interference in simple tasks: Data and theory. *Psychological Bulletin*, 116, 220—244.

Patrick, J., Spurgeon, P., and Shepherd, A. (1986). *A Guide to Task Analysis*. London: Taylor & francis.

Pellecchia, G. L., and Turvey, M. T. (2001). Cognitive activity shifts the attractors of bimanual rhythmic coordination. *Journal of Motor Behavior*, 33, 9—15.

Petersen, S. E., Robinson, D. L., and Morris, J. D. (1987). Contribution of the pulvinar to visual spatial orientation. *Neuropsychologia*, 25, 97—106.

Petride, H. L., and Mishkin, M. (1994). Behaviorism,cognitivism and the neuropsychology of memory. *American Scientist*, 82, 30—37.

Picard, R. W. (1997). *Affective Computing*. Cambridge, MA: MITPress.

Piechulla, W., Mayser, C., Gehrke, H., and König, W. (2003). Reducing driver's mental workload by means of an adaptive man-machine interface. *Transportation Research Part F*, 6, 233—248.

Pieraccini, R., Dayanidhi, K., Bloom, J., Dahan, J. G., and Phillips, M. (2003). A Multimodal Conversational Interface for a Concept Vehicle. *Proceedings of the 8th European Conference on Speech Communication and Technology—Eurospeech 2003,* September 1-4, 2003, Geneva, Switzerland, 2233—2236.

Pisoni, D. B. (1981). Speeded classification of natural and synthetic speech in a lexical decision task. *Journal of Acoustical Society of America*, 70, 98.

Pisoni, D. B., Nusbaum, H. C., Luce, P. A., and Slowiaczek, L. M. (1985). Speech perception, word recognition and the structure of the lexicon. *Speech Communication*, 4, 75—95.

Poock, G. K. (1980). *Experiments with voice input for command and control: using voice input to operate a distributed computer network*. Naval Postgraduate School Report. NP555-80-016, Monterey.

Poock, G. K. (1982). Voice recognition boosts command terminal throughput. *Speech Technology*, 1, 36—39.

Posner, M. I., Snyder, C. R. R., and Davidson, B. J. (1980). Attention and the detection of signals. *Journal of Experimental Psychology: General*, 109, 160—174.

Preece, J., Rogers, Y., and Sharp, H. C. (2002). *Beyond Human-computer Interaction*. Chichester: Wiley & Sons.

Price, P. J., Ostendorf, M., Shattuck-Hufnagel, S., and Fong, C. (1991). The use of prosody in syntactic disambiguation. *Journal of the Acoustical Society of America*, 90, 2956—2970.

Protopapas, A., and Lieberman, P. (1997). Fundamental frequency of phonation and perceived emotional stress. *Journal of Acoustical Society of America*, 101, 2267-2277.

Rahurkar, M. A., and Hansen, J. H. L. (2002). Frequency band analysis for stress detection using a teager energy operator based feature. *7th International Conference on Spoken Language Processing (ICSLP)*, September 16-20, 2002, Denver, Colorado. 2021—2024.

Rakers, G. (2001). Interation design process. In *User Interface Design for Electronic Appliances*, ed. B. T. K. Baumann, London: Taylor & Francis, 7—47.

Ranney, T. A., Mazzae, E., Garrott, R. and Goodman, M. (2000). *NHTSA Driver Distraction Research: Past, Present, and Future.* Driver Distraction Internet Forum, United States Department of Transportation, National Highway Traffic Safety Administration (NHTSA), Washington, DC.

Rasmussen, J. (1980). Models of mental strategies in process control. In *Human Detection and Diagnosis of System Failures*, eds. J. Rasmussen and W. B. Rouse, New York: Plenum, 241—258.

Rasmussen, J. (1985). The tole of hierarchical knowledge representation in decision making and system management. *IEEE Trans. Systems, Man, Cybernet*, SMC-15, 234—143.

Rasmussen, J. (1986). *Information Processing and Human-machine Interaction: An Approach to Cognitive Engineering.* New York: North Holland,.

Rasmussen, J. (1995). Cognitive system engineering approach to design of work support systems. *Widening Our Horizons - 31st Annual Conference of the Ergonomics Society of Australia*, June 15, . 19—28.

Ratcliff, R. (1985). Theoretical interpretations of the speed and accuracy of positive and negative responses. *Psychological Review*, 92, 212—225.

Ravden, S. J., and Johnson, G. I. (1989). *Evaluating Usability of Human-Computer Interfaces: A practical method.* Chichester: Ellis Horwood.

Redelmeier, D. A., and Tibshirani, R. J. (1997). Association between cellular calls and motor vehicle collisions. *New England Journal of Medicine*, 336, 453—458.

Reeves, B., and Nass, C. (1996). *The Media Equation: How people treat computer, and new media like real people and places.* New York: Cambridge University press.

Reid, G. B., and Nygren, T. E. (1988). The subjective workload assessment technique: A scaling procedure for measuring mental workload. In *Human Mental Workload*, eds. P.A. Hancock, N. and Meshkati, 189—218.

Reising, D. V. C., and Sanderson, P. M. (2002). Work domain analysis and sensors II: pasteurizer II case study. *International Journal of Human-Computer Studies*, 56, 597—637.

Rhyne, J. R., and Wolf, C. G. (1993). Recognition-based user interfaces. In *Advances in Human-Computer Interaction*, eds. H. R. Hartson and D. Hix, Norwood, New Jersey: Ablex, 191—212.

Richards, M. A. U., K. (1984a). How should people and computers speak to each other. *proceedings of INTERACT'84, 1ˢᵗ IFIP Conference Human-Computer Interaction*, London: Taylor & Francis. 62—67.

Richards, M. A., Underwood, K. (1984b). Talking to machines: how are people naturally inclined to speak? In *Contemporaty Ergonomics*, ed. B. Shackel, London: Taylor & Francis, 62—67.

Riley, M. A., Amazeen, E. L., Amazeen, P. G., Treffner, P. J., and Turvey, M. T. (1997). Effects of temporal scaling and attention on the asymmetric dynamics of bimanual coordination. *Motor Control*, 1, 263—283.

Ringle, M. D., Bruce, B. (1982). Conversation failure. In *Strategies for Natural Language Processing*, ed. W. Lehnert, and M. D. Ringle, NJ: Lawrence Erlbaum & Associates.

Ringle, M. D., and Halsted-Nussloch, R. H. (1989). Shaping User-Input: A Strategy for Natural Language Dialog Design. *Interacting with Computers*, 3, 227—244.

Robinson, C. R., and Eberts, R. E. (1987). Comparison of speech and pictorial displays in a cockpit environment. *Human Factors*, 29, 31—44.

Rodman, R. D. (1984). Research objectives in voice I/0. *Proceedings of Speech Tech 84*. April 2-4, 1984. New York. 225—230.

Roe, D. B., and Wilpon, J.G. (1993). Whether speech recognition: The next 25 years. *IEEE Communication Magazine,* (November), 54—62.

Roessler, R., and Lester, J.W. (1976). Voice predicts affect during psychotherapy. *Journal of Nervous and Mental Disease*, 163, 166—176.

Rokicki, S. R. (1995). Psychophysiological measures applied to operational test and evaluation. *Biological Psychology*, 40, 223—228.

Rood, G. M. (1998). *Human factors issues for the integration of alternative control technologies*. NATO report. RTO EN-3.

Rowell, M. (1999). Intelligent transportation, map databases and, standards. *Proceedings of Seminar on Integrated Solution for Land Vehicle Navigation*, Nuneaton, UK: Motor Industry Research Association (MIRA). 1—8.

Roy, S. D., Cort, A. and Rixon, G.A. (1983). The application of direct voice control to battlefield helicopters. *9th European Rotorcraft Forum, September13-15*, Stresia, Italy.

Rubio, A., Garcia, P., de la Torre A., Segura, J., Diaz, J., Benitez, M. C., Sánchez, V., Peinado, A. M., López, J. M., and Pérez, J. L. (1997). STACC: an automatic service for information access using continuouss speech recognition through telephone line. *Proceedings of the 5th European Conference on Speech Communication and Technology— Eurospeech 1997*, September 22-25, Rhodes, Greece, 1779—1782.

Ruiz, R., Legros, C., and Guell, A. (1990). Voice analysis to predict the psychological or physical state of a speaker. *Aviation and Space Environmental Medicine*, 61, 266-271.

Ruiz, R., Absil, E., Harmegnies, B., Legros, C., and Poch, D. (1996). Time- and spectrum-related variabilities in stressed speech under laboratory and real conditions. *Speech Communication*, 20, 111-129.

Rundicky, A. I., and Sakamoto, M. (1989). *Transcription conventions and evaluation technique for spoken language systems research.* Pittsburgh: Carnegie Mellon University, School of Computer Science. No. CMU-CS-89-194.

Rundicky, A. I. (1990). *The design of spoken language interfnces.* Pittsburgh: Carneaie Mellon University, School of Computer Science. No. CMU-CS-90-118.

Sakoe, H., and Chiba, S. (1971). Recognition of Continuously Spoken Words Based on Time- Normalization by Dynamic Programming. *The Journal of Acoustical Society of Japan*, 27.

Samuels, S. J. (1987). Factors that influence listening and reading comprehension. In *Comprhending Oral and Written language*, ed. H. S. J. Samuels, San Diego, CA: Academic Press., 295—325.

Sanders, G. A., Le, A. N., and Garofolo, J. S. (2002). Effects of word error rate in the DARPA communicator data during 2000 and 2001. *Proceedings of the 7th International Conference on Spoken Language Processing (ICSLP)*, September 16-20, 2002, Denver, Colorado. 277—280.

Sanders, S. S. and. McCormick, E. J. (1992). *Human factors in Engineering and Design.* 7th ed. Cambridge, England: McGraw-Hill, INC.

Schaefer, D. (2001). *Context-sensitive speech recognition in the air traffic control simulation.* EUROCONTROL Experimental Centre. EEC Note No. 02/2001.

Scherer, K. R. (1981). Speech and emotional states. In *Speech Evaluation in Psychiatry*, ed. J. K. Darby, New York: Grune & Stratton., 189—220.

Scherer, K. R. (1981). Vocal indicators of stress. In *Speech evaluation in psychiatry*, ed. J. K. Darby, New York: Grune & Stratton., 171—187.

Scherer, K. R. (1986). Vocal affect expression: A review and a model for future research. *Psychological Bulletin*, 99, 143—165.

Schiff, W., and Arnone, W. (1995). Perceiving and driving: Where parallel roads meet. In *Local Applications to the Ecological Approach to Human-Machine Systems*, eds. J. F. P. Hancock, J. Caird, and K. Vicente, Vol. 2, Mahwah, New Jersey: Erlbaum, 1—35.

Schmandt, C. (1994). *Voice Communication with Computers*. New York: Van Nostrand Reinhold.

Schraagen, J. M. C., Chipman, S. E., Shute, V., Ruisseau, J. I., Graff, N., Annett, J., Strub, M. H., and Sheppard, C. (2000). *State-of-the-art review of cognitive task analysis techniques*. NATO report. RTO-TR-24.

Schurick, J. M., Williges, B. H. and Maynard, J. F. (1985). User feedback requirements with automatic speech recognition. *Ergonomics*, 28, 1543—1555.

Sears, A., Feng, J., Oseitutu, K, and Karat, C.M. (2003). Hands-Free, Speech-Based Navigation: Difficulties, Consequences, and Solutions. *Human-Computer Interaction*, 18, 229—257.

Seyle, H. (1983). The stress concept: past, present and future. In *Stress Research: Issues for the Eighries*, ed. C. C. Cooper, New York: Wiley.

Shadish, W. R., Cook, T. D., and Campbell, D. T. (2002). *Experimental and Quasi-experimental Designs for Generalized Causal Inference*. Boston: Houghton Mifflin Company.

Shallice, T., and Warrington, E.K. (1970). Independent functioning of verbal memory stores: A neuropsychological study. *Quaterly Journal of Experimental Psychology*, 22, 261—173.

Sheniderman, B. (1992). *Designing the User Interface: Strategies for effective Human-Computer Interation. 2nd ed*. Massachusetts: Addison-Wesley.

Sheridan, T. B. (1997). Task analysis, task allocation and supervisory control. In *Handbook of Human-Computer Interaction*, eds. M. Helander, T. K. Landauer, P.Prabhu, North-Holland: Elsevier Science, 87—105.

Shimono, N. (2000). Development of a navigation support system for one person bridge operation (OPBO) and its evaluation. In *Transactions of The West Japan Society of Naval Architects*, 403—415.

Shneiderman, B. (2000). The Limits of Speech Recognition. *Communications of the ACM*, 43, 63—65.

Siewiorek, D., and Smailagic, A. (2002). User-Centered interdisciplinary design of wearable computer. In *Handbook of Human-Computer Interactions*, ed. A. Sears, J. A. and Jacko, New Jersey: Lawrence Erlbaum & Associates.

Siewiorek, D., Smailagic, A., and Hornyak, M. (2002). Multimodal contextual car-driver interface. *Proceedings of 4th IEEE International Conference on Multimodal Interface, October 14-16, 2002*, Pittsburgh, Peesylvania. 367—373.

Simpson, C. A., (1975). Occupational experience with a specifc phraseology: Group differences in intelligibility for synthesized and human speech. *Journal of the Acoustical Society of America*, 58(suppl. I), 57.

Simpson, C. A., and. Williams, D. H. (1980). Response time effects of alerting tone and semantic context for synthesized speech warnings. *Human Factors*, 22, 319—330.

Simpson, C. A., Coler, C. R. and Huff, E. M. (1982). Human factors of voice I/0 for aircraft cockpit controls and displays. *Workshop on Standardization for Speech I/0 Technology*. 159—166.

Simpson, C. A., McCauley, M. E., Roland E. F., Ruth, J. C. and Williges, B. H. (1985). System design for speech recognition and generation. *Human Factors*, 27, 115—141.

Simpson, C. A. (1985a). Pilot speech performance while talking to a speech recognizer and flying a competitive helicopter pursuit task. *4th Aerospace Behavioural Engineering Technical Conference, SEA Aerospace Congress and Exposition, October 14-17, Warrendale, PA.*

Simpson, C. A. (1985b). Speech variability effects on recognition accuracy associated with concurrent task performance by pilots. *Proceedings of the Third Symposium on Aviation Psychology*, April 22-25, 1985, Ohio State University, Columbia. 87—102.

Simpson, C. A. (1986). Speech variability effects on recognition accuracy associated with concurrent task performance by pilots. *Ergonomics*, 29, 1343—1357.

Sivak, M. (1995). The information that drivers use: is it indeed 90% visual? (keynote address). *6th International Conference on Vision in Vehicles*, September13-16, 1995, Derby, UK.

Slator, B. M., Anderson, M.P., and Conley, W. (1986). Pygmalion at the Interface. *Communications of the ACM*, 29, 599—604.

Smith, A. (1997). *Human-Computer Factors: A study of Users and Information Systems*. Cambridge, England: McGraw-Hill.

Smith, G. W. (1991). *Computers and Human Language*. Oxford, U.K: Oxford University Press.

South, A. J. (1999). Some characteristics of speech produced under high G-force and pressure breathing. *IEEE ICASSP - 99: International Conference on Acoustics, Speech, and System Performance*, March 15-19, 1999, Phoenix, Arizona, 2095—2098.

Spence, C., and Driver, J. (1997). On measureing selective attention to an expected sensory modality. *Perception & Psychophysics*, 59, 389—403.

Spence, C., Nicholis, M. E. R., and Driver, J. (2001). The cost of expecting events in the wrong sensory modality. *Perception & Psychophysics*, 63, 330—336.

Spiegel, M. F., Altom, M. J., Macchi, M. J., and Wallace, K. L. (1990). Comprehensive assessment of the telephone intelligibility of synthetic and natural speech. *Speech Communication*, 9, 279-291.

Stammers, R. B., and Shepherd, A. (1990). Task Analysis. In *Evaluation of Human Work - a practical ergonomics methodolgy*, eds. J. Wilson, E.N. Corlett, London: Taylor & Francis, 144—168.

Stanton, N. A., and Young, M.S. (1999). *A Guide to Methodology in Ergonomics - design for human use*. London: Taylor & Francis.

Stary, C. (2001). Handling user diversity in task-oriented design. In *Universal Access in HCI: Towards an information society for all*, ed. C. Stephanidis, NJ: Lawrence Erlbaum & Associates, 135—139.

Stedmon, A. W., and Gates, J. (1999). It's good to talk: Stress effects on articulation onset time. In *Engineering Psychology and Cognitive Ergonomics*, ed. D. Harris, Vol. 4, Ashgate, 277—283.

Steeneken, H. J. M. (1996). *Potentials of speech and language technology systems for military use: An application and technology oriented survey*. NATO, Defence Research Group. AC/243(Panel 3)TR/21.

Steeneken, H. J. M., and Hansen, J. H. L. (1999). Speech under stress conditions: overview of the effect of speech production and on system performance. *IEEE ICASSP - 99: Internationa. Conference on Acoustics, Speech, and System Performance*, March 1999, Phoenix, Arizona, 2079—2082.

Stein, N. L., and Trabasso, T. (1992). The organization of emotional experience: Creating links among emotion, thinking, language, and intentional action. *Cognition and Emotion*, 6, 225—244.

Sternberg, R. J. (1999). *Cognitive Psychology*. Harcourt College.

Stevens, A., Board, A., Allen, P. and Quimby, A. A. (1999). *Safety checklist for the assessment of in-vehicle information systems: A user manual*. Transport Research Laboratory. PA3536/00.

Stevens, A., Quimby, A., Board, A., Kersloot, T. and Burns, P. (2000). *Design guidelines for safety of in-vehicle information systems*. Transport Research Laboratory. PA 3721/01.

Stoffregen, T. A., and Bardy, B. G. (2001). On specification and the senses. *Behavioural and Brain Sciences*, 24, 195—261.

Stokes, A. F., and Wickens, C. D. (1988). Aviation Displays. In *Human Factors in Aviation*, eds. E. L. Weiner and D. C. Nagel, San Diego, CA: Academic Press, 387—431.

Story, M. F., Mueller, J. L., and Mace, R. L. (1998). *The Universal Design File: Designing for people of all ages and abilities*. Centre for Universal Design, School of Design, North Carolina State University.

Stough, P. (2001). *Aviation weather information*, NASA Langley Research Center.

Streeter, L. A., McDonald, N. H., Apple, W., Krauss, R. M., and Galotti, K. M. (1983). Acoustic and perceptual indicators of emotional stress. *Journal of Acoustical Society of America*, 73, 1354—1360.

Suh, N. P. (1990). *The Principles of Design*. Oxford University press, New York, Oxford.

Suhm, B., Myers, B., and Waibel, A. (2001). Multimodal error correction for speech user interface. *ACM Transaction on Computer-Human Interaction*, 8, 60—98.

Suhm, B. (2003). Improving Speech Interface Design by Understanding Limitations of Speech Interfaces. *Proceedings of the 8th European Conference on Speech Communication and Technology— Eurospeech 2003*, September 1-4, 2003, Geneva, Switzerland.

Suhm, B. (2003). Towards best practices for speech user interface design. *Proceedings of the 8ᵗʰ European Conference on Speech Communication and Technology—Eurospeech 2003,* September 1-4, 2003, Geneva, Switzerland.

Suzuki, J., and Ebukuro, R. (1980). Physical Distribution by Voice Input System. *NEC Technical Journal,* 132, 80—86.

Swail, C. (1997). Direct voice input for control of an avionic management system. *American Helicopter Society 53rd Annual Forum,* April 29-May 1, Virginia Beach, Virginia.

Takada, Y., and Shimoyama, O. (2001). Evaluation of driving-assistance Systems based on drivers' workload. *International Driving Symposium on Human Factors in Driver Assessment, Training and Vehicle Design,* August 14-17, 2001, Aspen, Colorado.

Tan, Y. H., and Thoen, W. (2002). Formal aspects of a generic model of trust for electronic commerce. *Decision Support Systems,* 33, 233—246.

Tannfors, L., and Lundgaard, R. (2004). Incomplete Design. In *M.Sc. Program in Interaction Design,* Chalmers University.

Taylor, M. M. (1986a). Direct voice input and its role in avionic systems. *International Conference on Speech Technology,* October 1986, Brighton, UK, 113—120.

Taylor, M. M. (1986b). Voice input applications in aerospace. In *Electronic Speech Recognition,* ed. G. Bristow, London: Colins, 322—337.

Taylor, M. M. (2001). *Visualisation of massive military datasets: Human factors, applications, and technologies.* NATO technical report. TRO-TR-030.

Terrier, P., and Cellier, J. M. (1999). Depth of processing and design-assessment of ecological interfaces: Task analysis. *International Journal of Human-Computer Studies,* 50, 287—307.

Thambiratnam, D., and Sridharan, S. (2000). Improving speech recognition accuracy for small vocabulary applications in adverse environment. *Internatiional Journal of Speech Technology,* 3, 109—117.

Thomas, C., and Bevan, N. (1996). *Usability Context Analysis: a practical guide.* Teddington, UK: Serco Usability Services,.

Tiger, L. (1992). *The Pursuit of Pleasure.* Boston: Little, Brown and Company,.

Tijerina, L., Parmer, E., and Goodman, M.J. (1998). Driver workload assessment of route guidance system destination entry while driving: A Test Track Study. *Proceedings of the 5ᵗʰ ITS World Congress*, October, Seoul, Korea.

Tijerina, L., Johnston, S., Parmer, E. and Winterbottom, M.D. (2000). *Driver distraction with route guidance systems*. NHTSA DOT HS 809—069.

Torenvliet, G. L., Jamieson, G. A., and Vicente, K. J. (2000). Making the most of ecological interface design: the role of individual differences. *Applied Ergonomics*, 31, 395—408.

Trbovich, P., and Harbluk, J. L. (2003). Cell Phone Communication and Driver Visual Behavior. *CHI 2003: The Impact of Cognitive Distraction, Conference on Human Factors and Computing Systems*. April 5-10, 2003, Florida, USA

Treisman, A. M. (1964). Verbal Cues, language, and meaning in selective attention. *American Journal of Psychology*, 77, 206—219.

Tsimhoni, O., and Smith, D., Green, P. (2004). Address entry while driving: Speech recognition versus a touch-screen heyboard. *Human Factors*, 46, 600—610.

Utsuki, N., and Okamura, N. (1976). *Relationship between emotional state and fundamental frequency of speech*. Japan: Japan Air Self Defence Force, Aeromedical Laboratory. Tech. Report 16.

Utter, D. (2000). *Passenger vehicle driver cell phone use results from the fall 2000 national occupant protection user survey*. Washington, DC: National Highway Traffic Safety Administration. DOT HAS 809/293.

Walker, M., Litman, D., Kamm, C., and Abella, A.(1997). PARADISE: A general framework for evaluating spoken dialogue agents. *Proceedings of the 35th Annual Meeting of the Association of Computational Linguistics (ACL/EACL 1997)*, July 11-12, 1997, Madrid, Spain, 271—280.

Wall, R. (2002). New NATO standards target UAV interoperability. *Aviation Week & Space Technology*, 157(15), 41.

van Bezooijen, R. (1990). *Sam Segmental* Test. Institute For Perception - TNO.

van Santen, J. P. H. (1993). Perceptual experiments for diagnostic testing of text-to-speech systems. *Computer Speech and Language*, 7, 49—100.

van Summers, W., Pisoni, D. B., Bernacki, R. H., Pedlow, R. I., and Stokes, M. A. (1988). Effects of noise on speech production: Acoustic and perceptual analyses. *Journal of the Acoustical Society of America*, 84, 917—928.

van Vianen, E., Thomas, B., and van Nieuwkasteele, M. (1996). A combined effort in the standardization of user interface testing. In *Usability Evaluation in Industry*, eds. P. W. Jordan, B. Thomas, B. A. Weerdmeester, I. L. McClelland, London: Taylor & Francis, 7—18.

Warnes, A. M., Fraser, D.A., Hawken, R.E. and Sievey, V. (1993). Elderly drivers and new road transport technology. In *Driving Future Vehicles*, eds. A. M. Parkes, and S. Franzen, London: Taylor & Francis, 99—117.

Warren, R. M. (1970). Perceptual restoration of missing speech sounds. *Science*, 167, 392—393.

Warren, R., and Wertheim, A. (1990). *Perception and the World of Self Motion*. Mahwah, New Jersey: Erlbaum.

Wastell, D. G., and Cooper, C.L. (1996). Stress and technological innovation: A comparative study of design practices and implementation strategies. *European Journal of Work and Organizational Psychology*, 5, 377—397.

Weegels, M. F. (2000). user's conceptions of voice-operated information services. *International Journal of Speech Technology*, 3, 75—82.

Weinstein, C. J. (1995). Military and government applications of human-machine communication by voice. *Proceedings of the National Academy of Science*, 92, 10011—10016.

Welch, J. R. (1977). *Automatic data-entry analysis*. Rome Air development Center. Tech. Rep. No. RADC TR-77-306.

Welch, J. R. (1980). Automatic speech recognition - putting it to work in industry. *Computer*, 14, 65—73.

Welford, A. T. (1978). Mental workload as a function of demand, capacity, strategy, and skill. *Ergonomics*, 21, 157—167.

Veltman, J. A., and Gaillard, A. W. K. (1996). Physiological indices of workload in a simulated flight task. *Biological Psychology*, 42, 323—342.

Verwey, W. B. (1989). Simple in-car route guidance information from another perspective: modality versus coding. *The First Vehicle Navigation and Information Systems Conference*, September 11-13, 1989, Toronto, Ontario, Canada.

Verwey, W. B. (1993). How can we prevent overload of the driver? In *Driving Future Vehicle*, ed. A. M. Parkes, and S. Franzen, London: Taylor & Francis, 235—244.

Whiteside, S. P. (1998). Simulated emotions: an acoustic study of voice and perturbation measures. *Proceedings of the 5th International Conference on Spoken Language Processing (ICSLP)*. November 30-December 4, 1998, Sydney, Australia 699–703.

Vicente, K. J. (1992). Ecological interface design: Theorical foundation. *IEEE Trans. on Systems, Man and Cybernetics,*, 22, 589—606.

Vicente, K. J., Christoffersen, K., and Pereklite, A. (1995). Supporting operator problem solving through ecological interface design. *IEEE Trans. on Systems. Man.and Cybernetics*, 25, 529—545.

Vicente, K. J. (1999). *Cognitive Work Analysis*. New Jersey: Lawrence Erlbaum & Associates.

Vicente, K. J. (2002). Ecological interface design: progress and challenges. *Human Factors*, 44, 62—78.

Wickens, C. D., and Vidulich, M. (1982). *S-C-R Compatibility & dual task performance in two complex information processing tasks: threat evaluation & fault diagnosis*. University of Illinois. Tech. Rep. No. EPL82-3/UNR-82-3.

Wickens, C. D., Sandry, D. L. and Vidulich, M. (1983). Compatibility and resource competition between modalities of input, central processing and output. *Human Factors*, 25, 227—248.

Wickens, C. D. (1991). Processing resources and attention. In *Multiple-Task Performance*, ed. D. Demos, London: Taylor & Francis, 3—34.

Wickens, C. D. Hollands, J. G. (2000). *Engineering Psychology and Human Performance*. New Jersey: Prentice Hall.

Wierwille, W. W. (1995). Development of an initial model relating driver influence visual demands to accident rate. *Third Annual Mid-Atlantic Human Factors Conference Proceedings. Blacksburg*, VA: Virginia Polytechnic Institute and State University.

Wieser, M. (1991). The Computer for the 21st Century. *Scientific American*, 94—104.

Williams, C. E., and Stevens, K.N. (1972). Emotions and speech: Some acoustic correlates. *Journal of the Acoustic Society of America*, 52, 1238—1250.

Williams, C. E., and Stevens, K.N. (1981). Vocal correlates of emotional statesstress. In *Speech evaluation in psychiatry*, ed. J. K. Darby, New York: NY Grune & Stratton.

Wilson, G. F., Purvis, B., Skelly, J., Fullenkamp, P., and Davis, I. (1987). Physiological data used to measure pilot workload in actual flight and simulatot conditions. *Proceedings of the 31ˢᵗ Annual Meeting of the Human Factors Society*, October 19-23, New York, USA. 779—783.

Wilson, G. F., and Fisher, F. (1991). The use of cardiac and eye blink measures to determine flight segment in F4 crews. *Aviation, Space and Environmental Medicine*, 62, 959—961.

Wilson, G. F., and Fisher, F. (1995). Cognitive task classification based upon topographic EEG data. *Biological Psychology*, 40, 239—250.

Wilson, G. F. and Russell, C. A. (1999). Operator Functional State Classification using Neural Networks with Combined Physiological and Performance Features. *Human Factors and Ergonomics Society 43ʳᵈ Annual Meeting*, Louston, Texas, USA. 1099—1102.

Wilson, G. F. (2002). Psychophysiological test methods and procedures. In *HFA Workshop: Psychophysiological Application to Human Factors, March 11-12, 2002*. Swedish Center for Human Factors in Aviation.

Wilson, G. F., Lambert, J.D., and Russell, C.A. (2002). Performance enhancement with real-time physiologically controoled adaptive aiding. In *HFA Workshop: Psychophysiological Application to Human Factors, March 11-12, 2002*. Swedish Center for Human Factors in Aviation.

Wilson, J., R., and Corlett, E.N. (1995). *Evaluation of human work*. London: Taylor & Francis.

Wilson, J. R. (2002). UAVs and the human factor. *Aerospace America*, 40, 54—57.

Violanti, J. M. (1997). Cellular phones and traffic accidents. *Public Health*, 111, 423—428.

Vloeberghs, C., Verlinde, P., Swail, C., Steeneken, H., van Leeuwen, D., Trancoso, I., South, A., Cupples, J., Anderson, T., and Hansen, J. (2000). *The impact of speech under "stress" on military speech technology*. NATO Research and Technology Organization. RTO-TR-10, AC/323(IST)TP/5.

Voiers, W. D. (1983). Evaluating processed speech using the Diagnostic Rhyme Test. *Speech Technology*, 1, 30—39.

Womack, B. D., Hansen, J. H. L. (1996). Classification of speech under stress using target driven features. *Speech Communication*, 20, 131—150.

Zajonc, R. (1997). Emotion. In *Handbook of Social Psychology,* 4th ed., eds. D. Gilbert, S. T. Fishe and G. Lindzey, Cambridge, England: McGraw-Hill., 591—631.

Zarembo, C. A. (1986). The usefulness of voice recognition on the floor of the NY Stock Exchange. *Proceedings of Speech Tech 86,* April 28-30 1986, New York. 69-72.

Zhou, G., Hansen, J.H.L., Kaiser, J.F. (1998). Linear and nonlinear speech feature analysis for stress classification. *Proceeding ICSLP 1998: The 5th International Conference on Spoken Language Processing.* November 30-December 4, 1998, Sydney, Australia. 883—886.

Zhou, G., Hansen, J. H. L., and Kaiser, J. F. (1999). Methods for stress classification: Nonlinear TEO and linear speech based feature. *IEEE ICASSP - 99: Internationa. Conference on Acoustics, Speech, and System Performance*, March 1999, Phoenix, Arizona, 2087—2090.

Zoltan-Ford, E. (1991). How to get people to say and type what computers can understand. *International Journal of Man-Machine Studies*, 34, 527—547.

Zue, V. (1997). Conversational interfaces: advances and challenges. *Proceedings of the 5thEuropean Conference on Speech Communication and Technology—Eurospeech 1997*, Rhodes, Greece. kn-9 - kn 18.

Index

abstraction hierarchy, 112-113, 159-160
ACC, 286-287
Accident, 8, 273, 277
ACT, 289
adaptive cruise control, 286
adaptive interface, 153-154
AFVs, 312
AH, 160
aircraft, 79-80, 299-320
Alternative Control Technology, 289
AMS, 315
anatomy, 28
ANN, 77-78, 175-177
anti-submarine warfare, 320
Army, 312, 314, 322
artificial neural network, 77
ASR (automatic speech recognition), 1, 5,
    7, 53, 67-68, 78-84, 90-94, 167-168,
    173-178, 182-208, 225-228, 232, 241,
    256, 291-292, 295-297, 299-303, 308-
    330
ASW, 320
ATC, 316-319
ATT, 252
Attention, 53-57, 98, 193
auditory distraction, 83
auditory feedback, 190-191
auditory presentation, 49, 221
auditory receptors, 34, 97
AUIS, 153, 155

autonomic system, 30
aviation application, 301
Avionics Management System, 315
axiomatic design theory, 133
axon, 28-29, 48
axon terminal, 28-29
biobehavioral, 63
BP (blood pressure), 75, 76
body fluids, 75
bottleneck, 54-57, 60, 98, 293
bottom-up, 37-38, 43-44, 47, 195
bottom-up perception, 37
brain, 9, 27-35, 39-55, 66-77, 86, 97-98,
    103, 122, 169, 170, 176, 211
brain damage, 31
Broca's area, 40
C2OTM, 312, 314
C3I, 309
CAI, 322, 323
CAVS, 262
cell assembly theory, 48
cell body, 28, 29, 30, 48
cellular phone, 272-281
CFG, 262, 263
CHAM model, 203
chemical transmitters, 28
CLID, 198
Cluster-Identification-Test, 198
CNI (Communication, Navigation and
    Identification), 308

CNS, 30, 76-77
cockpit, 3, 89, 194, 205, 300-309, 314
cognitive psychology, 9, 27, 101
cognitive science, 9, 14- 20, 48, 217
cognitive system engineering
    CSE, 96, 104
cognitive task analysis, 108
cognitive work analysis, 102, 110
Collision Avoidance, 252, 255
collision-warning system, 287
Command and Control system, 292
Command and Control on move, 312,
    314
command, control, communications and
    intelligent, 309
Computer Assisted Instruction, 322
concatenative synthesis, 177, 192
Concurrent system design methodology,
    260
consonants, 42
constraints, 2, 14, 16, 23, 65, 96, 100-
    124, 128, 152, 157-160, 172, 175,
    180, 191, 229, 255, 257, 288, 291,
    296, 302, 306
construct validity, 24-25
constructionist approach, 45
constructive perception, 37-38
context free grammar, , 262
context-in-use analysis, 144
control-centered approach, 124
cortex, 30-36, 39-40, 45, 47, 57, 76, 176
cortical connection, 36
Critical flicker frequency, 74
CTA, 108
CUA, 142, 144
CWA, 102
CWS, 287
Decay theory, 50
decision making, 9, 151, 260, 277, 310
dendrites, 28, 48
design methodologies, 13
design requirements, 118, 149, 164, 292
diagnostic rhyme test, 198
dialogue design, 95, 249, 256
dialogue generation system, 10
dialogue manager, 262
direct perception, 37
direct voice input, 289, 316, 325
discourse, 40, 180, 230

distributed cognitive theory, 104
distributed hierarchical system, 37
divided attention, 54, 61, 295
DM, 262
DRT, 198
dual-task performance, 57-58
DVI, 316
Ecological theory, 99
Ecological validity, 25
EEG (electroencephalogram), 32-33, 73,
    76-78, 93, 289
EID (ecological interface design), 101,
    105, 118, 125, 158, 159
EOG (electrooculogram), 75
EMG, 289
emotion, 48, 85-87, 93, 138, 177, 185,
    196, 236-237
encoding stage, 45, 305
environments, 2, 5, 54, 67, 77-78, 81,
    114, 126, 141-142, 154, 158, 161,
    170, 175, 228, 232, 253, 284, 291,
    295, 328
ergonomics, 5, 8-9, 17, 125, 190, 226,
    228, 231, 233, 245, 247, 257, 283,
    284, 328
ERP, 32-33, 76-78
error correction, 180, 199, 200-201, 204,
    207, 212, 221-222, 224, 233, 297, 303
error detection, 130, 191, 199
ethnographic approaches, 152
Ethnography, 152, 163-164
Event related potentials, 76
event-related potential, 32
EVMPI, 305, 306, 307
evolutionary principles, 211
Explicit memory, 47
external validity, 25
Eye blinks, 74-75
Eye/Voice Mission Planning Interface,
    306
FAA, 316
fatigue, 62-63, 67-68, 70, 72, 74, 81-82,
    161, 207, 282, 295, 302
Federal Aviation Administration, 316
fMRI, 32-33, 77
focused attention, 54-55, 60, 94, 220, 270
Forgetting, 49, 50
formant synthesis, 177, 192
formants, 42, 91

functional decay theory, 49

functional magnetic resonance imaging, 77

Fundamental frequency, 87

GA, 38

GCS, 310, 311

generic intelligent driver support, 255

G-force, 61, 290, 291, 295

GIDS, 255

Gisting, 318

global array, 38

GOMS (Goals, Operators, methods selection Roles), 96, 108-110, 117

GPS (Global Position System), 252, 256, 272

grammar, 39, 41-42, 44-45, 94, 169, 175, 184, 230, 244, 261-263, 268, 304, 327

grammatical rules, 40

graphical user interface, 241, 260

ground control system, 310

GUI, 260, 261

guidelines, 15-17, 21, 25-26, 143, 145, 152, 188, 241, 245-257, 279, 284, 303, 307, 323-324

Habits, 46

hands-free voice dialing, 276

HCI, 13, 109, 151, 158, 161, 163, 179, 182, 186, 284

hemisphere, 39, 40, 41, 43

hemispheres, 30

Heuristic evaluation, 132, 188, 245

Hierarchical Task Analysis, 107

HMM (Hidden Markov Model), 174-177

HR (heart rate), 73, 75-76, 89, 285, 287,

HRV (heart rate variability), 75-76, 285

HTA, 107

HUD (Head Up Display), 259-260, 278

human behavior, 2, 82, 99, 110, 186

human cognition, 9, 12, 21-22, 26-27, 97, 104, 115, 133, 168, 173, 186, 238, 239

human cognitive behavior, 9, 15, 21

human cognitive demands, 9

human factors, 2, 3, 5, 11, 9-12, 14, 64, 115, 144, 151, 157, 168, 179-180, 184, 186, 224-228, 231, 234, 237-238, 241, 245, 249, 254, 256, 270-273, 277-284, 291, 308, 315, 324

I Ching, 17-23, 136

ICE, 267

Implicit memory, 47

in-car entertainment, 256, 267

individuals behave, 170, 211

IP (information processing), 50, 57, 69, 71, 74-76, 96-99, 104, 106, 110, 118, 120, 124, 220-221, 268, 276, 290, 303

Intelligent Transport Systems, 252

Intelligent Vehicle Highway Systems, 252

interactive voice response, 174

inter-beat-interval, , 75

Interference, 46, 50, 60

interference theory, 49

Internal validity, 24

intonation, 40- 42, 169-170, 184, 195, 230-231, 246

ISO 13407, 142-143, 145

ISO DIS 9241-11, 125, 127, 138

ITS, 252-255, 271-273, 278, 281

IVCS (in-vehicle communication systems), 12, 251-272, 281, 283-285

IVHS, 252

IVR, 174

KBB, 160

key-word spotting system, 244

knowledge elicitation, 114

knowledge-based behavior, 160

language comprehension, 9, 40-41, 44, 193, 195, 236

language process, 28, 41, 50-51, 169, 177

LCD (Liquid Crystal display), 259-260

lexicon. Semantics, 40

long-term memory, 31, 47, 57, 97-98

LVCSR, 174

magnetic resonance imaging, 32-33

magnetonencephalogram, 32

market, 3- 5, 12, 124, 150, 164, 174, 227-229, 241, 249, 272, 277, 281

MCDS, 322

medium, 100, 182, 213-214, 256, 260, 268, 271, 298, 304, 312

MEG, 32-33, 77

memory system, 45-46, 48

mental workload, 8, 63, 69-77, 255, 287, 295, 303, 312

MHP, 110

minimalist hypothesis, 45

mobile command and control vehicle, 312
Model Human Processor, 110
modified rhyme test, 198
morphemes, 39, 169
motor output, 30
Motor theory, 44
MRI, 32-33
MRT, 198
multidisciplinary, 11, 17
multimedia, 105, 141, 213-215
multimodal, 11, 13, 36, 60, 141, 163, 167, 206, 208, 212-218, 220-222, 235, 258-259, 266, 271, 292, 311, 316, 324, 329
multiple resources theory, 59
multiple tasks performance, 53, 57
NASA-TLX, 71-72
natural language, 178, 214, 244
natural Language Understanding, 262
navigation system, 9, 277, 280
navigation systems, 8, 272-273, 277-281, 288, 301
Navy, 309-322
nervous system, 29- 31, 46, 74, 75, 77, 87
neuropsychology, 27-28, 49
neuroscience, 9, 27-31, 33, 46, 49-51, 98
NL, 178-179
NLU, 262
noise, 34, 61, 63, 65, 79-83, 88-89, 170, 173-175, 181, 200, 205, 230, 256, 283, 290-292, 295-296, 300, 302, 308, 313-315, 323
observation, 24, 76, 107-108, 114, 132, 144, 163-164
organization, 1, 2, 6, 13-14, 31, 35, 65, 86, 102, 104, 120, 141, 147, 149, 152, 242, 290
oxygen consumption, 32
parasympathetic, 30, 75
participatory design, 152
pattern analysis, 56
PCP, 150, 152
perception, 9, 14, 17, 21, 28, 31-48, 54, 59-60, 65, 68, 74, 79, 89, 97, 99, 101, 120, 122, 128, 148, 154, 157, 160, 169, 171, 176, 182, 193, 195-198, 216, 217, 247, 269, 276
performance constraints, 8

Performance evaluation, 180
performance measurement, 69, 70, 73, 92, 329
performance-resources function, 58
PET, 32, 33, 77
phonemes, 39, 40, 43, 169, 171, 177, 179, 184, 193, 195, 203, 236
Phonemic-restoration effect, 44
Phonetic refinement theory, 43
phonological loop, 48, 49
pitch, 41-42, 56, 81, 83, 87-90, 169, 185, 193, 196, 203
pleasure, 13, 138, 139, 140, 142, 148, 243
PNS, 30
POG, 306
point-of-gaze, 306
Positron emission tomography, 32-33
preattentively analysis, 56
Predictive validity, 25
PRF, 58
primary task, 57, 70, 71, 190, 256, 267, 270, 274-275, 297, 300-301, 304
product creation process, 150
prosody, 40, 169, 184, 193-196, 202, 236, 238, 246
PRP, 60
psychological refractory period, 60
Pursuit, 55
quasi-spatial processes, 39
questionnaires, 71, 108, 144, 247-248, 250
RBB (rule-based behavior), 112, 160
receptors, 29, 33-39, 43
recognition accuracy, 2, 10, 53, 67-68, 80-82, 84-85, 94, 168, 173, 175, 180-181, 185-186, 200-201, 204, 206, 221-232, 239, 246, 291, 296, 302, 309, 315, 317, 325, 327, 330
reference memory, 47
reinforcement, 46
reliability, 9-10, 68, 126-127, 135, 137, 168, 207-208, 227, 241, 285, 302, 322
reliability, 127, 140, 247, 285
requirements specification, 117, 144, 146
research methods, 22, 101
respiratory sinus arrhythmia, 75-76
Road Transport Informatics, 252
RSA, 75-76
Saccadic, 55

SACL, 69
Safety issue, 8
satisfaction, 13, 125-128, 133, 137-138, 142, 144, 146, 150, 154, 180, 187, 189, 226, 228, 230, 232, 241, 243-244, 247-249
SBB (skill based behavior), 160
Schemata, 46
S-C-R, 190, 304-305
Search-after-meaning theory, 45
secondary tasks, 71
selective-attention, 55
SENECa, 264
SENECs, 263
sensory input, 30, 40, 97, 169
sensory store, 97
Short-term memory, 47
SIAM, 187-188
single-energy arrays, 38
situation awareness, 8, 218, 259, 270, 289, 295, 301, 311-312, 323
slips, 202
SmartKom, 265-266
social-technical models, 152
socio-technical systems, 1, 7-12, 14
soft-system methodology, 152
spatial information, 49, 235
speech input, 3-8, 11, 167, 170, 190, 205-207, 212, 215, 226, 230, 259, 265, 268, 271, 286, 291, 296-300, 325
speech interface assessment method, 187
speech output, 59, 177, 192, 194, 235, 256, 259, 271
speech sounds, 41, 42, 44, 80, 173
speech synthesis, 10, 177, 192, 216, 240, 292, 293
speech understanding system, 10, 184, 296
speech user interface, 222, 238, 239
spinal cord, 30, 31, 35, 46
Spoken Language Dialogue Systems, 263
spoken-language-understanding, 178
SQUID, 32
SRK, 112, 113
SSM, 152
Stakeholders, 144, 147
Standard Segmental Test, 198
stimulus/central-processing/response, 304
storage stage, 45

stress, 41-42, 53, 61-69, 72-94, 170, 184-185, 195, 201, 203, 206, 221, 229, 231, 237, 277, 282, 289-292, 295, 296, 300, 302, 309, 313, 317, 325, 327-328, 330
Stress Arousal Checklist, 69
stressors, 62-64, 66, 80, 84-92, 182, 302
subjective evaluation, 71-73, 92, 329
SUI, 238, 241
SWAT, 71-72
syllables, 40, 88, 193, 230
sympathetic, 30, 75-76
syntax, 39, 44-45, 84-85, 177, 182, 184, 190, 200, 205, 206, 240, 303, 308, 317, 324, 327
Syntax, 205, 213
synthesis system, 10
synthetic speech, 1, 42, 167-168, 192, 194-199, 205, 237, 240, 254, 268, 305
task analysis, 70, 96, 102, 106-108, 110-122, 114, 157, 162, 182, 187, 317, 325
task-orientated, 2, 12, 25, 244
technical centered approach, 123
technical evaluation, 244
Technical evaluation, 243
telecommunication, 225, 234, 237, 249
telephone voice services, 225
TEO, 91
text-to-speech synthesis, 177, 192
TICS, 252
Time On Task, 232
timing, 20, 22-23, 41-42, 60, 71, 77, 121, 170, 196, 202, 254, 277, 279
top-down, 38, 43-44, 47, 54-55, 120
top-down perception, 37
TOT, 232
TRACE model, 44
Transport Information and Control Systems, 252
Trust, 137
TTS, 185, 194-198, 225, 228, 240, 262
UAVs, 310-311
UCD, 13-14, 124, 131, 141-144, 147, 150-156, 164, 187, 249
universal access, 160-161, 163
Unmanned Aerial Vehicles, 310
usability evaluation, 13-14, 116, 128, 131, 133, 142, 146, 149, 151, 187, 188, 241-249

usage-centered approach, 124, 155, 157-
    158, 162
user analysis, 8, 144, 150
User error, 202
User Words On Task, 232
user-centered design, 13, 124-125, 142,
    162, 182, 288
utility, 13, 76, 127
UWOT, 232
validity, 16, 22-25, 38, 68, 91-93, 135,
    199, 278, 285
Validity, 23, 248
vehicles, 79, 192, 251-254, 268, 272, 283,
    287, 310-314, 318
vibration, 61, 63, 79-81, 191, 200, 230,
    286, 290-291, 295, 302, 313, 314
VICO, 258-259
visual cues, 42, 171, 305
visual feedback, 82, 190-191, 207, 235,
    299, 307
visually guided action, 39

visual-spatial recognition, 39
visuo-spatial sketch pad, 49
vocabulary, 8, 68, 81, 84-85, 94, 173-175,
    177, 181-184, 190, 200, 205-206, 209,
    221, 224, 230, 232, 239, 242, 244,
    256-257, 265, 282, 290-291, 297, 299,
    308-309, 322-324, 327
Vowel, 42
WER, 232
Wernicke's aphasia, 40
Wizard-of-Oz, 145, 183, 209, 221, 259,
    285
WM, 193
Word Error Rate, 232
working memory, 21, 47-49, 59, 78, 98,
    170, 190-193, 196, 239-240, 268
workload, 9, 53, 58, 61-79, 84, 87-94, 98,
    195, 201, 203, 206, 220, 268-271,
    277, 285-298, 301-308, 311-315, 323,
    328